De Gruyter Proceedings in Mathematics

Painlevé Equations and Related Topics

Proceedings of the International Conference,
Saint Petersburg, Russia, June 17–23, 2011

edited by
Alexander D. Bruno
Alexander B. Batkhin

De Gruyter

Mathematics Subject Classification 2010: Primary: 34M55, 34M35, 34M56, 33E17, 34G56, 35B06, 35Q40, 37K10; Secondary: 34E05, 34M03, 30F35, 11F03, 14H45, 14H55, 30F10, 33C60, 33E05, 34M40, 35Q15, 32G08, 32G15, 32G34, 33C15, 33E10, 33E99, 34G20, 34M50, 35C15, 37J35, 39A14, 45F05, 76L05.

ISBN 978-3-11-027558-2
e-ISBN 978-3-11-027566-7

Library of Congress Cataloging-in-Publication Data

A CIP catalog record for this book has been applied for at the Library of Congress.

Bibliographic information published by the Deutsche Nationalbibliothek

The Deutsche Nationalbibliothek lists this publication in the Deutsche Nationalbibliografie; detailed bibliographic data are available in the internet at http://dnb.dnb.de.

Typesetting: PTP-Berlin Protago-TEX-Production GmbH, www.ptp-berlin.eu
Printing and binding: Hubert & Co. GmbH & Co. KG, Göttingen
♾ Printed on acid-free paper

Printed in Germany

www.degruyter.com

Preface

Painlevé equations are six equations P_l of the form

$$y'' = \frac{Q_l(x, y, y')}{R_l(x, y)}, \quad l = 1, \dots, 6,$$

where Q_l and R_l are polynomials. They have solutions without movable critical points (the Painlevé property), and all singularities of solutions are concentrated in singular points $x = \text{const}$ of the equations. In the result, singularities of solutions at these points are very complicated, especially at the infinity singular point $x = \infty$.

Wach Painlevé equation P_l has several sequences of equations P_l^m of higher orders with $m = 1, 2, \dots$ and $P_l^1 = P_l$ having the Painlevé property and forming infinite number of equations of finite number of hierarchies. Recently these hierarchies were intensively studied, but in this volume they are practically not discussed.

There are many applications the Painlevé equations, some of which are presented at various chapters in this book. For applications, discrete Painlevé equations were constructed, which are also presented.

The **main problems** for Painlevé equations and their generalizations are:

1. finding asymptotic forms and asymptotic expansions of solutions near singular points of equations;
2. their interpretation, including questions of convergence;
3. computing formulas of the connection of asymptotic forms of one solution near different points;
4. test of the Painlevé property, which in some cases can be done by calculating asymptotic expansions of solutions near nonsingular points of equations.

All these problems are presented in this book.

To solve these problems, we have two main approaches:

1. the method of isomonodromy deformations (the so-called Riemann–Hilbert method);
2. Power Geometry.

An essential part of this book is a competition of these two approaches. Each of them has its own advantages and disadvantages. The symbiosis of these two methds could perhaps bring about the solution of all the mentioned problems.

Power Geometry allows us to find asymptotic forms and asymptotic expansions of solutions to equations of all classes: algebraic, ordinary differential, and partial differential. But to date the connection formulas have not been found with its help. On the other hand, Power Geometry allows us to test the Painlevé property algorithmically. The main disadvantage of Power Geometry is that it is generally unknown; it is, for example, used by authors in only five chapters of this volume.

The book is divided into 5 parts.

Part I describes methods for finding asymptotic expansions of such solutions to one ordinary differential equation (ODE), for which the orders of the first derivative and the second derivative differ from the order of the solution by -1 and -2, respectivly. Convergence of the power expansions is also looked at. Part I also gives applications of these methods for Painlevé equations P_3, P_5, and P_6.

Part II describes methods for finding asymptotic expansions of such solutions to one ODE, for which orders of the first and the second derivatives differ from the order of the solution by γ and 2γ, respectively, where γ is a real number and $\gamma \neq -1$. Part II also gives detailed applications for all the Painlevé equations, P_1–P_6. In particular, periodic and elliptic asymptotic expansions are computed.

Part III describes applications of the method of isomonodromy deformations to several problems in ODE, including Painlevé equations. This approach allows us to compute some asymptotic forms (found in Part I), to prove convergence of some asymptotic expansions (found in Part I), and to calculate some new connection formulas.

Part IV contains chapters dealing with the selection of equations with the Painlevé property (the Painlevé test) in several classes of ODE. If the equations of the class are sums of terms with unknown coefficients, then the coefficients can be found by methods of Plane Power Geometry (see Part I), as was done for equations P_5 and P_6 (see references in Part I and in the Comments).

The chapters in Part V describe different methods and approaches to ODEs and applications of some of them to Painlevé equations. Two chapters consider the discrete Painlevé equations.

The Comments section contains some editorial remarks regarding some of the chapters contained in this volume.

This book presents a whole spectrum of studying mathematical problems connected with Painlevé equation, and all the methods (old and new) for the solution of ODEs. This book will be of interest and useful for mathematicians and physicists, from students to professors.

Moscow, March 2012

Alexander B. Batkhin
Alexander D. Bruno

Alexander D. Bruno, Professor and Head, Department of Singular Problems, Keldysh Institute of Applied Mathematics (KIAM), Russian Academy of Sciences, Moscow, Russia; Department of Theory of Dynamical Systems, Lomonosov Moscow State University, Moscow, Russia. E-mail: abruno@keldysh.ru Web: http://en.wikipedia.org/wiki/Alexander_Dmitrievich_Bruno

Alexander B. Batkhin, Associate Professor. Department of Singular Problems. Keldysh Institute of Applied Mathematics (KIAM), Russian Academy of Sciences, Moscow, Russia. E-mail: batkhin@gmail.com

Contents

Preface v

I Plane Power Geometry

1 *Alexander D. Bruno*
Plane Power Geometry for One ODE and P_1–P_6 3

1.1 Statement of the Problem 3
1.2 Computation of Truncated Equations 4
1.3 Computation of Expansions of Solutions to the Initial Equation (1.1) . 6
1.4 Extension of the Class of Solutions 7
1.5 Solution of Truncated Equations 7
1.6 Types of Expansions 10
1.7 Painlevé Equations P_l 11

2 *Alexander B. Batkhin and Natalia V. Batkhina*
New Simple Exact Solutions to Equation P_6 13

2.1 Introduction 13
2.1.1 Power Geometry Essentials 13
2.1.2 Matching "Heads" and "Tails" of Expansions 14
2.2 Constructing the Template of an Exact Solution 15
2.3 Results 17
2.3.1 Known Exact Solutions to P_6 17
2.3.2 Computed Solutions 17
2.3.3 Generalization of Computed Solutions 20

3 *Irina V. Goryuchkina*
Convergence of a Formal Solution to an ODE 23

3.1 The General Case 23
3.2 The Case of Rational Power Exponents 24

3.3 The Case of Complex Power Exponents ... 25
3.4 On Solutions of the Sixth Painlevé Equation ... 25

4 *Irina V. Goryuchkina*
Asymptotic Expansions and Forms of Solutions to P_6 27
4.1 Asymptotic Expansions near Singular Points of the Equation ... 27
4.2 Asymptotic Expansions near a Regular Point of the Equation ... 30
4.3 Boutroux-Type Elliptic Asymptotic Forms ... 30

5 *Anastasia V. Parusnikova*
Asymptotic Expansions of Solutions to P_5 33
5.1 Introduction ... 33
5.2 Asymptotic Expansions of Solutions near Infinity ... 35
5.3 Asymptotic Expansions of Solutions near Zero ... 35
5.4 Asymptotic Expansions of Solutions in the Neighborhood of the Nonsingular Point of an Equation ... 37

II Space Power Geometry

6 *Alexander D. Bruno*
Space Power Geometry for one ODE and P_1–P_4, P_6 41
6.1 Space Power Geometry ... 41
6.2 Asymptotic Forms of Solutions to Painlevé Equations P_1–P_4, P_6 ... 44
6.2.1 Equation P_1 ... 44
6.2.2 Equation P_2 ... 45
6.2.3 Equation P_3 for $cd \neq 0$ ... 46
6.2.4 Equation P_3 for $c = 0$ and $ad \neq 0$ ... 47
6.2.5 Equation P_3 for $c = d = 0$ and $ab \neq 0$ ... 48
6.2.6 Equation P_4 ... 49
6.2.7 Equation P_6 ... 50

7 *Alexander D. Bruno and Anastasia V. Parusnikova*
Elliptic and Periodic Asymptotic Forms of Solutions to P_5 53
7.1 The Fifth Painlevé Equation ... 53
7.2 The case $\delta \neq 0$ ... 54
7.2.1 General Properties of the P_5 Equation ... 54
7.2.2 The First Family of Elliptic Asymptotic Forms ... 55

7.2.3 The First Family of Periodic Asymptotic Forms 57
7.2.4 The Second Family of Periodic Asymptotic Forms 58

7.3 The Case $\delta = 0,\ \gamma \neq 0$ 59
7.3.1 General Properties 59
7.3.2 The Second Family of Elliptic Asymptotic Forms 60
7.3.3 The Third Family of Periodic Asymptotic Forms 62
7.3.4 The Fourth Family of Periodic Asymptotic Forms 63

7.4 The Results Obtained 64

8 *Alexander D. Bruno*
Regular Asymptotic Expansions of Solutions to One ODE and P_1–P_5 67

8.1 Introduction 67

8.2 Finding Asymptotic Forms 68

8.3 Computation of Expansions (8.2) 69

8.4 Equation P_1 71

8.5 Equation P_2 73
8.5.1 Elliptic Asymptotic Forms, Face $\Gamma_3^{(2)}$ 73
8.5.2 Periodic Asymptotic Forms, Face $\Gamma_4^{(2)}$ 74

8.6 Equation P_3 75
8.6.1 Case $cd \neq 0$ 75
8.6.2 Case $c = 0, ad \neq 0$ 76
8.6.3 Case $c = d = 0, ab \neq 0$ 77

8.7 Equation P_4 77
8.7.1 Elliptic Asymptotic Forms, Face $\Gamma_3^{(2)}$ 78
8.7.2 Periodic Asymptotic Forms, Face $\Gamma_4^{(2)}$ 78

8.8 Equation P_5 79
8.8.1 Case $d \neq 0$, Elliptic Asymptotic Forms, Face $\Gamma_1^{(2)}$ 79
8.8.2 Case $d \neq 0$, Periodic Asymptotic Forms, Face $\Gamma_2^{(2)}$ 81
8.8.3 Case $d = 0, c \neq 0$, Elliptic Asymptotic Forms, Face $\tilde{\Gamma}_1^{(2)}$ ... 81
8.8.4 Case $d = 0, c \neq 0$, Periodic Asymptotic Forms, Face $\tilde{\Gamma}_2^{(2)}$ ... 81

III Isomondromy Deformations

9 *Dmitry V. Artamonov*
Isomonodromic Deformations on Riemann Surfaces 85

9.1 Introduction 85

9.2 The Space of Parameters $\tilde{T}$ 86
9.3 The Description of Bundles with Connections on a Riemann Surface . 86
9.4 Isomonodromic Deformations 87

10 *Mikhail V. Babich*
On Birational Darboux Coordinates of Isomonodromic Deformation Equations Phase Space 91

11 *Yuliya P. Bibilo and Renat R. Gontsov*
On the Malgrange Isomonodromic Deformations of Nonresonant Irregular Systems 95
11.1 Introduction 95
11.2 The Malgrange Isomonodromic Deformation of the Pair (E^0, ∇^0) ... 96
11.3 Specificity of Meromorphic 2×2-Connections 98

12 *Davide Guzzetti*
Critical behavior of P_6 Functions from the Isomonodromy Deformations Approach 101
12.1 Introduction 101
12.2 Behavior of $y(x)$ 102
12.3 Parameterization in Terms of Monodromy Data 104

13 *Alexander Ya. Kazakov*
Isomonodromy Deformation of the Heun Class Equation 107
13.1 Introduction 107
13.2 Gauge Transforms of Linear Differential Equations 108
13.3 Gauge Transforms of the Hypergeometric Class Equations 111
13.4 Gauge Transform of Heun Class Equations 112
13.4.1 Formulation of the Problem 112
13.4.2 Initial System of Equations and Equation *Heunc*2 113
13.4.3 Parameters of the Transformed Equation 114
13.5 Conclusion 116

14 *Vladimir P. Leksin*
Isomonodromy Deformations and Hypergeometric-Type Systems 117
14.1 Schlesinger Families of Fuchsian Systems 117
14.2 Schlesinger Systems 118

14.3 Upper-Triangular Schlesinger Systems 118
14.4 Jordan–Pochhammer Systems 120
14.5 The Basic Result 121

15 *Dmitrii P. Novikov*
A Monodromy Problem Connected with P_6 123
15.1 Preliminaries I 123
15.2 Preliminaries II 124
15.3 Main Result 125
15.4 Example 126

16 *Yousuke Ohyama*
Monodromy Evolving Deformations and Confluent Halphen's Systems 129
16.1 Introduction 129
16.2 Quadratic Systems and Nonassociative Algebras 131
16.3 Monodromy Evolving Deformations 133
16.4 Halphen's Confluent Systems and Monodromy Evolving Deformations 135

17 *Yoshikatsu Sasaki*
On the Gauge Transformation of the Sixth Painlevé Equation 137
17.1 Linearizations of the Sixth Painlevé Equation 137
17.1.1 LODE L_{VI} 138
17.1.2 LODE $\bar{L}_{\mathrm{VI}}$ 139
17.1.3 LODE $\tilde{L}_{\mathrm{VI}}$ 140
17.1.4 Schlesinger System with Symmetric Gauge 142
17.1.5 Schlesinger System with Asymmetric Gauge 143
17.2 Schlesinger Transformation $\tilde{L}_{\mathrm{VI}} \to L_{\mathrm{VI}}$. 143

18 *Ilya V. Vyugin*
Expansions for Solutions of the Schlesinger Equation at a Singular Point 151
18.1 Introduction 151
18.2 Schlesinger Equation and Isomonodromic Deformations 154
18.3 Sketch of the Proof 155

IV Painlevé Property

19 *Pantelis A. Damianou*
Painleve Analysis of Lotka–Volterra Equations 161

20 *Evgenii Gricuk and Valerii Gromak*
Painlevé Test and Briot–Bouquet Systems 165

21 *Valerii Gromak*
Solutions of the Chazy System 167

22 *Yasin Adjabi and Arezki Kessi*
Third-Order Ordinary Differential Equations with the Painlevé Test 171

22.1 Introduction 171

22.2 Simplified Equation 172

22.3 Reduced Equations 174

22.3.1 Leading Order $k = -1$ 174

22.3.2 Leading Order $k = -2$ 181

22.3.3 Leading Order $k = -3$ 182

22.3.4 Leading Order $k = -4$ 183

23 *Ivan P. Martynov, Vyacheslav A. Pronko and Tatsyana K. Andreeva*
Analytic Properties of Solutions of a Class of Third-Order Equations with an Irrational Right-Hand Side 185

V Other Aspects

24 *Yurii V. Brezhnev*
The Sixth Painlevé Transcendent and Uniformizable Orbifolds 193

24.1 Algebraic Solutions of $\mathcal{P}_6$ and Uniformization Theory 193

24.2 On the General Solution to Equation (24.1) 194

24.3 Calculus: Abelian Integrals and Affine (Analytic) Connections 195

25 *Yurii V. Brezhnev*
On Uniformizable Representation for Abelian Integrals 199

25.1 Introduction 199

25.2 Schwarz Equation and Equations on Tori 200

25.3 Holomorphic Elliptic Integrals and Hypergeometric Functions 201
25.3.1 Lemniscate 201
25.3.2 Equi-Anharmonic Curve 203

25.4 Abelian Integrals for Genus $g > 1$ 204
25.4.1 Higher Genera. Examples 205

26 *Rustem N. Garifullin*
Phase Shift for a Special Solution to the Korteweg–de Vries Equation in the Whitham Zone 209

26.1 Introduction 209

26.2 Evaluation of the Phase Shift 210

27 *Valentina A. Golubeva*
Fuchsian Reduction of Differential Equations 213

27.1 Fuchsian Reduction 215
27.1.1 Two Simple Examples 215
27.1.2 A More Complex Example 216

27.2 Two Applications: Astronomy and Relativity Theory 217
27.2.1 Astronomy. A Model of Gaseous Stars 217
27.2.2 Relativity. Gowdy Space-Time 218

27.3 Fuchsian Systems for Feynman Integrals 220

28 *Kohei Iwaki*
The Voros Coefficient and the Parametric Stokes Phenomenon for the Second Painlevé Equation 225

28.1 Introduction 225

28.2 Connection Formula for the Parametric Stokes Phenomenon 226

28.3 Derivation of the Connection Formulas Through the Analysis of the Voros Coefficient of (P_2) 228

29 *Alexander Ya. Kazakov*
Integral Symmetry and the Deformed Hypergeometric Equation 231

30 *Alexander Ya. Kazakov and Sergey Yu. Slavyanov*
Integral Symmetries for Confluent Heun Equations and Symmetries of Painlevé Equation P_5 237

31 *Dmitry Korotkin and Peter Zograf*
From the Tau Function of Painlevé P_6 Equation to Moduli Spaces 241

32 *Yousuke Ohyama*
On particular Solutions of q-Painlevé Equations and q-Hypergeometric Equations 247
32.1 Introduction 247
32.2 q-Difference Equation of the Hypergeometric Type 247
32.3 Hypergeometric Solutions of the q-Painlevé Equations 250

33 *Sergey Yu. Slavyanov*
Derivation of Painlevé Equations by Antiquantization 253

34 *Kouichi Takemura*
Integral Transformation of Heun's Equation and Apparent Singularity 257
34.1 Heun's Equation and Integral Transformation 257
34.2 Apparent Singularity and Integral Representation of Solutions 258
34.3 Elliptical Representation of Heun's Equation and Integral Transformation 259

35 *Vladimir Tsegel'nik*
Painlevé Analysis of Solutions to Some Nonlinear Differential Equations and their Systems Associated with Models of the Random-Matrix Type 263
35.1 Introduction 263
35.2 Model of the Random-Matrix Type with Airy Kernal 264
35.3 System of Differential Equations Associated with the Dyson Process . 264
35.4 Solutions of the Traveling-Wave Form of a Partial Differential Equation 265

36 *Pavlos Xenitidis*
Reductions on the Lattice and Painlevé Equations P_2, P_5, P_6 267
36.1 Introduction 267
36.2 Symmetries of the ABS Equations 268
36.3 Reduction on the Lattice and Discrete Painlevé Equations 269
36.4 Continuous Symmetry Reductions 269

Comments 271

Part I

Plane Power Geometry

Chapter 1

Plane Power Geometry for One ODE and P_1–P_6

Alexander D. Bruno

Abstract. Power Geometry (PG) is the third level of Mathematical Analysis (the first is Classic Analysis, the second is Functional Analysis, and the third is Nonlinear Analysis $\equiv$ PG). It gives algorithms for calculation of asymptotic forms and asymptotic expansions of solutions to one equation (algebraic, ordinary differential, and partial differential) and to a system of such equations. Here we consider such asymptotic expansions of solutions to single ordinary differential equation (ODE) which can be calculated by algorithms of planar Power Geometry, and we point out some applications to Painlevé equations.

Keywords. Order of a Solution, First Approximation of a Solution, Truncated Equation, Asymptotic Expansion.

2010 Mathematics Subject Classification. Primary: 34M55; Secondary: 34E05.

1.1 Statement of the Problem

Let x be independent and y be dependent variables, $x, y \in \mathbb{C}$. A *differential monomial* $a(x, y)$ is a product of an ordinary monomial $cx^{r_1}y^{r_2}$, where $c = \text{const} \in \mathbb{C}$, $(r_1, r_2) \in \mathbb{R}^2$, and a finite number of derivatives of the form $d^l y/dx^l$, $l \in \mathbb{N}$. A sum of differential monomials $f(x, y) = \sum a_i(x, y)$ is called the *differential sum* [1, 2].

The First Problem. Let a differential equation be given

$$f(x, y) = 0, \tag{1.1}$$

where $f(x, y)$ is a differential sum. As $x \to 0$, or as $x \to \infty$, for solutions $y = \varphi(x)$ to the equation (1.1), find all expansions of the form

$$y = c_r x^r + \sum c_s x^s, \quad c_r = \text{const} \in \mathbb{C}, \quad c_r \neq 0, \tag{1.2}$$

where c_s are polynomials in $\log x$, and power exponents $r, s \in \mathbb{C}$, $\omega\text{Re}\, r > \omega\text{Re}\, s$. Here and below $\omega = -1$ if $x \to 0$, and $\omega = 1$ if $x \to \infty$.

This work was supported by RFBR, Grant Nr. 11-01-00023.

The procedure to compute expansions (1.2) consists of three steps:

(1) computation of truncated equations

$$\hat{f}_j^{(d)}(x, y) = 0; \tag{1.3}$$

(2) computation of the first approximations $y = c_r x^r$, $c_r \neq 0$ as solutions to the truncated equation (1.3);

(3) computation of further terms $c_s x^s$ of the expansion (1.2) as solutions to the linear equations

$$\mathcal{L}(x) c_s x^s + \theta_s(x) = 0, \tag{1.4}$$

where operator $\mathcal{L}(x) = \delta \hat{f}_j^{(d)} / \delta y$ on $y = c_r x^r$ and $\delta f / \delta y$ means the first variation or the Freshe/Gatou derivation. If $f(x, y)$ is a polynomial then $\delta f / \delta y = \partial f / \partial y$. The equations (1.4) can be solved successively along increasing $-\omega \operatorname{Re} s$.

1.2 Computation of Truncated Equations

To each differential monomial $a(x, y)$, we assign its (vector) *power exponent* $Q(a) = (q_1, q_2) \in \mathbb{R}^2$ by the following rules:

$$Q(c x^{r_1} y^{r_2}) = (r_1, r_2), \quad Q(d^l y / dx^l) = (-l, 1); \tag{1.5}$$

when differential monomials are multiplied, their power exponents must be added as vectors $Q(a_1 a_2) = Q(a_1) + Q(a_2)$.

The set $\mathbf{S}(f)$ of power exponents $Q(a_i)$ of all differential monomials $a_i(x, y)$ present in the differential sum

$$f(x, y) = \sum a_i(x, y) \text{ over } Q_i \in \mathbf{S}$$

is called the *support of the sum* $f(x, y)$. Obviously, $\mathbf{S}(f) \in \mathbb{R}^2$. The convex hull $\Gamma(f)$ of the support $\mathbf{S}(f)$ is called the *polygon of the sum* $f(x, y)$. The boundary $\partial\Gamma(f)$ of the polygon $\Gamma(f)$ consists of the vertices $\Gamma_j^{(0)}$ and the edges $\Gamma_j^{(1)}$. They are called *(generalized) faces* $\Gamma_j^{(d)}$, where the upper index indicates the dimension of the face and the lower one is its number. Each face $\Gamma_j^{(d)}$ corresponds to the *truncated sum*

$$\hat{f}_j^{(d)}(x, y) = \sum a_i(x, y) \quad \text{over} \quad Q(a_i) \in \Gamma_j^{(d)} \cap \mathbf{S}(f).$$

Example. Consider the third Painlevé equation P_3, multiplied by its common divisor,

$$f(x, y) \stackrel{def}{=} -xyy'' + xy'^2 - yy' + ay^3 + by + cxy^4 + dx = 0, \tag{1.6}$$

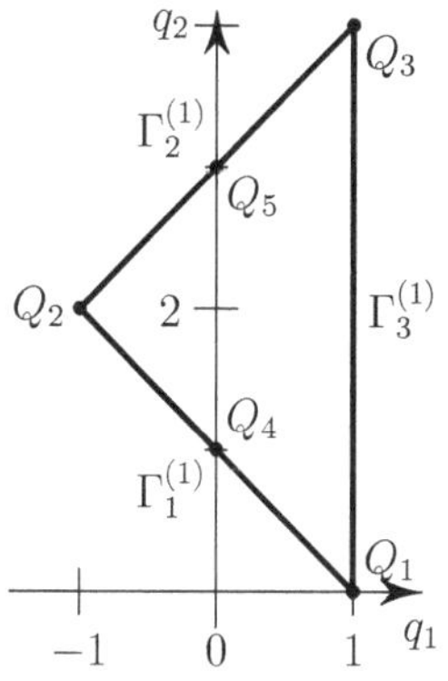

Figure 1.1. Support S and polygon Γ of the equation (1.6).

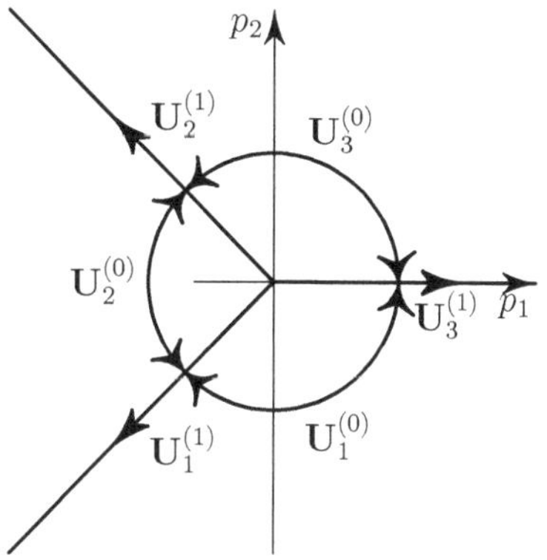

Figure 1.2. Normal cones $U_j^{(d)}$ for the polygon Γ of Figure 1.1.

assuming the complex parameters $a, b, c, d \neq 0$. Here the first three differential monomials have the same power exponent $Q_2 = (-1, 2)$, then $Q_5 = (0, 3)$, $Q_4 = (0, 1)$, $Q_3 = (1, 4)$, $Q_1 = (1, 0)$. They are shown in Figure 1.1 in coordinates q_1, q_2. Their convex hull $\Gamma(f)$ is the triangle with three vertices $\Gamma_1^{(0)} = Q_1$, $\Gamma_2^{(0)} = Q_2$, $\Gamma_3^{(0)} = Q_3$, and with three edges $\Gamma_1^{(1)}, \Gamma_2^{(1)}, \Gamma_3^{(1)}$. The vertex $\Gamma_2^{(0)} = Q_2$ corresponds to the truncation

$$\hat{f}_2^{(0)}(x, y) = -xyy'' + xy'^2 - yy',$$

and the edge $\Gamma_1^{(1)}$ corresponds to the truncation

$$\hat{f}_1^{(1)}(x, y) = \hat{f}_2^{(0)}(x, y) + by + dx. \tag{1.7}$$

■

Let the plane $\mathbb{R}_*^2$ be dual to the plane $\mathbb{R}^2$ such that for $P = (p_1, p_2) \in \mathbb{R}_*^2$ and $Q = (q_1, q_2) \in \mathbb{R}^2$ the scalar product $\langle P, Q \rangle \stackrel{\text{def}}{=} p_1 q_1 + p_2 q_2$ is defined. Each face $\Gamma_j^{(d)}$ in $\mathbb{R}_*^2$ corresponds to its own *normal cone* $\mathbf{U}_j^{(d)}$, formed by the outward normal vectors P to the face $\Gamma_j^{(d)}$. For the edge $\Gamma_j^{(1)}$, the normal cone $\mathbf{U}_j^{(1)}$ is the ray orthogonal to the edge $\Gamma_j^{(1)}$ and directed outward the polygon $\Gamma(f)$. For the vertex $\Gamma_j^{(0)}$, the normal cone $\mathbf{U}_j^{(0)}$ is the open sector (angle) in the plane $\mathbb{R}_*^2$ with the vertex at the origin $P = 0$ and limited by the rays which are the normal cones of the edges adjacent to the vertex $\Gamma_j^{(0)}$.

Example. For equation (1.4), the normal cones $\mathbf{U}_j^{(d)}$ of the faces $\Gamma_j^{(d)}$ are shown in Figure 1.2. ■

Thus, to each face $\Gamma_j^{(d)}$ there correspond the normal cone $\mathbf{U}_j^{(d)}$ in the plane $\mathbb{R}_*^2$ and the truncated equation $\hat{f}_j^{(d)}(x, y) = 0$.

Theorem 1.1 ([1]). *If the expansion* (1.2) *satisfies the equation* (1.1), *and* $\omega(1, \operatorname{Re}\, r) \in \mathbf{U}_j^{(d)}$, *then the truncation* $y = c_r x^r$ *of the solution* (1.2) *is the solution to the truncated equation* $\hat{f}_j^{(d)}(x, y) = 0$.

As truncated equations are quasi-homogeneous, it is not difficult to find their power solutions. Hence, to find all truncated solutions $y = c_r x^r$ to equation (1.1), we need to compute the support $\mathbf{S}(f)$, the polygon $\Gamma(f)$, all its faces $\Gamma_j^{(d)}$, and their normal cones $\mathbf{U}_j^{(d)}$. Then for each truncated equation $\hat{f}_j^{(d)}(x, y) = 0$, we need to find all its solutions $y = c_r x^r$ which have one of the vectors $\pm(1, \operatorname{Re}\, r)$ lying in the normal cone $\mathbf{U}_j^{(d)}$.

1.3 Computation of Expansions of Solutions to the Initial Equation (1.1)

If the solution to a truncated equation is $y = c_r x^r$, then the operator

$$\mathcal{L} = \delta \hat{f} / \delta y \text{ on } y = c_r x^r$$

is the Euler one:

$$\mathcal{L} = x^\beta \sum_{l=0}^{n} b_l x^l \frac{d^l}{dx^l},$$

where $b_l = \text{const} \in \mathbb{C}$. It has the *characteristic polynomial*

$$\nu(k) = x^{-\beta-k} \mathcal{L}(x) x^k. \tag{1.8}$$

Roots k_j of the polynomial $\nu(k)$ are *critical numbers* if $\omega \operatorname{Re} k_j \leqslant \omega \operatorname{Re} r$.

If the truncated solution $y = c_r x^r$ has no critical numbers, then the initial equation (1.1) has the unique solution of the form (1.2), where c_s are constants. Such solutions belong to **Type 1** (*power expansions*). For them $|\operatorname{Im}\, s / \operatorname{Re}\, s| <$ const. If the order $\varkappa$ of the characteristic polynomial $\nu(k)$ equals to the maximal order n of the derivatives in equation (1.1), then the expansion (1.2) converges [1–4]. If $\varkappa < n$, then the power expansion (1.2) can diverge and can be continued by an exponential expansion of Type 6 (see below) as a solution to equation (1.1).

If the truncated solution $y = c_r x^r$ has critical numbers, then we obtain solutions (1.2) of **Type 2,** where c_r is a constant, but c_s are polynomials in $\log x$ (*power-logarithmic expansions*).

1.4 Extension of the Class of Solutions

We will consider such a generalization of the power function cx^r which preserves its main properties. The real number

$$p_\omega(\varphi(x)) = \omega \varlimsup_{x^\omega \to \infty} \frac{\log|\varphi(x)|}{\omega \log|x|},$$

where $\arg x = \text{const} \in [0, 2\pi)$, is called the *order* of the function $\varphi(x)$ on the ray when $x \to 0$ or $x \to \infty$. The order $p_\omega(\varphi)$ is not defined on the ray $\arg x = \text{const}$, where the limit point $x = 0$ or $x = \infty$ is a point of accumulation of poles of the function $\varphi(x)$.

The Second Problem. As $x \to 0$ or as $x \to \infty$, for solutions $y = \varphi(x)$ to equation (1.1), find all expansions of the form

$$y = c_r(x)x^r + \sum c_s(x)x^s, \quad r, s \in \mathbb{C}, \omega \operatorname{Re} r > \omega \operatorname{Re} s, \tag{1.9}$$

where $c_r(x)$ and $c_s(x)$ are either constants or functions with property $p_\pm(c_r(x)) = p_\pm(c_s(x)) = 0$ and $p_\pm(d^l c_r(x)/dx^l) = \pm(d^l c_s(x)/dx^l) = -l$ for $l = 1, \dots, n$, where n is the maximal order of derivations in the initial equation (1.1).

Due to the rule (1.5), the order of a differential monomial $a_i(x, y)$ on a solution (1.9) is the same as for solution (1.2). Thus the plane Power Geometry works, and Theorem 1.1 is true for the solutions (1.9). Now the truncated equations $\hat{f}_j^{(d)}(x, y) = 0$ are the same as before, but the class of their desired solutions $y = c_r(x)x^r$ is wider.

1.5 Solution of Truncated Equations

We can simplify a truncated equation $\hat{f}_j^{(d)} = 0$ by means of *power transformations*

$$\begin{aligned} \log x &= \alpha_{11} \log u + \alpha_{12} \log v, \\ \log y &= \alpha_{21} \log u + \alpha_{22} \log v, \end{aligned} \qquad \alpha = \begin{pmatrix} \alpha_{11} & \alpha_{12} \\ \alpha_{21} & \alpha_{22} \end{pmatrix}, \quad \alpha_{ij} \in \mathbb{R}, \ \det \alpha \neq 0,$$

which induce the affine geometry in the plane of power exponents $Q = (q_1, q_2)$: $x^{q_1} y^{q_2} = u^{\tilde{q}_1} v^{\tilde{q}_2}$, where $(\tilde{q}_1, \tilde{q}_2) = (q_1, q_2)\alpha$. Moreover, we can further simplify the equations by means of the *logarithmic transformation*

$$\xi = \ln x$$

and by the change

$$z = x^i.$$

Using them, we can find solutions to a truncated equation in the form $c_r(x)x^r$, $r =$ const, where

Case a: $c_r = \text{const} \in \mathbb{C}$;

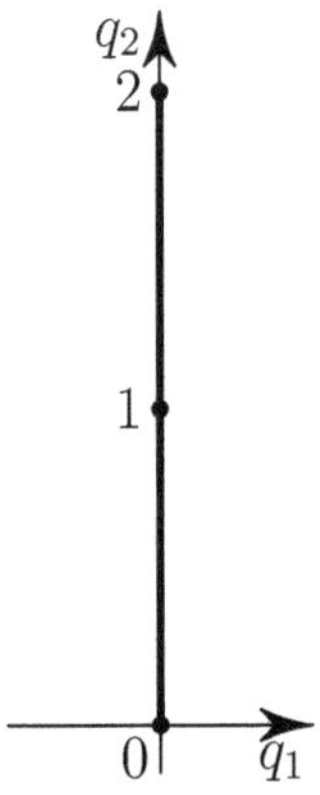

Figure 1.3. Support and polygon for the equation (1.10).

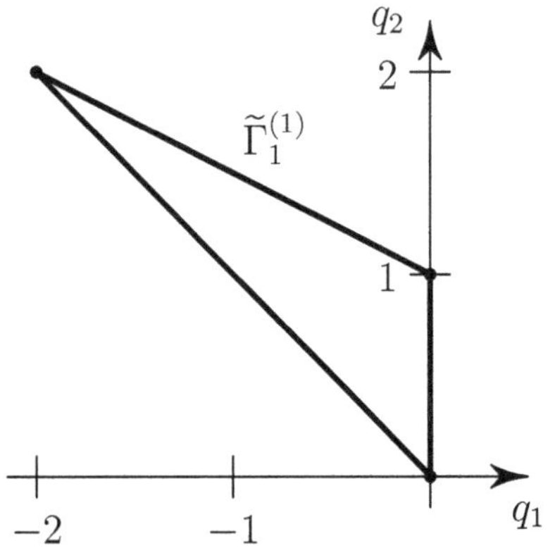

Figure 1.4. Support and polygon of the equation, obtained from (1.10) after logarithmic transformation.

Case b: $c_r(x)$ is a power expansion in $\log x$;
Case c: $c_r(x)$ is a power expansion in x^i.

Example. For the truncated equation (1.7) corresponding to the edge $\Gamma_1^{(1)}$ with the normal vector $-(1,1)$, we make the power transformation

$$\begin{aligned} \ln x &= \ln u \\ \ln y &= \ln u + \ln v, \end{aligned} \qquad \alpha = \begin{pmatrix} 1 & 0 \\ 1 & 1 \end{pmatrix},$$

i.e., $x = u$, $y = uv$. Since $y' = xv' + v$, $y'' = xv'' + 2v'$, then, canceling x and collecting similar terms, the equation (1.7) takes the form

$$-x^2 v v'' + x^2 v'^2 - x v v' + bv + d = 0. \tag{1.10}$$

If $b \neq 0$ it has the constant solution $v = -d/b$ (Case a). Its support consists of three points $\widetilde{Q}_1 = (0,2)$, $\widetilde{Q}_2 = (0,1)$, $\widetilde{Q}_3 = 0$ placed in the axis $\tilde{q}_1 = 0$ (see Figure 1.3).

Now we make the logarithmic transformation $\xi = \log x$. Since $v' = \dot{v}/x$, $v'' = (\ddot{v} - \dot{v})/x^2$, where $\dot{} = d/d\xi$, then, collecting similar terms, equation (1.10) takes the form $-v\ddot{v} + \dot{v}^2 + bv + d = 0$. Its support and polygon are shown in Figure 1.4.

Applying the technique described above to this equation, we obtain the expansion of its solutions $v(\xi) = -(b/2)\xi^2 + \tilde{c}\xi + \sum_{k=0}^{\infty} c_k \xi^{-k}$, where $\tilde{c}$ is an arbitrary constant, and the constants c_k are uniquely determined. In the original variables, we obtain the family of nonpower asymptotic forms of the solutions to the original equation (1.6)

$$y \sim x\left[-(b/2)(\log x)^2 + \tilde{c}\log x + \sum_{k=0}^{\infty} c_k (\log x)^{-k}\right], \quad x \to 0 \text{ (Case b)}.$$

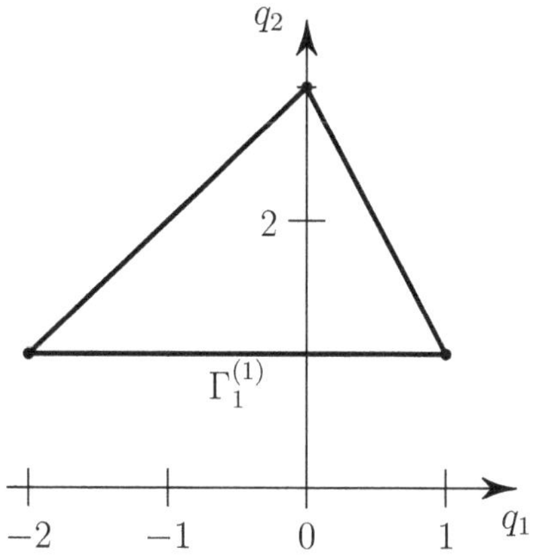

Figure 1.5. Support $S(f)$ and polygon $\Gamma(f)$ for the equation P_2 with $a = 0$.

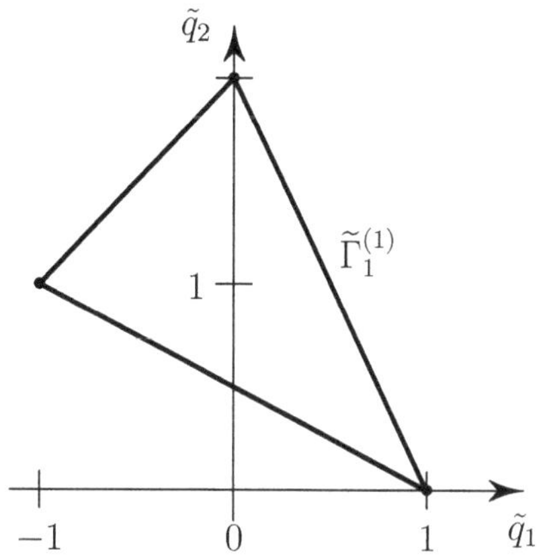

Figure 1.6. Support and polygon of equation (1.11).

According to Example 3 of [5], if $b \neq 0$, the expansion $v(\xi)$ is indeed the polynomial $v(\xi) = -b(\xi + \tilde{c})^2 - d/(2b)$.

After substitution $t = x^i$, equation (1.10) takes the form

$$v\left(t^2\frac{d^2v}{dt^2} + t\frac{dv}{dt}\right) - t^2\left(\frac{dv}{dt}\right)^2 + bv + d = 0.$$

Support of the equation is the same as for (1.10) (Figure 1.3). If $b = 0$ and $d = 4$, it has the solution

$$v = t + t^{-1} = x^i + x^{-i} = e^{i\log x} + e^{-i\log x} = 2\cos\log x \quad \text{(Case c).} \qquad \blacksquare$$

If the truncated equation corresponds to a horizontal edge, then we make the logarithmic transformation

$$\eta = \log y$$

and obtain solutions to the truncated equation in the form

$$y = C\exp\varphi(x) \quad \textbf{(Case d),}$$

where $\varphi'(x)$ is a power expansion in x and C is an arbitrary constant.

Example. The second Painlevé equation P_2

$$f(x, y) \stackrel{\text{def}}{=} -y'' + 2y^3 + xy + a = 0$$

with parameter $a = 0$ has the support $\mathbf{S}(f)$, and the polygon $\Gamma(f)$ shown in Figure 1.5. $\Gamma(f)$ has the horizontal edge $\Gamma_1^{(1)}$ with the truncated equation

$$\hat{f}_1^{(1)}(x,y) \stackrel{\text{def}}{=} -y'' + xy = 0.$$

After transformations $\log y = \eta$ and $\eta' = \zeta$ we obtain the equation

$$h(x,\zeta) \stackrel{\text{def}}{=} -(\zeta' + \zeta^2) + x = 0. \tag{1.11}$$

Its support and polygon are shown in Figure 1.6. The edge $\widetilde{\Gamma}_1^{(1)}$ corresponds to the truncated equation

$$\hat{h}_1^{(1)}(x,\zeta) = -\zeta^2 + x = 0$$

with two solutions $\zeta = \pm\sqrt{x}$. They correspond to two solutions to equation (1.11):

$$\zeta = (-1)^l x^{1/2} - \frac{1}{4x} + \sum_{k=2}^{\infty} c_{kl} x^{(1-3k)/2} \stackrel{\text{def}}{=} \varphi'(x), \quad l = 1,2,$$

where c_{kl} are uniquely determined constants. Returning to initial variables, we obtain the truncated solutions of Case d. ■

1.6 Types of Expansions

Expansions (1.9) are obtained via successive solution of equations (1.4). If the truncated solution belong to Case b, then we obtain solutions (1.9) of **Type 3,** where $c_r(x)$ and $c_s(x)$ are power series in decreasing powers of $\log x$ (*complicated expansions*). The truncated solution $y = c_r(x)x^r$ has its characteristic polynomial $\lambda(k)$. Absence of critical numbers is sufficient for existence and uniqueness of the expansion [5].

If the truncated solution belong to Case c, then we obtain solution (1.9) of **Type 4,** where r and s are real, but $c_r(x)$ and $c_s(x)$ are series in powers of x^i, and $c_r(x)$ contains finite number of such terms (*semiexotic expansions*), or **Type 5,** where r, s and $c_s(x)$ are as in type 4, but $c_r(x)$ is a sum of infinitely many powers of x^i and they are bounded from one side (*exotic expansions*).

Types 4 and 5 differ from Types 1 and 2 by the form of $c_r(x)$ and by the absence of the restriction $|\text{Im } s/\text{Re } s| < \text{const}$. Expansions of Types 1–5 and the algorithms of their computation were described in [1,2]. Now we introduce a new type:

Type 6. *Exponential expansions*

$$y = b_0(x) + Ce^{\varphi(x)} + \sum_{k=2}^{\infty} b_k(x)C^k e^{k\varphi(x)}, \tag{1.12}$$

where $b_0(x)$, $b_k(x)$ and $\varphi'(x)$ are power expansions and C is an arbitrary constant. To the initial part $b_0(x) + Ce^{\varphi(x)}$ there corresponds its characteristic polynomial $\mu(k)$. Absence of critical numbers k_j is enough for teh existence of the expansion (1.12). The expansions $\varphi(x)$ and $b_k(x)$ are unique [6].

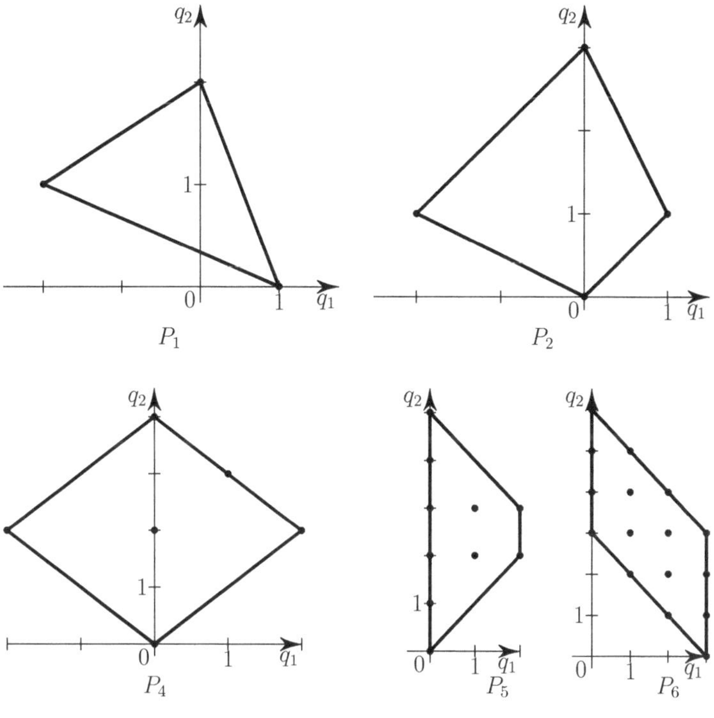

Figure 1.7. Supports and polygons for equations P_1, P_2, P_4–P_6.

1.7 Painlevé Equations P_l

Supports and polygons of Painlevé equations P_1, P_2, P_4, P_5, P_6 are shown for generic cases in Figure 1.7. Expansions of solutions of Type 1 exist for all P_1–P_6, of Types 2–5 only for P_3, P_5, P_6, and of Type 6 only for P_1–P_5, and only near infinity [2,5,7–12]. For some other applications of Power Geometry, see [13–19].

Bibliography

[1] A. D. Bruno, Asymptotics and expansions of solutions to an ordinary differential equation, *Russian Mathem. Surveys*, **59**(3): 429–480, 2004 (= *UMN* **59**(3): 31–80, 2004).

[2] A. D. Bruno and I. V. Goryuchkina, Asymptotic expansions of solutions to the sixth Painleve equation, *Transactions of Moscow Math. Soc.*, **71**: 1–105, 2010 (= *Tr. MMO*, **71**: 6–118, 2010).

[3] A. D. Bruno and I. V. Goryuchkina, On the convergence of a formal solution to an ordinary differential equation, *Doklady Mathematics*, **81**(3): 358–361, 2010 (= *DAN* **432**(2): 151–154, 2010).

[4] I. V. Goryuchkina, Convergence of formal solution to an ODE, *Chapter 3, this volume*.

[5] A. D. Bruno, *On complicated expansions of solutions to ODE*, preprint of KIAM Nr. 15, Moscow, 2011, 26 p. (in Russian); available from the Keldysh Institute of Applied Mathematics of RAS: `http://library.keldysh.ru/preprints/Default.asp?lg=e`.

[6] A. D. Bruno, Exponential expansions of solutions to an ordinary differential equation, *Doklady Mathem.*, **85**(2), 2012 (= *DAN* **443**(5), 2012).

[7] A. D. Bruno and V. Yu. Petrovich, *Singularities of solutions to the first Painlevé equation*, preprint of KIAM Nr. 75, Moscow, 2004, 17 p. (in Russian); available from the Keldysh Institute of Applied Mathematics of RAS: `http://library.keldysh.ru/preprints/Default.asp?lg=e`.

[8] A. D. Bruno and A. V. Gridnev, *Power and exponential expansions of solutions to the third Painlevé equation*, Preprint of KIAM Nr. 51, Moscow, 2003, 32 pp. (in Russian); available from the Keldysh Institute of Applied Mathematics of RAS: `http://library.keldysh.ru/preprints/Default.asp?lg=e`.

[9] A. D. Bruno and A. V. Gridnev, *Nonpower expansions of solutions to the third Painleve equation*, Preprint of KIAM Nr. 10, Moscow, 2010, 18 p. (in Russian); available from the Keldysh Institute of Applied Mathematics of RAS: `http://library.keldysh.ru/preprints/Default.asp?lg=e`.

[10] A. D. Bruno, Power Geometry as a new calculus, *Analysis and Applications*, in: H. G. W. Begehr, R. P. Gilbert and M. W. Wong (eds.), *ISAAC 2001*, pp. 51–71, Kluwer Academic Publishers Dordrecht/Boston/London, 2003.

[11] A.V. Parusnikova, Asymptotic expansions of solutions to P_5, *Chapter 5, this volume.*

[12] I. V. Goruchkina, Asymptotic expansions and forms of solutions to P_6, *Chapter 4, this volume.*

[13] A. D. Bruno and N. A. Kudryashov, Expansions of solutions to the equation P_1^2 by algorithms of Power Geometry, *Ukrainian Mathematical Bulletin*, **6**(3): 311–337, 2009.

[14] A. D. Bruno, *Power Geometry in Algebraic and Differential Equations,* Elsevier Science, Amsterdam, 2000, 385 p. (= Fizmatlit, Moscow, 1998, 288 p.).

[15] A. D. Bruno and T. V. Shadrina, Axisymmetric boundary layer on a needle, *Transactions of Moscow Math. Soc.,* **68**: 201–259, 2007 (= *Tr. MMO*, **68**: 224–287, 2007).

[16] A. D. Bruno, *Power Geometry as a new Mathematics*, preprint of KIAM Nr. 28, Moscow, 2007, 26 p.; available from the Keldysh Institute of Applied Mathematics of RAS: `http://library.keldysh.ru/preprints/Default.asp?lg=e`.

[17] A.D. Bruno, Power Geometry in nonlinear partial differential equations, *Ukrainean Mathem. Bulletin,* **5**(1): 32–45, 2008.

[18] A. D. Bruno and V. F. Edneral, Algorithmic analysis of local integrability, *Doklady Mathematics,* **79**(1): 48–52, 2009 (= *DAN,* **424**(3): 299–303, 2009).

[19] A. D. Bruno, Analysis of singularities and of integrability of ODE's by algorithms of Power Geometry, *Banach Center Publications*, **94**: 83–98, 2011.

Author Information

Alexander D. Bruno, Russian Academy of Sciences, Keldysh Institute of Applied Mathematics (KIAM), Moscow, Russia.
E-mail: `abruno@keldysh.ru`

Chapter 2

New Simple Exact Solutions to Equation P_6

Alexander B. Batkhin and Natalia V. Batkhina

Abstract. We compute exact solutions to the sixth Painlevé equation (P_6) in the form of finite sums of power functions with rational power exponents. Our method essentially uses algorithms of Power Geometry for computing power expansions of solutions to an ordinary differential equation and computer algebra. New exact solutions are obtained.

Keywords. Sixth Painlevé Equation, Power Geometry, Exact Solution, Algebraic Solution.

2010 Mathematics Subject Classification. 34M55.

2.1 Introduction

Let us recall the basic definitions and results of Power Geometry to simplify the explanation of the method for constructing an exact solution to differential equations.

2.1.1 Power Geometry Essentials

Let x be independent and y be dependent variables, $x, y \in \mathbb{C}$. A *differential monomial* $a(x, y)$ is a product of an ordinary monomial $cx^{r_1}y^{r_2}$, where $c = \text{const} \in \mathbb{C}$, $(r_1, r_2) \in \mathbb{R}^2$, and a finite number of derivatives of the form $d^l y/dx^l$, $l \in \mathbb{N}$. A sum of differential monomials

$$f(x, y) = \sum a_i(x, y) \tag{2.1}$$

is called the *differential sum*. Let a differential equation be given

$$f(x, y) = 0, \tag{2.2}$$

where $f(x, y)$ is a differential sum. To each differential monomial $a(x, y)$, we assign its (vector) *power exponent* $Q(a) = (q_1, q_2) \in \mathbb{R}^2$ by the following rules:

$$Q\left(cx^{r_1}y^{r_2}\right) = (r_1, r_2); \quad Q\left(d^l y/dx^l\right) = (-l, 1);$$

when differential monomials are multiplied, their power exponents must be added as vectors $Q(a_1a_2) = Q(a_1) + Q(a_2)$. The set $\mathbf{S}(f)$ of power exponents $Q(a_i)$ of all

This work was partially supported by RFBR, Grant Nr. 11-01-00023.

differential monomials $a_i(x, y)$ present in the differential sum is called the *support of the sum* $f(x, y)$. Obviously, $\mathbf{S}(f) \in \mathbb{R}^2$.

The convex hull $\Gamma(f)$ of the support $\mathbf{S}(f)$ is called the *Newton–Bruno polygon* of the sum $f(x, y)$. The boundary $\partial\Gamma(f)$ of the polygon $\Gamma(f)$ consists of the faces $\Gamma_j^{(2)}$, the edges $\Gamma_j^{(1)}$ and the vertices $\Gamma_j^{(0)}$. They are called (generalized) *faces* $\Gamma_j^{(d)}$ where the upper index indicates the dimension of the face and the lower one is its number.

Each face $\Gamma_j^{(d)}$ corresponds to the *truncated sum*

$$\hat{f}_j^{(d)}(x, y) = \sum a_i(x, y) \text{ over } Q(a_i) \in \Gamma_j^{(d)} \cap \mathbf{S}(f). \tag{2.3}$$

Let the plane $\mathbb{R}_*^2$ be dual to the plane $\mathbb{R}^2$ such that for $P = (p_1, p_2) \in \mathbb{R}_*^2$ and $Q = (q_1, q_2) \in \mathbb{R}^2$, the scalar product

$$\langle P, Q \rangle \stackrel{\text{def}}{=} p_1 q_1 + p_2 q_2$$

is defined. Each face $\Gamma_j^{(d)}$ in $\mathbb{R}_*^2$ corresponds to its own *normal cone* $\mathbf{U}_j^{(d)}$ formed by the outward normal vectors P to the face $\Gamma_j^{(d)}$.

Each face $\Gamma_j^{(d)}$ corresponds to the normal cone $\mathbf{U}_j^{(d)}$ in the plane $\mathbb{R}_*^2$ and to the truncated equation

$$\hat{f}_j^{(d)}(x, y) = 0. \tag{2.4}$$

Theorem 2.1 ([1, Chap. VI, Theorem 1.1]). *If the power expansion* $y = c_r x^r + \sum c_s x^s$, $c_r = \text{const} \in \mathbb{C}$, $c_r \neq 0$, *satisfies the equation* (2.2), *and* $\omega(1, r) \in \mathbf{U}_j^{(d)}$, *where* $\omega = -1$, *if* $x \to 0$ *or* $\omega = 1$, *if* $x \to \infty$ *then the truncation* $y = c_r x^r$ *of the power expansion is the solution to the truncated equation* (2.4).

As truncated equations are quasi–homogeneous, it is not difficult to find its power solutions. Hence, to find all truncated solutions $y = c_r x^r$ to the equation (2.2), we need to compute the support $\mathbf{S}(f)$, the polygon $\Gamma(f)$, all its faces $\Gamma_j^{(d)}$, and their normal cones $\mathbf{U}_j^{(d)}$. For each truncated equation $\hat{f}_j^{(d)}(x, y) = 0$, we need to find all its solutions $y = c_r x^r$ which have one of the vectors $\pm(1, r)$ lying in the normal cone $\mathbf{U}_j^{(d)}$.

2.1.2 Matching "Heads" and "Tails" of Expansions

The modification of the method for computing exact solutions, proposed in [3] for finding exact solutions to N. Kovalewski equations, is used.

This method is based on the fitting of two power series expansions near the origin and at infinity and getting conditions on the coefficients of expansions in the form of a system of algebraic equations. It is possible to get the exact solution in the form of finite sum of power functions with rational degrees by solving this system of equations.

The modification of the method consists of the fact that, using the form of asymptotic expansions of solutions to the equation P_6 at the origin and at infinity, the template of an exact solution with undefined coefficients is composed. After substituting such a solution into the equation P_6 one can obtain the system of algebraic equations for undefined coefficients of exact solution and parameters of the equation. The system obtained is solved with the help of a computer algebra system using a Gröbner basis [4].

2.2 Constructing the Template of an Exact Solution

It is well known fact that Painlevé equations in general and P_6 in particular have a lot of different applications especially in mathematical physics. P_6 is

$$y'' = \left(\frac{1}{y} + \frac{1}{y-1} + \frac{1}{y-x}\right)\frac{(y')^2}{2} - \left(\frac{1}{x} + \frac{1}{x-1} + \frac{1}{y-x}\right)y' + \frac{y(y-1)(y-x)}{x^2(x-1)^2}\left[\alpha + \beta\frac{x}{y^2} + \gamma\frac{x-1}{(y-1)^2} + \delta\frac{x(x-1)}{(y-x)^2}\right],$$

where α, β, γ, δ are complex parameters, x and y are complex independent and dependent variables correspondingly, $y' = dy/dx$.

P_6 is invariant under the finite group of Bäcklund transformations of order 24 with 3 generators. Let $y(x) = y(x;\alpha,\beta,\gamma,\delta)$ be a solution of P_6 for definite values of equation's parameters. The transformation $T_j : y \to y_j$ determines the new solution

(a) $y_1(x;-\beta,-\alpha,\gamma,\delta) = 1/y(1/x)$;

(b) $y_2(x;-\beta,-\gamma,\alpha,\delta) = 1 - 1/y(1/(1-x))$;

(c) $y_3(x;-\beta,-\alpha,1/2-\delta,1/2-\gamma) = x/y(x)$.

Knowledge of the exact solutions of the equation P_6 for definite values of equation's parameters α, β, γ, δ is valuable for applications.

Some elementary exact solutions in the form of algebraic functions were obtained with the help of Bäcklund transformations (see [2, § 48]).

The choice of the form of the exact solution is determined with the help of Power Geometry algorithms (see [5, Ch. 1]). One can write the equation P_6 as the sum of differential monomials

$$\begin{aligned} f(x,y) \stackrel{\text{def}}{=} {} & 2y''x^2(x-1)^2y(y-1)(y-x) - (y')^2\big[\,x^2(x-1)^2(y-1)(y-x) \\ & + x^2(x-1)^2y(y-x) + x^2(x-1)^2y(y-1)\,\big] \\ & + 2y'\big[\,x(x-1)^2y(y-1)(y-x) + x^2(x-1)y(y-1)(y-x) \\ & + x^2(x-1)^2y(y-1)\,\big] - \big[\,2\alpha y^2(y-1)^2(y-x)^2 \\ & + 2\beta x(y-1)^2(y-x)^2 + 2\gamma(x-1)y^2(y-x)^2 \\ & + 2\delta x(x-1)y^2(y-1)^2\,\big] = 0, \end{aligned}$$

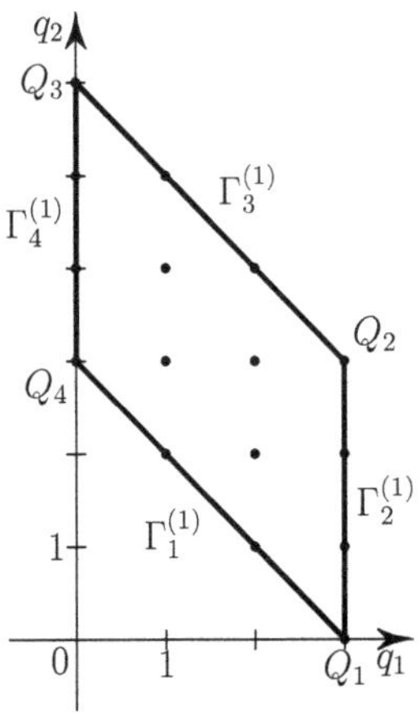

Figure 2.1. Support and polygon for P_6

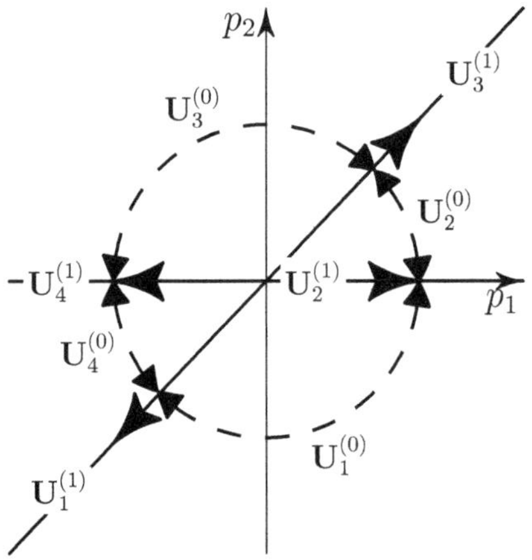

Figure 2.2. Normal cones for P_6

then compute the support $\mathbf{S}(f)$, the polygon $\Gamma(f)$ of this sum, and normal cones (see Figures 2.1 and 2.2, taken from [5]).

The power expansion of a solution near the origin is a power series of increasing power exponents, and the power expansion of a solution near infinity is a power series of decreasing power exponents. Therefore the matching condition may be fulfilled if the degree of the initial expansion term of the expansion near the origin are less then the degree of the initial expansion term of the expansion near infinity.

The initial expansion terms of expansions corresponding to the edges $\Gamma_4^{(1)}$ and $\Gamma_3^{(1)}$ have power exponents equal to 0 and 1, respectively. The asymptotic expansions corresponding to the edges Q_4 and Q_2 were computed in Theorem 2.2.1 and in § 5.1 [5] and are of the form

$$\mathcal{A}_0 : c_r x^r + \sum c_s x^s, \; 0 < \operatorname{Re} r < 1, \; s \in \{r + lr + m(1-r)\}, \tag{2.5}$$

$$\mathcal{A}_\infty : c_r x^r + \sum c_s x^s, \; 0 < \operatorname{Re} r < 1, \; s \in \{r - lr - m(1-r)\}, \tag{2.6}$$

where $l, m \geqslant 0, l + m > 0, l, m \in \mathbb{Z}$.

Thus there are only four pairs of power expansions at the origin and at infinity that are suitable for the matching procedure:

(a) the edge $\Gamma_4^{(1)}$ and the edge $\Gamma_3^{(1)}$;

(b) the vertex Q_4 and the vertex Q_2;

(c) the vertex Q_4 and the edge $\Gamma_3^{(1)}$;

(d) the edge $\Gamma_4^{(1)}$ and the vertex Q_2.

Therefore the exact solutions should be sought in the form

$$y(x) = \sum_{k=0}^{l} a_k x^{k/l}, \quad l \in \mathbb{N}. \tag{2.7}$$

2.3 Results

2.3.1 Known Exact Solutions to P_6

There is an elementary solution to P_6 of the form $y(x) = \varkappa x^n$ [2, § 48]. Algebraic solutions that determine algebraic curves with different values of genus were computed by Dubrovin and Mazzocco (2000) [6], Gromak et al. (2002) [2], Hitchin (2003) [7], and Boalch (2005, 2006) [8]. For the full list of algebraic solutions and their classification, see [9].

2.3.2 Computed Solutions

If all $a_k = 0$ for $k = 1, \dots, l-1$, then (2.7) gives the linear solution of the equation P_6. The following table summarizes the families of linear solutions for different values of parameters of the equation P_6.

Sol. No	Parameters α	β	γ	δ	Free params
$ax+1$	$1/2$	$-(a+1)^2/2$	0	$a - a^2/2$	a
$ax+1-a$	γ/a^2	$-1/2$	γ	$1/2 - \gamma(1-1/a)^2$	a, γ
$x+b$	$1/2$	$-(b+1)^2/2$	$(b-1)^2/2$	$1/2$	b

Here we present the list of solutions of the form (2.7) for $l = 2, \dots, 6$. There is an elementary solution $y = ax^n$, $n \in \mathbb{C}$ to the equation P_6 for $\alpha = \beta = 0$, $\gamma = n^2/2$, $\delta = n(2-n)/2$; therefore we list only those solutions of the form (2.7) for which at least more than one coefficient $a_k \neq 0$. The value of the denominator of power in template solution (2.7) is shown in the upper left corner of the table.

$l=2$	Parameters				Coefficients		
Sol. Nr.	α	β	γ	δ	a_0	a_1	a_2
2-1	$1/2$	0	$-5/8$	$1/8$	$1/2$	$\pm 1/2$	0
2-2	$1/2$	0	$9/8$	$3/8$	0	$\pm 1/2$	$1/2$
2-3	$1/32$	$-25/32$	$9/32$	$15/32$	-2	± 3	0
2-4	$1/32$	$-25/32$	$1/32$	$7/32$	0	± 3	-2

One parameter family is: $y(x) = a_2(1 \mp \sqrt{x})^2 \pm \sqrt{x}$ for P_6 parameters values

$$\alpha = \frac{1}{8a_2^2}, \beta = -\frac{(4a_2-1)^2}{8a2^2}, \gamma = \frac{(a_2-1)^2}{8a_2^2}, \delta = \frac{3a_2^2+2a_2-1}{8a_2^2}.$$

Pairs of solutions (2-1, 2-2) and (2-3, 2-4) are connected by the Bäcklund transformation

$$T_4 : y_4(x;\alpha,\beta,-\delta+1/2,-\gamma+1/2) = xy(1/x;\alpha,\beta,\gamma,\delta).$$

$l=3$	Parameters				Coefficients			
Sol. Nr.	α	β	γ	δ	a_0	a_1	a_2	a_3
3-1	$1/18$	$-2/9$	0	$1/2$	0	-1	-1	0
3-2	$2/9$	$-1/18$	$1/2$	0	0	$1/2$	$1/2$	0
3-3	$2/9$	$-1/18$	$1/2$	0	0	$-(1+z_0)/2$	$z_0/2$	0
3-4	$1/2$	0	$2/9$	$-7/18$	$1/3$	$1/3$	$1/3$	0
3-5	$1/2$	0	$2/9$	$-7/18$	$1/3$	$-(1+z_0)/3$	$z_0/3$	0
3-6	$1/2$	0	$8/9$	$5/18$	0	$1/3$	$1/3$	$1/3$
3-7	$1/2$	0	$8/9$	$5/18$	0	$-(1+z_0)/3$	$z_0/3$	$1/3$
3-8	$1/18$	$-8/9$	0	$1/2$	1	1	1	1
3-9	$1/18$	$-8/9$	0	$1/2$	1	$-1-z_0$	z_0	1

Here z_0 is the root of the equation $z^2+z+1=0$. Solutions 3-4 and 3-6 are connected by the Bäcklund transformation T_4 (see above).

$l=4$	Parameters				Coefficients				
Sol. No	α	β	γ	δ	a_0	a_1	a_2	a_3	a_4
4-1	$9/32$	$-1/32$	$1/2$	0	0	$1/3$	$1/3$	$1/3$	0
4-2	$9/32$	$-1/32$	$1/2$	0	0	$\pm 1/3$	0	$\pm 1/3$	0
4-3	$1/32$	$-9/32$	0	$1/2$	0	$\mp i$	1	$\pm i$	0
4-4	$1/2$	0	$9/32$	$-9/32$	$1/4$	$\pm 1/4$	$1/4$	$\pm 1/4$	0
4-5	$1/2$	0	$9/32$	$-9/32$	$1/4$	$\mp i/4$	$-1/4$	$\pm i/4$	0
4-6	$1/2$	0	$25/32$	$7/32$	0	$\pm 1/4$	$1/4$	$\pm 1/4$	$1/4$

Solutions 4-4 and 4-6 are connected by the Bäcklund transformation T_4 as mentioned above.

$l = 5$	Parameters				Coefficients				
Sol. No	α	β	γ	δ	a_0	a_1	a_2	a_3	a_4
5-1	$8/25$	$-1/50$	$1/2$	0	0	$1/4$	$1/4$	$1/4$	$1/4$
5-2	$8/25$	$-1/50$	$1/2$	0	0	A_1	$z_0^3/4$	$z_0^2/4$	$z_0/4$
5-3	$1/50$	$-8/25$	0	$1/2$	0	-1	-1	-1	-1
5-4	$1/50$	$-8/25$	0	$1/2$	0	B_1	z_1^3	z_1^2	z_1
5-5	$1/2$	0	$8/25$	$-11/50$	$1/5$	$1/5$	$1/5$	$1/5$	$1/5$
5-6	$1/2$	0	$8/25$	$-11/50$	$1/5$	C_1	$z_0^3/5$	$z_0^2/5$	$z_0/5$

Here z_0 is the root of the equation $z^4+z^3+z^2+z+1=0$, $A_1=-1/4-a_2-a_3-a_4$, $C_1=-1/5-a_2-a_3-a_4$ and z_1 is the root of the equation $z^4-z^3+z^2-z+1=0$, $B_1=1-a_2-a_3-a_4$, $a_5=0$.

$l = 6$	Parameters			Coefficients				
No	α	γ	δ	a_1	a_2	a_3	a_4	a_5
6-1	$1/72$	0	$1/2$	± 1	1	± 1	1	± 1
6-2	$1/72$	0	$1/2$	$-(1+z_0)$	$-z_0$	1	$1+z_0$	z_0
6-3	$1/72$	0	$1/2$	$1-z_1$	z_1	-1	$1-z_1$	z_1
6-4	$1/2$	$25/72$	$-13/72$	$\pm 1/6$	$1/6$	$\pm 1/6$	$1/6$	$\pm 1/6$
6-5	$1/2$	$25/72$	$-13/72$	$-(1+z_0)/6$	$z_0/6$	$1/6$	$-(1+z_0)/6$	$z_0/6$
6-6	$1/2$	$25/72$	$-13/72$	$(1-z_1)/6$	$-(z_1)/6$	$-1/6$	$(z_1-1)/6$	$z_1/6$

Here z_0 is the root of the equation z^2+z+1, z_1 is the root of the equation z^2-z+1, $a_6 = 0$ for all found solutions, $a_0 = 0$, $\beta = -25/72$ for solutions 6-1–6-3 and $a_0 = 1/6$, $\beta = 0$ for solutions 6-4–6-6.

Solution 3-1 can be obtained from elementary solutions $y = x^{4/3}$ or $y = x^{1/3}$ by the Bäcklund transformations

$$T_8 : y_8(x; -\delta + 1/2, -\gamma, -\beta, -\alpha + 1/2) = \frac{x\big(y(x;\alpha,\beta,\gamma,\delta) - 1\big)}{y(x;\alpha,\beta,\gamma,\delta) - x}$$

and

$$T_9 : y_9(x; \gamma, \delta - 1/2, \alpha, \beta + 1/2) = \frac{y(x;\alpha,\beta,\gamma,\delta) - x}{y(x;\alpha,\beta,\gamma,\delta) - 1},$$

respectively.

The same transformations connect the solution 5-3 with elementary solutions $y = x^{6/5}$ and $y = x^{1/5}$, respectively. It is possible that some other solutions among those mentioned above may be obtained from the known elementary solutions of the equation P_6 by the Bäcklund transformations, but the authors known nothing about it.

Theorem 2.2. *All the above computed exact solutions to* P_6 *can be written in the form* $P(x, y) = 0$*, where* $P(x, y)$ *is a polynomial which defines an algebraic curve of genus equals zero.*

2.3.3 Generalization of Computed Solutions

Solutions 3-4, 3-8, 4-4 and 5-5 can be written as the sum of a finite geometrical progression and can then be generalized for the case of any power exponents. The direct substitution shows that the function

$$y = b\frac{x-1}{x^b-1}$$

is the exact solution of the equation P_6 for $\alpha = 1/2$, $\beta = 0$, $\gamma = (1-b)^2/2$, $\delta = -(2+b)^2/2$.

Acknowledgments. The authors express their gratitude to Professor A. D. Bruno for the fruitful ideas, without which this work would not be held.

Bibliography

[1] A. D. Bruno, *Power Geometry in Algebraic and Differential Equations,* Elsevier, Amsterdam, 2000, 385 p. = Fizmatlit, Moscow, 1998, 288 p.

[2] V. I. Gromak, I. Laine, S. Shimomura, *Painlevé Differential Equations in Complex Plane,* Walter de Gruyter, Berlin, New York, 2002.

[3] A. D. Bruno, I. N. Gashenenko, Simple exact solutions to the N. Kowalewski equations, *Doklady Mathematics,* **74**(1): 536–539, 2006 (= *DAN,* **409**(4): 439–442, 2006).

[4] D. Cox, J. Little, D. O'Shea, *Ideals, Varieties and Algorithms: an introduction in computational algebraic geometry and commutative algebra*, Springer, 2007.

[5] A. D. Bruno, I. V. Goryuchkina, Asymptotic expansions of solutions to the sixth Painlevé equation, *Transactions of Moscow Math. Soc.,* **71**: 1–105, 2010 (= *Tr. MMO,* **71**: 6–118, 2010).

[6] B. Dubrovin, M. Mazzocco, Monodromy of certain Painlevé transcendets and reflection groups, *Invent. Math.,* **141**(1): 55–147, 2000.

[7] N. J. Hitchin, A lecture on the octahedron, *Bull. London Math. Soc.,* **35**(5): 577–600, 2003.

[8] P. Boalch, The fifty-two icosahedral solutions to Painlevé VI, *J. Reine Angew. Math.,* **596**: 183–214, 2006.

[9] O. Lisovyy, Yu. Tykhyy, Algebraic solutions of the sixth Painlevé equation. `http://arXiv.org/abs/0809.4873v2`.

Author Information

Alexander B. Batkhin, Keldysh Institute of Applied Mathematics of RAS, Miusskaya Sq. 4, Moscow, Russia.
E-mail: batkhin@gmail.com

Natalia V. Batkhina, Volzhsky Institute of Humanities branch of Volgograd State University, 40 let Pobedy, 11, Volzhsky, Volgograd Region, Russia.
E-mail: batkhina@vgi.volsu.ru

Chapter 3

Convergence of a Formal Solution to an ODE

Irina V. Goryuchkina

Abstract. In 2004 Professor Bruno formulated a theorem on convergence of a power series solution to an ordinary differential equation. We proved this theorem in two cases: for rational power exponents in the expansion, and for complex and irrational (but not rational) power exponents with one complex or irrational generatrix of the set of power exponents in the expansion. In the proofs we used Power Geometry, the majorant method, and a new trick. These proofs can be generalized on the general case of this theorem. These results are new.

Keywords. Formal Solutions, Convergence of Series, Majorant Method.

2010 Mathematics Subject Classification. Primary: 34M55; Secondary: 34E05.

3.1 The General Case

We consider an ordinary differential equation of the form

$$f\left(x, y, y', \dots, y^{(n)}\right) = 0, \tag{3.1}$$

where $f(x, y, y', \dots, y^{(n)})$ is a polynomial of its variables.

Let for $|x| \to 0$, $\arg x \in (-\pi, \pi)$ equation (3.1) have a formal solution of the form

$$y = \sum c_s x^s, \quad s \in \mathbf{K} \subset \mathbb{C}, \tag{3.2}$$

where power exponents s are complex, $\operatorname{Re} s$ increase, the number of power exponents s with the same real parts $\operatorname{Re} s$ is finite, and coefficients c_s are complex constants. We enumerate power exponents s in the order of increasing real parts $\operatorname{Re} s : \operatorname{Re} s_0 \leq \operatorname{Re} s_1 \leq \operatorname{Re} s_2 \leq \dots$.

By the substitution

$$y = \sum_{s=s_0}^{s_m} c_s x^s + u, \tag{3.3}$$

where $m \in \mathbb{Z}$, $m \geq 0$, $\operatorname{Re} s_m \geq n$, and power exponents s and coefficients c_s are from (3.2), we reduce equation (3.1) to the form

$$f_1(x, u) \stackrel{def}{=} \mathcal{L}(x)u + g(x, u, u', \dots, u^{(n)}) = 0, \tag{3.4}$$

This work was supported by RFBR, Grant Nr. 11-01-00023.

where the linear differential operator $\mathcal{L}(x)$ has form

$$\mathcal{L}(x) = x^{v} \sum_{l=1}^{n} a_l x^l \frac{d^l}{dx^l}, \tag{3.5}$$

$\mathcal{L}(x) \not\equiv 0$, $v \in \mathbb{C}$, a_l are complex constants. The function g contains terms independent of $u, u', \dots, u^{(n)}$, linear terms in $u, u', \dots, u^{(n)}$ of the form $c\, x^{v_1+l}\, u^{(l)}$ with $\operatorname{Re} v_1 > \operatorname{Re} v$, $l \le n$, $c = \text{const} \in \mathbb{C}$, and nonlinear terms in $u, u', \dots, u^{(n)}$.

The linear differential operator $\mathcal{L}(x)$ has eigenvalues $\lambda_1, \dots, \lambda_n$. We order these eigenvalues by increasing real parts $\operatorname{Re}\lambda_1 \le \dots \le \operatorname{Re}\lambda_n$. We suppose that $\operatorname{Re} s_m \ge \operatorname{Re}\lambda_n$ in substitution (3.3). Then equation (3.4) has unique solution of the form

$$u = \sum_{s=s_m+1}^{\infty} c_s x^s, \tag{3.6}$$

where power exponents $s \in \mathbb{C}$, $\operatorname{Re} s$ increase, complex coefficients c_s are uniquely determined constants.

Theorem 3.1 ([1]). *If in equation* (3.4), *which we obtain after substitution* (3.3) *into equation* (3.1), *the order of the highest derivative in $\mathcal{L}(x)u$ is equal to the order of the highest derivative in the sum f_1, then the series* (3.6) *converges for sufficiently small $|x|$ and* $\arg x \in (-\pi, \pi)$.

3.2 The Case of Rational Power Exponents

Let for $x \to 0$ equation (3.1) have the formal solution

$$y = \sum_{s=s_0}^{\infty} c_s x^s, \tag{3.7}$$

where $s \in \mathbb{Z}$, $s \ge s_0 > -\infty$, s increase, c_s are constant coefficients.

After substitution (3.3) with $s \in \mathbb{Z}$, $s_m \ge n$, $s_m \ge s_0$, $s_m \ge \operatorname{Re}\lambda_n$, equation (3.1) takes the form (3.4). Equation (3.4) contains integer power exponents of x and nonnegative integer powers of $u, u', \dots, u^{(n)}$.

For $x \to 0$, equation (3.4) has unique solution of the form

$$u = \sum_{s=s_m+1}^{\infty} c_s x^s, \tag{3.8}$$

where power exponents $s \in \mathbb{N}$, s increase, complex coefficients c_s are uniquely determined constants.

Theorem 3.2. *If in equation* (3.4), *which we obtain by means of substitution* (3.3) *with* $s \in \mathbb{Z}$, $s_m \geq n$, $s_m \geq s_0$, $s_m \geq \operatorname{Re}\lambda_n$ *from equation* (3.1), *the order of the highest derivative in* $\mathcal{L}(x)u$ *is equal to the order of the highest derivative in the sum* f_1, *then the series* (3.8) *converges for sufficiently small* $|x|$.

Here we consider the case when expansion (3.8) contains only integer power exponents. The case of rational power exponents with a finite common denominator m is reduced to this case by the substitution $z = x^{1/m}$.

The constraint on $\arg x$ is necessary only for expansions (3.6) with irrational or complex power exponents and can be dropped in the case under study.

The proof of Theorem 3.2 was published in [2] and [3].

3.3 The Case of Complex Power Exponents

We consider formal solution of the form (3.2), where

$$\mathbf{K} = \{s_0 + m_1 r_1 + m_2 r_2,\ m_1, m_2 \in \mathbb{Z},\ m_1, m_2 \geq 0\}, \tag{3.9}$$

$$\begin{gathered} s_0 \in \mathbb{C} \setminus \mathbb{Q},\ \ r_1 = \langle R_1,\ (1, s_0)\rangle,\ \ r_2 = \langle R_2,\ (1, s_0)\rangle, \\ R_1 = (\alpha_1, \beta_1),\ \ R_2 = (\alpha_2, \beta_2),\ \ R_1,\ R_2 \in \mathbb{Z}^2. \end{gathered} \tag{3.10}$$

After substitution (3.3) with $s \in \mathbf{K}$, equation (3.1) takes the form (3.4). The function f_1 contains complex power exponents of x and nonnegative integer power exponents $u, u', \dots, u^{(n)}$.

For $|x| \to 0$, $\arg(x) \in (-\pi, \pi)$, equation (3.4) has a unique solution of the form (3.6) with power exponents $s \in \mathbb{C}$.

Theorem 3.3. *If in equation* (3.4), *which we obtain after substitution* (3.3) *from the equation* (3.1), *the order of the highest derivative in* $\mathcal{L}(x)u$ *is equal to the order of the highest derivative in the sum* f_1, *then the series* (3.6) *with* s *as defined by* (3.9) *and* (3.10) *converges for sufficiently small* $|x|$ *and* $\arg(x) \in (-\pi, \pi)$.

The proof of Theorem 3.3 was published in [4]. A similar theorem is true for irrational s_0, r_1, r_2 in (3.10).

3.4 On Solutions of the Sixth Painlevé Equation

According to Theorems 3.2 and 3.3, all power expansions of solutions to the sixth Painlevé equation near its three singular points are convergent. Near the regular point all expansions are power expansions and form 17 families. They converge according to theorem 3.2, and in some cases according to the Cauchy theorem.

Bibliography

[1] A. D. Bruno, Asymptotic forms and expansions of solutions to an ordinary differential equation, *Russ. Math. Surv.*, **59**(3): 429–480, 2004 (= *UMN*, **59**(3) : 31–80, 2004).

[2] A. D. Bruno and I. V. Goryuchkina, Asymptotic expansions of solutions to the sixth Painlevé equation, *Transactions Moscow Math Soc.*, **71**: 1–105, 2010 (= *Tr. MMO*, **71**: 6–118, 2010).

[3] A. D. Bruno and I. V. Goryuchkina, On the convergence of a formal solution to an ordinary differential equation, *Doklady Mathem.*, **81**(3): 358–361, 2010 (= *DAN*, **432**(2): 151–154, 2010).

[4] I. V. Goryuchkina, *Theorem on convergence of a formal solution to ODE*, Preprint No. 25 Keldysh Inst. Appl. Math., Moscow, 2011 (in Russian); available from the Keldysh Institute of Applied Mathematics of RAS at `http://library.keldysh.ru/preprints/Default.asp?lg=e`.

Author Information

Irina V. Goryuchkina, Keldysh Institute of Applied Mathematics of RAS, Miusskaya Sq. 4, 125047, Moscow, Russia.
E-mail: igoryuchkina@gmail.com

Chapter 4

Asymptotic Expansions and Forms of Solutions to P_6

Irina V. Goryuchkina

Abstract. Using algorithms of Power Geometry, we have found asymptotic expansions and asymptotic forms of solutions to the sixth Painlevé equation for all values of its four complex parameters near its three singular points $x = 0, 1, \infty$ and near its regular points $x = x_0 \neq 0, 1, \infty$. There are five types of asymptotic expansions of solutions: power, power-logarithmic, complicated, exotic, and half-exotic. Near all singular points of the equation they form 117 families. Most of these expansions are new. Near the regular points of the equation there are 17 families of power expansions, eight of which are new. In addition, we show that near the singular points of the sixth Painlevé equation the Boutroux-type elliptic asymptotic forms are absent, in contrary to other Painlevé equations.

Keywords. Painlevé Equations, Asymptotic Forms, Asymptotic Expansions.

2010 Mathematics Subject Classification. Primary: 34M55; Secondary: 34E05.

4.1 Asymptotic Expansions near Singular Points of the Equation

The sixth Painlevé equation is

$$\begin{aligned} y'' =& \frac{(y')^2}{2}\left(\frac{1}{y}+\frac{1}{y-1}+\frac{1}{y-x}\right)-y'\left(\frac{1}{x}+\frac{1}{x-1}+\frac{1}{y-x}\right) \\ &+\frac{y(y-1)(y-x)}{x^2(x-1)^2}\left[a+b\frac{x}{y^2}+c\frac{x-1}{(y-1)^2}+d\frac{x(x-1)}{(y-x)^2}\right], \end{aligned} \tag{4.1}$$

where a, b, c, and d are complex parameters, x and y are complex variables, and $y' = dy/dx$. This equation has three singular points $x = 0$, $x = \infty$, and $x = 1$, and three main symmetries that allow to transfer expansions of solutions near one of its singular points to expansions of the solutions near the same or a different one of singular points. These symmetries are associated with three changes of variables:

$$1)\ x = z,\ y = z/w;\quad 2)\ x = 1/z,\ y = 1/w;\quad 3)\ x = 1-z,\ y = 1-w. \tag{4.2}$$

This work was supported by RFBR, Grant Nr. 11-01-00023.

At first we obtained 21 families of such asymptotic expansions of solutions to equation (4.1) near the singular point $x = 0$, for which the order of the first term is less than unity. We refer to these expansions as *basic expansions*. All other asymptotic expansions of solutions to equation (4.1) near its three singular points $x = 0$, $x = 1$, and $x = \infty$ were obtained from the basic expansions using symmetries (4.2). Altogether the expansions form 117 families [1]. At $x = 0$ we seek asymptotic expansions of the form

$$y = c_r x^r + \sum_s c_s x^s, \tag{4.3}$$

where power exponents r and s are complex, $\operatorname{Re} s > \operatorname{Re} r$, $\operatorname{Re} s$ increase. They are one of three types: power, power-logarithmic or complicated. Expansions of these types have a finite number of terms s with the same value $\operatorname{Re} s$. Coefficients c_r and c_s are varied according to their type:

Type 1. c_r and c_s are constant (*power expansions*);

Type 2. c_r is constant, c_s are polynomials in $\log x$ (*power-logarithmic expansions*);

Type 3. c_r and c_s are series in decreasing powers of $\log x$ (*complicated expansions*).

At $x = 0$ we also consider expansions of the form

$$y = \sum_\rho c_\rho x^\rho + \sum_s c_s x^s, \tag{4.4}$$

where power exponents ρ and s are complex, all $\operatorname{Re} \rho$ are the same, $\operatorname{Re} \rho < \operatorname{Re} s$, $\operatorname{Re} s$ increase, the first sum contains more than one term, and complex coefficients c_ρ and c_s are polynomials in $\log x$. We differentiate into two types of the expansions (4.4).

Type 4. The first sum in (4.4) contains a finite number of terms. Coefficients c_ρ are constants for extreme values of $\operatorname{Im} \rho$. The number of power exponents s with the same real parts $\operatorname{Re} s$ are finite (*half-exotic expansions*);

Type 5. The first sum in (4.4) contains infinite number of terms, but $\operatorname{Im} \rho$ are bounded either below or above. Coefficient c_ρ is constant for the extreme value of $\operatorname{Im} \rho$ (*exotic expansions*).

In addition, we suppose that $\arg x$ is bounded from two sides. Here we give examples for each type of expansions.

Theorem 4.1. *For $x \to 0$ there exists the family of power expansions of solutions to equation* (4.1) *with two parameters c_r and r, and constant coefficients, which has form*

$$\mathcal{A}_0 : y = c_r x^r + \sum_s c_s x^s, \tag{4.5}$$

where complex power exponent r is arbitrary with $\operatorname{Re} r \in (0, 1)$, *complex power exponents $s \in \{r + lr + m(1 - r),\ l, m \geq 0;\ l + m > 0;\ l, m \in \mathbb{Z}\}$; complex coefficient*

c_r is an arbitrary nonzero constant, and other complex coefficients c_s are uniquely determined constants. Family $\mathcal{A}_0$ exists for all values of parameters of equation (4.1).

The family $\mathcal{A}_0$ was known.

Theorem 4.2. *For $x \to 0$ and $a = c \neq 0$, $2\sqrt{2a} \in \mathbb{Z}\backslash\{0\}$ there exists the family of power-logarithmic expansions of solutions to equation* (4.1) *of the form*

$$\mathcal{B}_2: \quad y = 2 + \sum_{s=1}^{\infty} c_s(\log x)x^s, \tag{4.6}$$

where coefficients $c_s(\log x)$ are polynomials in $\log x$.

Theorem 4.3. *For $x \to 0$ and $a \neq c \neq 0$ there exists the family of complicated expansions of solutions to equation* (4.1) *of the form*

$$\mathcal{B}_3: \quad y = \psi_0 + \sum_{\sigma=1}^{\infty} \psi_\sigma x^\sigma, \tag{4.7}$$

where

$$\psi_0 = \frac{2}{c-a}\frac{1}{\log^2 x} + \frac{c_{-3}}{\log^3 x} + \sum_{s=4}^{\infty}\frac{c_{-s}}{\log^s x} = \frac{2(c-a)}{(c-a)^2(\log x + C_0)^2 - 2a},$$

where coefficients c_{-3} and C_0 are arbitrary constants, other coefficients c_{-s} are uniquely determined constants, and ψ_σ are a series in decreasing powers in $\log x$.

Theorem 4.4. *For $x \to 0$ there exists the family of half-exotic expansions of solutions to equation* (4.1) *with two parameters c_ρ and ρ, and constant coefficients, which has the form*

$$\mathcal{H}_0: \quad y = c_\rho x^\rho + c_1 x + c_{2-\rho}x^{2-\rho} + \sum_s c_s x^s + \cdots, \tag{4.8}$$

where $\rho - 1$ is a purely imaginary arbitrary constant, s runs over the set $\{l + k(\rho - 1);\ l,k \in \mathbb{Z};\ l \geq 2, |k| \leq l\}$, complex coefficient c_ρ is an arbitrary constant, and other complex coefficients c_1, $c_{2-\rho}$ and c_s are uniquely determined constants.

Theorem 4.5. *For $x \to 0$ there exists the family of exotic expansions of solutions to equation* (4.1) *with two parameters C_1 and ρ, and constant coefficients, which has the form*

$$\begin{aligned}\mathcal{B}_0^\tau: \quad y &= \frac{\rho^2}{\beta\cos^2[\log(C_1 x)^\gamma] + \alpha\sin^2[\log(C_1 x)^\gamma]} + \sum_{\operatorname{Re} s \geq 1} c_s x^s \\ &= x^\rho\left(c_\rho + \sum_{k=1}^{\infty}\tilde{c}_k x^{k\rho}\right) + \sum_{\operatorname{Re} s\geq 1} c_s x^s,\end{aligned} \tag{4.9}$$

where ρ *is a purely imaginary nonzero arbitrary constant,* $s \in \{\rho + l\rho + m(1-\rho);\ l, m \geq 0;\ l + m > 0;\ l, m \in \mathbb{Z}\}$, $\tau = \operatorname{sgn}(\operatorname{Im}\rho)$, $\alpha + \beta = (\rho^2 - 2c + 2a)/(2a)$, $\alpha\beta = \rho^2/(2a)$, $2\gamma = i\rho$, *complex coefficients* $\tilde{c}_k$ *and* c_s *are uniquely determined constants.*

4.2 Asymptotic Expansions near a Regular Point of the Equation

In the neighborhood of a regular point $x = x_0 \neq 0, 1, \infty$ of the sixth Painlevé equation (4.1) there are 17 families of power expansions of its solutions for all values of its four parameters [2]. The expansions are Laurent or Taylor series. Of these, one family of expansions has the pole of the second order, two families of expansions have poles of the first order, and the others are families of Taylor expansions. Eight families of expansions are new (cf. [3]).

4.3 Boutroux-Type Elliptic Asymptotic Forms

Theorem 4.6. *Near the singular points* $x = 0, 1, \infty$ *of the sixth Painlevé equation, the Boutroux-type elliptic asymptotic form of solutions are absent* [4] *in contrast to other Painlevé equations* [5–7].

Bibliography

[1] A. D. Bruno, I. V. Goryuchkina, Asymptotic expansions of solutions to the sixth Painlevé equation, *Transactions Moscow Math Soc.*, **71**: 1–105, 2010 (= *Tr. MMO,* **71**: 6–118, 2010).

[2] A. D. Bruno, I. V. Goryuchkina, All expansions of solutions to the sixth Painlevé equation near its nonsingular point, *Doklady Mathem.*, **79**(3): 397–402, 2009 (= *DAN,* **426**(5): 586–591, 2009).

[3] V. I. Gromak, I. Laine, S. Shimomura, *Painlevé Differential Equations in Complex Plane,* Walter de Gruyter, Berlin, New York, 2002.

[4] I. V. Goryuchkina, *Three-dimensional analysis of asymptotic forms of the solutions to the sixth Painlevé equation*, Preprint No. 56, Keldysh Inst. Appl. Math. Moscow (2010), 24 pp. (in Russian); available from from the Keldysh Institute of Applied Mathematics of RAS: `http://library.keldysh.ru/preprints/Default.asp?lg=e`..

[5] A. D. Bruno, I. V. Goryuchkina, Boutroux asymptotic forms of solutions to Painleé equations and power geometry, *Doklady Mathem.*, **78**(2): 681–685, 2008 (= *DAN,* **422**(2): 157–160, 2008).

[6] A. D. Bruno, I. V. Goryuchkina, Asymptotic forms of solutions to the third Painlevé equation, *Doklady Mathem.*, **78**(2): 765–768, 2008 (= *DAN,* **422**(6): 729–731, 2008).

[7] A. D. Bruno, I. V. Goryuchkina, Asymptotic forms of solutions to the fourth Painlevé equation, *Doklady Mathem.,* **78** (3): 868–873 (= *DAN*, **423**(4): 443–448, 2008, 2008).

Author Information

Irina V. Goryuchkina, Keldysh Institute of Applied Mathematics of RAS, Miusskaya Sq. 4, Moscow, Russia.
E-mail: igoryuchkina@gmail.com

Chapter 5

Asymptotic Expansions of Solutions to P_5

Anastasia V. Parusnikova

Abstract. By means of Power Geometry we obtained all asymptotic expansions of solutions to the equation P_5 of the following five types: power, power-logarithmic, complicated, exotic, and half-exotic, for all values of 4 complex parameters of the equation. They form 16 and 30 families in the neighborhood of singular points $z = \infty$ and $z = 0$, respectively. There are 10 families in the neighborhood of a nonsingular point. Over 20 families are new.

Keywords. Painlevé Equations, Asymptotic Expansion.

2010 Mathematics Subject Classification. Primary: 34M55; Secondary: 34E05.

5.1 Introduction

We consider the fifth Painlevé equation

$$w'' = \left(\frac{1}{2w} + \frac{1}{w-1}\right)(w')^2 - \frac{w'}{z} + \frac{(w-1)^2}{z^2}\left(\alpha w + \frac{\beta}{w}\right) + \frac{\gamma w}{z} + \frac{\delta w(w+1)}{w-1}, \tag{5.1}$$

where $\alpha, \beta, \gamma, \delta$ are complex parameters, z is an independent complex variable, and w is a dependent one. The fifth Painlevé equation (5.1) has two singular points: $z = 0$ and $z = \infty$. The aim of the present chapter is to find all the asymptotic expansions of solutions to the fifth Painlevé equation. We obtain these expansions using the methods of Power Geometry [1, 2]. We are looking for the expansions of the following form near the singular points of the equation:

$$w = c_r(z)z^r + \sum_{s\in\mathbf{K}} c_s(z)z^s, \tag{5.2}$$

where $c_r(z), c_s(z), r, s \in \mathbb{C}$, $\mathbf{K} \subset \{s \mid \operatorname{Re} s > \operatorname{Re} r\}$ for the expansions in the neighborhood of $z = 0$ and $\mathbf{K} \subset \{s \mid \operatorname{Re} s < \operatorname{Re} r\}$ for the expansions in the neighborhood of $z = \infty$; the set $\mathbf{K}$ is countable.

This work was supported by RFBR, Grant Nr. 11-01-00023.

We obtain the expansions (5.2) of the following five types:

Type 1: $c_r(z)$ and $c_s(z)$ are constant (*power expansions*);

Type 2: $c_r(z)$ is constant, $c_s(z)$ are polynomials in $\log z$ (*power-logarithmic*);

Type 3: $c_r(z)$ and $c_s(z)$ are power series in $\log z$ (*complicated expansions*);

Type 4: $r,\ s \in \mathbb{R}$, $c_r(z)$ and $c_s(z)$ are series in z^i, and c_r is a sum of countable number of terms, the set of power exponents of z^i in c_r is bounded from either above or below (*exotic expansions*);

Type 5: $r,\ s \in \mathbb{R}$, $c_r(z)$ is a finite sum of powers of z^i with complex coefficients, and $c_s(z)$ are a power series over z^i (*half-exotic*).

The apparatus of Power Geometry [1,2] permits us to work with ordinary differential equations of the n-th order which have the form $f(z, w, w', w'', \ldots, w^{(n)}) = 0$, where f is a polynomial in its variables (we call the left part of the equation a differential sum). To transform equation (5.1) into a differential sum we multiply it by $z^2 w(w-1)$. The next step of the equation's exploration is the construction of its polygon $\Gamma(f)$. The polygon $\Gamma(f)$ for equation (5.1) is shown in Figure 5.1. The edges $\Gamma_j^{(1)},\ j = 1, 2, 3$ give the expansions in the neighborhood of $z = \infty$, the edge $\Gamma_1^{(1)}$ gives the expansions near $z = 0$.

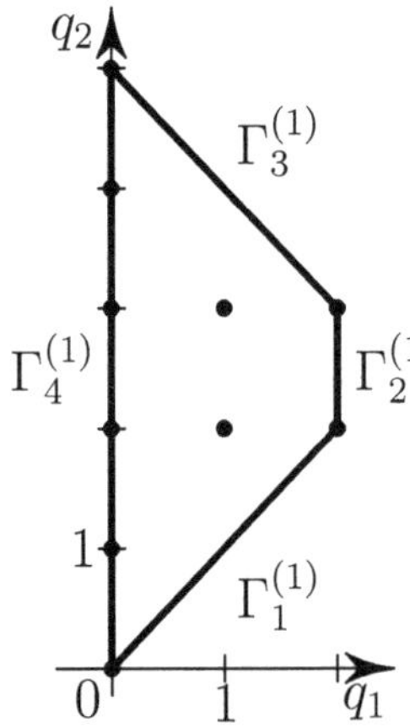

Figure 5.1. Support and polygon for equation (5.1) multiplied by $z^2 w(w-1)$.

Equation (5.1) has the symmetry

$$(z, w, \alpha, \beta, \gamma, \delta) = (\tilde{z}, 1/\tilde{w}, -\tilde{\beta}, -\tilde{\alpha}, -\tilde{\gamma}, \tilde{\delta}) \tag{5.3}$$

and has 3 special solutions:

$I_1 :\ w = 0$ for $\beta = 0$,
$I_2 :\ w = 1$ for $\delta = 0$,
$I_3 :\ w = \infty$ for $\alpha = 0$.

5.2 Asymptotic Expansions of Solutions near Infinity [3]

Theorem 5.1. *In the neighborhood of $z = \infty$ there exist 10 asymptotic expansions of solutions to equation* (5.1) *(power expansions):*

If $\alpha\beta\delta \neq 0$, there exist the following 5 expansions of solutions

$$\mathcal{D}_k : w = (-1)^k \sqrt{\frac{\beta}{\delta}}\frac{1}{z} + \left(-\frac{2\beta}{\delta} + (-1)^k \frac{\gamma}{2\delta}\sqrt{\frac{\beta}{\delta}}\right)\frac{1}{z^2} + \sum_{s=3}^{\infty}\frac{c_{sk}}{z^s}.$$

$$\mathcal{E}_1 : w = -1 + \frac{2\gamma}{\delta z} + \sum_{s=2}^{\infty}\frac{c_s}{z^s}.$$

$$\mathcal{F}_k : w = (-1)^k \sqrt{-\frac{\delta}{\alpha}}z + 2 + (-1)^k \frac{1}{2}\frac{\gamma}{\sqrt{-\alpha\delta}} + \sum_{s=1}^{\infty}\frac{c_{s,k}}{z^s},$$

where c_s, c_{sk} are uniquely determined complex constants, $k = 1, 2$.

If $\alpha\beta\gamma \neq 0$, $\delta = 0$ there exist the following 5 expansions of solutions

$$\mathcal{D}_k : w = (-1)^k \sqrt{-\frac{\beta}{\gamma}\frac{1}{\sqrt{z}}} + \frac{\beta}{\gamma}\frac{1}{z} + \sum_{s=3}^{\infty}\frac{c_{s,k}}{z^{s/2}}.$$

$$\mathcal{E}_2 : w = 1.$$

$$\mathcal{F}_k : w = (-1)^k \sqrt{-\frac{\gamma}{\alpha}}\sqrt{z} + 1 + \sum_{s=1}^{\infty}\frac{c_{s,k}}{z^{s/2}},$$

where c_s, c_{sk} are uniquely-determined complex constants, $k = 3, 4$. For every expansion except $\mathcal{E}_2$ there exist two corresponding exponential additions of the form $b(z)Ce^{\varphi(z)}$, where C is an arbitrary constant, $\varphi'(z)$, $b(z)$ are power series. Moreover, every exponential addition can be continued to the exponential expansion of the following type:

$$w = \sum_{k=0}^{\infty} b_k(z)C^k e^{k\varphi(z)}, \tag{5.4}$$

where C is an arbitrary constant, $\varphi'(z)$, $b_k(z)$ are power series.

If $\beta = 0$ there exist two families of expansions $\mathcal{U}_1$, $\mathcal{U}_2$ of the type (5.4) *with $b_0 \equiv 0$. If $\alpha = 0$ there exist two families $\mathcal{V}_1$, $\mathcal{V}_2$, which can be obtained from $\mathcal{U}_1$, $\mathcal{U}_2$ by the symmetry* (5.3)*. If $\gamma = \delta = 0$ there exist one family for $\alpha = 0$ and one more for $\beta = 0$.*

5.3 Asymptotic Expansions of Solutions near Zero [4]

The following truncated equation corresponds to the edge $\Gamma_4^{(1)}$

$$-z^2 w(w-1)w'' + z^2\left(\frac{3}{2}w - \frac{1}{2}\right)(w')^2 - zw(w-1)w' + (w-1)^3(\alpha w^2 + \beta) = 0. \tag{5.5}$$

As the differential operator (the first variation of (5.5)) vanishes only if $w = 1$ thus $w = 1$ is its special solution. We substitute $w = 1+\tilde{w}$ into (5.1) and obtain an equation $g(z,\tilde{w}) = 0$ with polygon $\tilde{\Gamma} = \Gamma(g)$ having edges $G_j^{(1)}$, $j = 1,2,3,4$ and 4 vertices (see Figure 5.2). As we consider the case $z \to 0$, the appropriate truncated solutions can be given by equations corresponding to vertex $(0,2)$ and edges $G_3^{(1)}, G_4^{(1)}$.

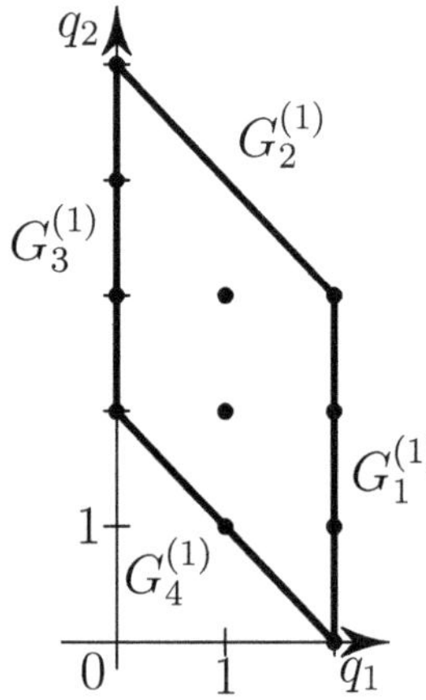

Figure 5.2. Support and polygon for equation (5.1) after substitution $w = 1 + \tilde{w}$ and multiplication by $z^2(\tilde{w} + 1)\tilde{w}$.

Theorem 5.2. *In the neighborhood of $z = 0$ there exist the following 8 families of expansions, corresponding to the edge $G_4^{(1)}$:*

$$\mathcal{H}_1 : y = 1 - \frac{2\delta}{\gamma}x + \sum_{s \in \mathbf{K}} c_s x^s,$$

where $a = \left(\operatorname{sgn} \operatorname{Re} \frac{\gamma}{\sqrt{-2\delta}}\right) \frac{\gamma}{\sqrt{-2\delta}}$, $\mathbf{K} = \{s : s = l + m + ma,\ l, m \in \mathbb{Z},\ l, m \geq 0,\ l + m > 0\}$, c_{a+1} is an arbitrary constant,exists if $\gamma\delta \neq 0$, $\gamma^2/2\delta = -p^2$, $p \in \mathbb{R} \setminus \mathbb{N}$. If $a \in \mathbb{Q}$ the expansion converges according to Theorem 1.7.2 from [2]*;*

$$\mathcal{H}_2 : y = 1 - \frac{2\delta}{\gamma}x + \sum_{s=1}^{\infty} c_s x^s,$$

where c_s, $1 \leq s \leq a$ are constant and c_s, $s \geq a + 2$ are polynomials in $\log z$ with uniquely-determined coefficients, $c_{a+1} = A \log x + C$, where C is an arbitrary constant, exists if $\gamma\delta \neq 0$, $\gamma^2/2\delta = -n^2, n \in \mathbb{N}$.

$$\mathcal{H}_3 : y = 1 - \frac{\delta}{\gamma}x - \frac{\gamma}{2}(\ln x + C)^2 x + \sum_{p=2}^{\infty} \varphi_p x^p,$$

where C is an arbitrary constant, φ_p are a series in decreasing powers of $\log x$ *with uniquely-determined coefficients;*

$$\mathcal{H}_1^\tau : y = 1 + \left(-\frac{2\delta}{\gamma} + Cx^{i\tau\gamma/\sqrt{2\delta}}\right)x + \sum_{\operatorname{Re} s>1} c_s x^s, \gamma^2/\delta \in \mathbb{R}_+,$$
$$\tau = \pm 1, \quad C \neq 0;$$

$$\mathcal{H}_4 : y = 1 + \left(c_r x^{ir} - \frac{\gamma}{r^2} + \frac{\gamma^2 - 2\delta r^2}{4c_r r^4} x^{-ir}\right)x + \sum_{\operatorname{Re} s>1} c_s x^s,$$
$$r \in \mathbb{R} \setminus \{0\}, \quad c_r \in \mathbb{C},$$

r and c_r are arbitrary constants.

The families $\mathcal{H}_1$, $\mathcal{H}_1^\tau$, $\mathcal{H}_2$, and $\mathcal{H}_3$ are one-parametric, and the family $\mathcal{H}_4$ is two-parametric.

If $\gamma \neq 0$, $\delta = 0$ there exist the families $\mathcal{H}_3$ and $\mathcal{H}_4$ (we should substitute $\delta = 0$ in the corresponding formulae).

If $\gamma = 0$, $\delta \neq 0$, there exist two families of expansions

$$\mathcal{H}_j^{(1)} : y = 1 + (-1)^j \sqrt{-2\delta}\,(\ln x + C)\,x + \sum_{p=2}^{\infty} \varphi_p x^p, \; j = 5, 6,$$

where C is an arbitrary constant, φ_p are series in decreasing powers of $\log x$ *with uniquely determined coefficients; and the family $\mathcal{H}_4$ (we should substitute $\gamma = 0$ in the corresponding equations).*

Theorem 5.3. *In the neighborhood of $z = 0$ there exist the following 21 families of asymptotic expansions corresponding to the edge $G_3^{(1)}$ and to the vertex $G_3^{(0)} =$ $(0,3)$ obtained from the corresponding families of expansions of solutions to the sixth Painlevé equation: $\mathcal{B}_i$, $i = 1, 1', 2, 2', 3, 4, 5, 6, 8, 9, 10$, $\mathcal{B}_j^\tau$, $j = 0, 1, 2, 6, 7$, $\mathcal{A}_1$. The family $\mathcal{A}_2$ is obtained by the symmetry* (5.3) *from family $\mathcal{A}_1$. Families $\mathcal{B}_1$ and $\mathcal{B}_2$ are of Type 1, but $\mathcal{B}_{1'}$ and $\mathcal{B}_{2'}$ are of Type 2.*

5.4 Asymptotic Expansions of Solutions in the Neighborhood of the Nonsingular Point of an Equation [5]

To explore the expansions near the nonsingular points $z = z_0$, $z_0 \neq 0$, $z_0 \neq \infty$ of the equation, we perform the transformation $z = t + z_0$ which permits us to apply to the transformed equation the algorithms from Power Geometry described above.

Theorem 5.4. *In the neighborhood of the nonsingular point $z = z_0$ of equation* (5.1) *there exist 10 families of asymptotic expansions of its solutions. The new one is*

$$\mathcal{O}_8 : w = 1 - \frac{\gamma}{2z_0}(z - z_0)^2 + \sum_{s=4}^{\infty} c_s (z - z_0)^s,$$

where c_4 is an arbitrary constant, other c_s are uniquely determined. The expansions exist when $\gamma \neq 0$, $\delta = 0$. One of the 10 families is two-parameter, the rest of the families are one-parameter.

Theorem 5.5. *The expansions of all 10 families converge in the deleted neighborhood of $z = z_0$.*

Some of these results are not new – they can be found in [6, 7].

Bibliography

[1] A. D. Bruno, Asymptotics and expansions of solutions to an ordinary differential equation, *Russian Mathem. Surveys,* **59**(3): 429–480, 2004 (= *UMN*, **59**(3): 31–80, 2004).

[2] A. D. Bruno, I. V. Goryuchkina, Asymptotic expansions of solutions to the sixth Painlevé equation, *Transactions Moscow Math Soc.,* **71**: 1–105, 2010 (= *Trudy MMO*, **71**: 6–118, 2010).

[3] A. D. Bruno, A. V. Parusnikova, *Asymptotic expansions of solutions to the fifth Painleve equation,* Preprint Nr.. 39, Keldysh Inst. Appl. Math., Moscow, 2010 (in Russian); available from the Keldysh Institute of Applied Mathematics of RAS: `http://library.keldysh.ru/preprints/Default.asp?lg=e`.

[4] A. D. Bruno, A. V. Parusnikova, Local expansions of solutions to the fifth Painleve equation, *Doklady Mathematics,* **83**(3), 2011 (= *DAN*, **438**(4): 439–443, 2011).

[5] A. D. Bruno, A. V. Parusnikova, Expansions of solutions to the fifth Painleve equation near its nonsingular point, *Doklady Mathemaics,* **85**(1), 2012 (= *DAN*, **442**(5): 583–588, 2012).

[6] V. I. Gromak, I. Laine, S. Shimomura, *Painlevé Differential Equations in Complex Plane,* Walter de Gruyter, Berlin, New York, 2002.

[7] E. S.Karulina, Expansions of solutions to the fifth Painlevé equation in a neigbourhood of a nonsingular point, *Journal of Mathematical Sciences,* **145**(5): 5252–5259, 2007.

Author Information

Anastasia V. Parusnikova, Moscow State Institute of Electronics and Mathematics (Technical University), Bolshoy Trehsvyatitelskiy per. 3, 109028, Moscow, Russia.
E-mail: `parus-a@mail.ru`

Part II

Space Power Geometry

Chapter 6

Space Power Geometry for one ODE and P_1–P_4, P_6

Alexander D. Bruno

Abstract. In this chapter we explain algorithms of Space Power Geometry (SPG) for one ordinary differential equation (ODE). The algorithms allow calculation of asymptotic forms of solutions to an ODE, which cannot be calculated by algorithms of the Plane PG. We apply this approach to Painlevé equations P_l. It appears that equations P_1–P_4 have elliptic asymptotic forms of solutions, equations P_2, P_4 also have periodic asymptotic forms, and all their families are two-parameter. Solutions to equation P_6 have neither elliptic nor periodic asymptotic forms.

Keywords. Elliptic, Periodic, Asymptotic Forms of Solutions, Painlevé Equations.

2010 Mathematics Subject Classification. Primary: 34M55; Secondary: 34E05.

6.1 Space Power Geometry

Let x be independent and y be dependent variables, $x, y \in \mathbb{C}$. A *differential monomial* $a(x, y)$ is a product of an ordinary monomial $cx^{r_1}y^{r_2}$, where $c = \text{const} \in \mathbb{C}$, $(r_1, r_2) \in \mathbb{R}^2$, and a finite number of derivatives of the form $d^l y/dx^l$, $l \in \mathbb{N}$. The sum of differential monomials $f(x, y) = \sum a_i(x, y)$ is called the *differential sum*.

In [1] it was shown that as $x \to 0$ ($\omega = -1$) or as $x \to \infty$ ($\omega = 1$) solutions $y = \varphi(x)$ to the ODE $f(x, y) = 0$, where $f(x, y)$ is a differential sum, can be found by means of algorithms of Plane PG if

$$p_\omega(\varphi(x)) - l = p_\omega\big(d^l\varphi/dx^l\big), \quad l = 1, \ldots, n,$$

where the order $p_\omega(\varphi(x))$ was introduced in Section 4 of [1] and n is the maximal order of derivatives in $f(x, y)$. Here we introduce algorithms which allow us to calculate solutions $y = \varphi(x)$ with the property

$$p_\omega\big(\varphi(x)\big) + l\gamma_\omega = p_\omega\big(d^l\varphi/dx^l\big), \quad l = 1, \ldots, n,$$

where $\gamma_\omega \in \mathbb{R}$, $\omega = \pm 1$.

This work was supported by RFBR, Grant Nr. 11-01-00023.

Lemma 6.1. *If*

$$p_\omega(\varphi(x)) = -\gamma_\omega + p_\omega(\varphi'(x)),$$

then $\omega + \omega\gamma_\omega \geqslant 0$.

Note, that in Plane PG we had $\gamma_\omega = -1$, i. e. $\omega + \omega\gamma_\omega = 0$. So, interesting new possibilities correspond to $\omega + \omega\gamma_\omega > 0$.

We consider the ODE

$$f(x,\ y) = \sum_i a_i(x, y) = 0, \tag{6.1}$$

where $f(x,\ y)$ is a differential sum. To each differential monomial $a_i(x, y)$, we assign its (vector) *power exponent* $\mathbf{Q}(a_i) = (q_1, q_2, q_3) \in \mathbb{R}^3$ by the following rules:

$$\mathbf{Q}\left(cx^{r_1}y^{r_2}\right) = (r_1,\ r_2,\ 0);\quad \mathbf{Q}\left(d^l y/dx^l\right) = (0,\ 1,\ l);$$

the power exponent of the product of differential monomials is the sum of power exponents of factors $\mathbf{Q}(a_1 a_2) = \mathbf{Q}(a_1) + \mathbf{Q}(a_2)$.

The set $\tilde{\mathbf{S}}(f)$ of power exponents $\mathbf{Q}(a_i)$ of all differential monomials $a_i(x, y)$ presented in the differential sum $f(x, y)$ is called the *space support of the sum* $f(x, y)$. Obviously, $\tilde{\mathbf{S}}(f) \subset \mathbb{R}^3$. The convex hull $\mathbf{\Gamma}(f)$ of the support $\tilde{\mathbf{S}}(f)$ is called the *polyhedron of the sum* $f(x, y)$. The boundary $\partial\mathbf{\Gamma}(f)$ of the polyhedron $\mathbf{\Gamma}(f)$ consists of the vertices $\mathbf{\Gamma}_j^{(0)}$, the edges $\mathbf{\Gamma}_j^{(1)}$ and the faces $\mathbf{\Gamma}_j^{(2)}$. They are called (generalized) *faces* $\mathbf{\Gamma}_j^{(d)}$, where the upper index indicates the dimension of the face, and the lower one is its number. Each face $\mathbf{\Gamma}_j^{(d)}$ corresponds to the *space truncated sum*

$$\check{f}_j^{(d)}(x, y) = \sum a_i(x, y) \text{ over } \mathbf{Q}(a_i) \in \mathbf{\Gamma}_j^{(d)} \cap \tilde{\mathbf{S}}(f). \tag{6.2}$$

Example. Consider the third Painlevé equation P_3, multiplied by its common divisor,

$$f(x, y) \stackrel{\text{def}}{=} -xyy'' + xy'^2 - yy' + ay^3 + by + cxy^4 + dx = 0, \tag{6.3}$$

assuming the complex parameters $a,\ b,\ c,\ d \neq 0$. Here the first two differential monomials have the same power exponent $\mathbf{Q}_1 = (1, 2, 2)$, then $\mathbf{Q}_5 = (0, 2, 1)$, $\mathbf{Q}_3 = (0, 3, 0)$, $\mathbf{Q}_4 = (0, 1, 0)$, $\mathbf{Q}_2 = (1, 4, 0)$, $\mathbf{Q}_6 = (1, 0, 0)$. They are shown in Figure 6.1 in coordinates q_1, q_2, q_3. Their convex hull $\mathbf{\Gamma}(f)$ has 5 two-dimensional faces $\mathbf{\Gamma}_j^{(2)}$. The far face $\mathbf{\Gamma}_1^{(2)}$, spanned by $\mathbf{Q}_1, \mathbf{Q}_2, \mathbf{Q}_6$, corresponds to the truncation

$$\check{f}_1^{(2)} = -xyy'' + xy'^2 + cxy^4 + dx = x(-yy'' + y'^2 + cy^4 + d). \tag{6.4}$$

■

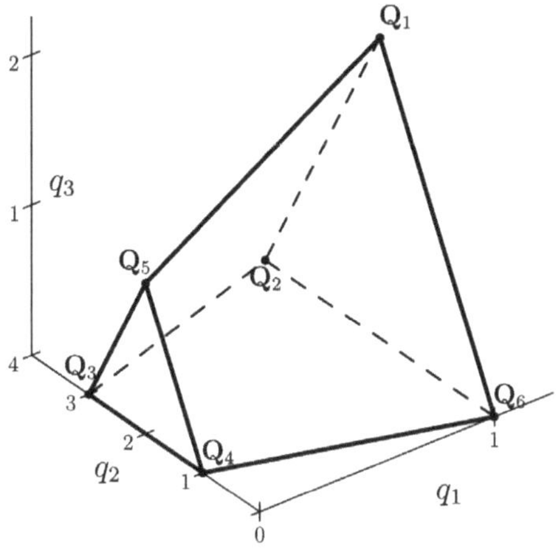

Figure 6.1. Three-dimensional support $\tilde{\mathbf{S}}$ and polyhedron $\mathbf{\Gamma}$ for equation (6.4).

Thus, each face $\mathbf{\Gamma}_j^{(d)}$ corresponds to the truncated equation

$$\check{f}_j^{(d)}(x, y) = 0. \tag{6.5}$$

Let $\mathbf{N}_j = (n_1, n_2, n_3)$ be the external normal to two-dimensional face $\mathbf{\Gamma}_j^{(2)}$. We will consider only normals with $n_1 \neq 0$, so we can assume that $n_1 = \omega$. If the face $\mathbf{\Gamma}_j^{(2)}$ has the normal $\mathbf{N}_j = (1, 0, 0)$ then the corresponding truncation $\check{f}_j^{(2)}(x, y) = x^{q_1} g(y)$, where the differential sum $g(y)$ contains y and its derivatives but does not contain x. In that case the full sum $f(x, y)$ can be written as $f(x, y) = x^{q_1} g(y) + x^{q_1 - \gamma} h(x, y)$, where $\gamma > 0$ and $h(x, y)$ is a differential sum.

Example. In (6.3) and (6.4) $q_1 = 1$, $\gamma = 1$,

$$g(y) = -yy'' + y'^2 + cy^4 + d, \quad h(x, y) = -yy' + ay^3 + by. \qquad \blacksquare$$

Remark 6.2. If $y(x)$ is a solution to the equation $g(y) = 0$ with the property

$$0 < \varepsilon < |y(x)|, |y'(x)|, \ldots, |y^{(n)}| < \varepsilon^{-1}, \tag{6.6}$$

then $y(x)$ can be the asymptotic form of the solutions to the full equation (6.1). Here ε is a small number.

Let the power transformation of variables $x, y \to u, v$:

$$y = x^{\alpha} v, \; u = x^{\beta}, \tag{6.7}$$

transforms $f(x, y)$ into $f^*(u, v)$.

Theorem 6.3. *Let the face $\mathbf{\Gamma}_1^{(2)}$ of $\mathbf{\Gamma}(f)$ have the normal $\mathbf{N}_j = (\omega, n_2, n_3)$ with $\omega + n_3 > 0$, then the power transformation (6.7) with $\alpha = n_2$, $\beta = 1 + \omega n_3$ transforms the truncation $\check{f}_j^{(2)}(x, y)$ of $f(x, y)$ into the truncation*

$$\check{f}_j^{*(2)}(u, v) = u^{q_1} g(v) \tag{6.8}$$

of $f^*(u, v)$, corresponding to the face $\mathbf{\Gamma}_j^{*(2)}$ of $\mathbf{\Gamma}(f^*)$ with the normal $\mathbf{N}_j^* = (1, 0, 0)$. Here $\check{f}_j^{*(2)}(u, v)$ equals to $\check{f}_j^{(2)}(x, y)$ after substitution

$$\beta^l \, u^{[\alpha + l(\beta - 1)]/\beta} \, d^l v / du^l \tag{6.9}$$

instead of $y^{(l)} = d^l y / dx^l$.

So, if $v = \varphi(u)$ is a solution to the equation $g(v) = 0$ and $|\varphi(u)|$ is bounded from zero and infinity as $|y|$ in (6.6), then the initial equation $f(x, y) = 0$ can have a solution with asymptotic form

$$y \sim x^\alpha \varphi(x^\beta), \quad x \to \infty. \tag{6.10}$$

6.2 Asymptotic Forms of Solutions to Painlevé Equations P_1–P_4, P_6

6.2.1 Equation P_1

$$f(x, y) \stackrel{\text{def}}{=} -y'' + 3y^2 + x = 0. \tag{6.11}$$

The support $\tilde{\mathbf{S}}(f)$ consists of three points $\mathbf{Q}_1 = (0, 1, 2)$, $\mathbf{Q}_2 = (0, 2, 0)$, $\mathbf{Q}_3 = (1, 0, 0)$. They lie in a plane with normal $\mathbf{N} = (1, 1/2, 1/4)$; see the space support $\tilde{\mathbf{S}}(f)$ and its convex hull $\mathbf{\Gamma}(f)$ in Figure 6.2. Here $\mathbf{\Gamma}(f) = \mathbf{\Gamma}_1^{(2)}$ and $f = \check{f}_1^{(2)}$, i.e., the truncated equation (6.11) coincides with the initial equation (6.11).

We make the power transformation (6.7) with $\alpha = 1/2$, $\beta = 5/4$. According to Theorem 6.3 and equation (6.9), equation (6.11) gives an equation with the truncation

$$\begin{aligned}\check{f}_1^{*(2)}(u, v) &= -(5/4)^2 u^{(1/2 + 2/4)(4/5)} \ddot{v} + 3u^{4/5} v^2 + u^{4/5} \\ &= u^{4/5} \left[-(5/4)^2 \ddot{v} + 3v^2 + 1 \right] \stackrel{\text{def}}{=} u^{4/5} g(v).\end{aligned}$$

Here and below, $\dot{v} = dv/du$. The equation $g(v) = 0$ has the first integral

$$\dot{v}^2 = 2(4/5)^2 (v^3 + v + C_0), \tag{6.12}$$

where here and below C_0 is an arbitrary constant. If the discriminant $\Delta = -4 - 27C_0^2$ of the polynomial $P(v) = v^3 + v + C_0$ is not zero, then solutions to equation (6.12) form a two–parameter family of elliptic functions

$$v = \varphi(u) \stackrel{\text{def}}{=} 2\wp\left(\frac{4}{5} u + C_1; -1, -\frac{1}{2} C_0 \right),$$

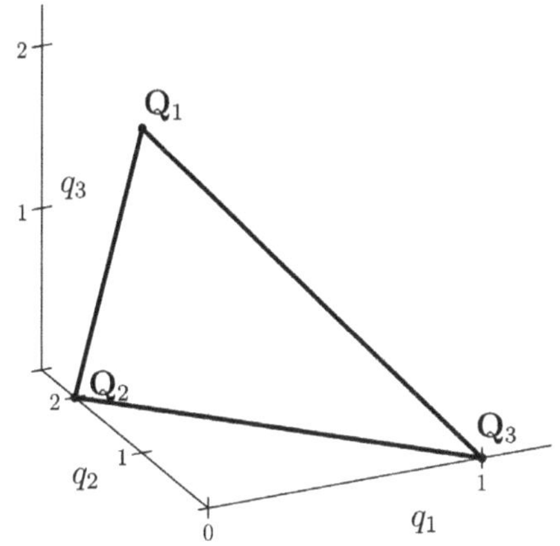

Figure 6.2. Support and polyhedron for equation (6.11).

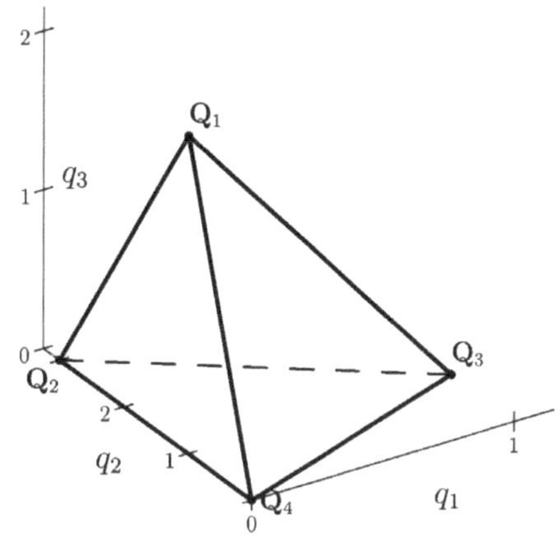

Figure 6.3. Support and polyhedron for equation (6.13).

where $\wp(u; g_2, g_3)$ is the Weierstrass elliptic function [2, Chap. 13] and C_1 is an arbitrary constant. Thus, the equation P_1 has a two–parameter family of solutions with asymptotic forms (6.10), i.e., $y = x^{1/2}\varphi(x^{5/4})$. These asymptotic forms were found by Boutroux [3] and were studied in many other publications.

If $\Delta = 0$, i. e. $C_0 = \mp 2i/(3\sqrt{3})$, then the polynomial $P(v)$ has double root and

$$v = \pm\frac{i}{\sqrt{3}}\frac{1+10\varkappa+\varkappa^2}{(1-\varkappa)^2}, \qquad \varkappa = C_1 \exp\sigma\sqrt{\pm 2i\sqrt{3}}\,\frac{4u}{5}, \qquad \sigma = \pm 1.$$

6.2.2 Equation P_2

$$f(x,y) \stackrel{\text{def}}{=} -y'' + 2y^3 + xy + a = 0, \tag{6.13}$$

where a is the complex parameter. If $a \neq 0$ the support $\tilde{\mathbf{S}}(f)$ consists of four points $\mathbf{Q}_1 = (0,1,2)$, $\mathbf{Q}_2 = (0,3,0)$, $\mathbf{Q}_3 = (1,1,0)$, $\mathbf{Q}_4 = 0$; see Figure 6.3. Their convex hull $\mathbf{\Gamma}(f)$ is a tetrahedron. It has 4 faces with external normals $\mathbf{N}_1 = (0,0,-1)$, $\mathbf{N}_2 = (-1,0,0)$, $\mathbf{N}_3 = (1,1/2,1/2)$, $\mathbf{N}_4 = (1,-1,1/2)$. Two of them $\mathbf{N}_3$ and $\mathbf{N}_4$ have $n_1 + n_3 > 0$.

First we consider the face $\mathbf{\Gamma}_3^{(2)}$ with the truncated equation

$$\check{f}_3^{(2)}(x,y) \stackrel{\text{def}}{=} -y'' + 2y^3 + xy = 0. \tag{6.14}$$

After transformation (6.7) with $\alpha = 1/2$, $\beta = 3/2$, using (6.9), we obtain the truncated equation

$$\begin{aligned}\check{f}_3^{*(2)}(u,v) &= -(3/2)^2 u^{(1/2+2\cdot 1/2)\cdot 2/3}\ddot{v} + 2uv^3 + uv\\ &= u\left[-(3/2)^2\ddot{v} + 2v^3 + v\right] \stackrel{\text{def}}{=} u\,g(v) = 0.\end{aligned}$$

Equation $g(v) = 0$ has the first integral

$$\dot{v}^2 = \frac{4}{9}\left(v^4 + v^2 + C_0\right). \tag{6.15}$$

If the discriminant $\Delta = 16C_0(4C_0 - 1)^2$ of the polynomial $P(v) = v^4 + v^2 + C_0$ is different from zero, i.e., $C_0 \neq 0$ and $C_0 \neq 1/4$, equation (6.15) has a two–parameter family of elliptic solutions

$$v = \varphi(u) \stackrel{\text{def}}{=} \alpha_4\left[1 + \frac{\sqrt{1-4C_0}}{\alpha_4^2 + 1/6 - 2\xi}\right], \quad \xi = \wp\left(\frac{2}{3}u + C_1; C_0 + \frac{1}{12}, \frac{C_0 - 36}{6^3}\right),$$

$\alpha_4 = \sqrt{(\sqrt{1-4C_0} - 1)/2}$ is a root of the polynomial $P(v)$ (see [2, Sect. 13.5], [3–5], [6, Chap. 7]). So the equation P_2 has a two–parameter family of solutions with asymptotic forms (6.10), i.e., $y \sim x^{1/2}\varphi(x^{3/2})$.

If $C_0 = 0$, then $v = -\sinh^{-1}\left(\frac{2}{3}u + C_1\right)$. If $C_0 = 1/4$, then $v = i(1+\varkappa)/(1-\varkappa)$, $\varkappa = C_1 \pm i\sqrt{2}u$.

Now we consider the face $\mathbf{\Gamma}_4^{(2)}$ with the truncated equation

$$\check{f}_4^{(2)}(x, y) \stackrel{\text{def}}{=} -y'' + xy + a = 0. \tag{6.16}$$

After transformation (6.7) with $\alpha = -1$, $\beta = 3/2$, using (6.9), we obtain the truncated equation

$$\check{f}_4^{*(2)}(u, v) \stackrel{\text{def}}{=} -(3/2)^2 u^{(-1+2\cdot 1/2)\cdot 2/3}\ddot{v} + v + a \stackrel{\text{def}}{=} g(v) = 0.$$

It has the first integral

$$\dot{v}^2 = (4/9)(v^2 + 2av + C_0). \tag{6.17}$$

Equation (6.17) has a two–parameter family of periodic solutions $v = \varphi(u)$ with the complex period $3\pi i$, which give asymptotic forms (6.10), i.e., $y \sim \varphi(x^{3/2})/x$, for solutions to equation P_2.

6.2.3 Equation P_3 for $cd \neq 0$

See (6.3) and Figure 6.1. The polyhedron $\mathbf{\Gamma}(f)$ has 5 faces $\mathbf{\Gamma}_j^{(2)}$ with external normals $\mathbf{N}_j$:

$$\mathbf{N}_1 = (1,0,0), \quad \mathbf{N}_2 = (-1,0,0), \quad \mathbf{N}_3 = (0,0,-1),$$
$$\mathbf{N}_4 = (-1,1,1), \quad \mathbf{N}_1 = (-1,-1,1).$$

Only one face $\mathbf{\Gamma}_1^{(2)}$ with truncated equation (6.4) has external normal $\mathbf{N} = (n_1, n_2, n_3)$ with $n_1 + n_3 > 0$. It is $\mathbf{N}_1 = (1,0,0)$, so $\alpha = 0$, $\beta = 1$, and we do not need to make the transformation (6.7). Equation $\check{f}_1^{(2)} = 0$ from (6.4) has the first integral

$$y'^2 = cy^4 + C_0 y^2 - d \stackrel{\text{def}}{=} P(y). \tag{6.18}$$

The discriminant of $P(y)$ is $\Delta = -16cd\left(4cd + C_0^2\right)^2$.

If $\Delta \neq 0$, i.e., $C_0^2 \neq -4cd$, then solutions to equation (6.18) are elliptic functions

$$y = \varphi(x) \stackrel{\text{def}}{=} \alpha_4\left[1 + \frac{\sqrt{C_0^2 + 4cd}}{3\sqrt{C_0^2 + 4cd} - 2C_0 - 12\xi}\right],$$

$$\xi = \wp\left(u + C_1; \frac{C_0^2 - 12cd}{12}, -\frac{C_0^3 + 36C_0cd}{6^3}\right),$$

$\alpha_4 = \sqrt{\left(\sqrt{C_0^2+4cd}-C_0\right)/(2c)}$ is a root of the polynomial $P(y)$ (see [2, Sect. 13.5], [7]). They give a two-parameter family of asymptotic forms $y \sim \varphi(x)$ for solutions to equation (6.3).

If $\Delta = 0$, i.e., $C_0 = \pm 2i\sqrt{cd}$, then $y = B(1+\varkappa)/(1-\varkappa)$, $\varkappa = C_1 \exp 2B\sigma\sqrt{c}x$, $B = \sqrt{\mp\sqrt{d/c}}$.

6.2.4 Equation P_3 for $c = 0$ and $ad \neq 0$

See (6.3) and Figure 6.4. Among faces $\mathbf{\Gamma}_j^{(2)}$ only one face $\mathbf{\Gamma}_1^{(2)}$ with the truncated equation

$$\check{f}_1^{(2)} \stackrel{\text{def}}{=} -xyy'' + xy'^2 + ay^3 + dx = 0 \tag{6.19}$$

has the external normal $\mathbf{N} = (n_1, n_2, n_3)$ with $n_1 \neq 0$ and $n_1 + n_3 > 0$. It is $\mathbf{N}_1 = (1, 1/3, -1/3)$, so $\alpha = 1/3$, $\beta = 2/3$ in the power transformation (6.7). According to Theorem 6.3, equation (6.19) takes the form

$$\check{f}_1^{*(2)}(u, v) \stackrel{\text{def}}{=} x\left(-\left(\frac{2}{3}\right)^2 v\ddot{v} + \left(\frac{2}{3}\right)^2 \dot{v}^2 + av^3 + d\right) \stackrel{\text{def}}{=} xg(v).$$

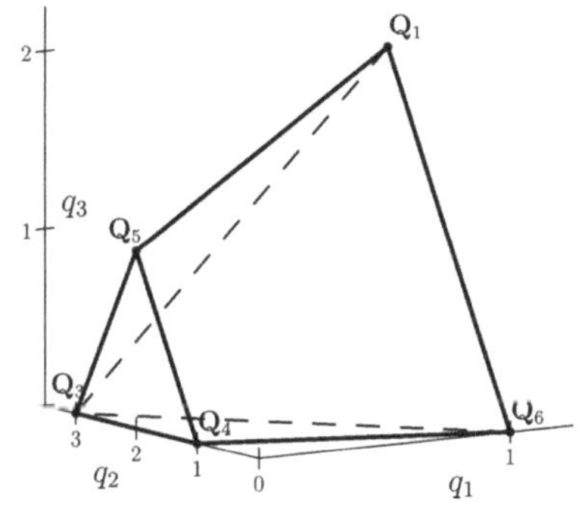

Figure 6.4. Support and polyhedron for equation (6.3) with $c = 0$.

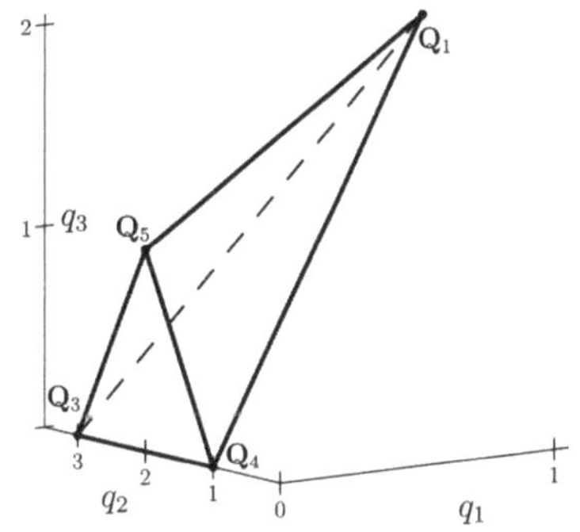

Figure 6.5. Support and polyhedron for equation (6.3) with $c = d = 0$.

The equation $g(v) = 0$ has the first integral

$$\dot{v}^2 = \left(\frac{3}{2}\right)^2 \left(2av^3 + C_0 v^2 - d\right) \stackrel{\text{def}}{=} \left(\frac{3}{2}\right)^2 P(v).$$

The discriminant of the polynomial $P(v)$ is $\Delta = 4d\left(C_0^3 - 27a^2 d\right)$. If $\Delta \neq 0$, i. e. $C_0^3 \neq 27a^2 d$, then the last equation has the elliptic solutions

$$v = \varphi(u) \stackrel{\text{def}}{=} \frac{2}{a}\wp\left(\frac{3}{2}u + C_1; \frac{C_0^2}{12}, \frac{a^2 d}{4} - \frac{C_0^3}{6^3}\right) - \frac{C_0}{3a}$$

(see [2, Sect. 13.5], [6, Chap. 16 and Appendix C]). Thus, solutions to the initial equation (6.3) have the asymptotic forms $y \sim \varphi\left(x^{2/3}\right)$.

6.2.5 Equation P_3 for $c = d = 0$ and $ab \neq 0$

See (6.3) and Figure 6.5. Among faces $\mathbf{\Gamma}_j^{(2)}$ only one face $\mathbf{\Gamma}_1^{(2)}$ has the external normal $\mathbf{N} = (n_1, n_2, n_3)$ with $n_1 \neq 0$ and $n_1 + n_3 > 0$. Namely, $\mathbf{N}_1 = (1, 0, -1/2)$. The corresponding truncated equation

$$\check{f}_1^{(2)} \stackrel{\text{def}}{=} -xyy'' + xy'^2 + ay^3 + by = 0 \tag{6.20}$$

according to Theorem 6.3 after the power transformation $y = v$, $u = x^{1/2}$ takes the form

$$\check{f}_1^{*(2)}(u, v) \stackrel{\text{def}}{=} -\frac{1}{4}v\ddot{v} + \frac{1}{4}\dot{v}^2 + av^3 + bv \stackrel{\text{def}}{=} g(v).$$

The equation $g(v) = 0$ has the first integral

$$\dot{v}^2 = 4\left(2av^3 + 2C_0 v^2 - 2bv\right) \stackrel{\text{def}}{=} 4P(v).$$

The discriminant of the polynomial $P(v)$ is $\Delta = 16b^2\left(C_0^2 + 4ab\right)$. If $\Delta \neq 0$, i.e., $C_0^2 \neq -4ab$, then the last equation has elliptic solutions

$$y = \varphi(u) \stackrel{\text{def}}{=} \frac{2}{a}\wp\left(2u + C_1; \frac{C_0^2}{3}, \frac{a^2 b}{2} - \frac{C_0^3}{27}\right) - \frac{2C_0}{3a}.$$

Thus, solutions to the initial equation (6.3) have asymptotic forms $y \sim \varphi(x^{1/2})$.

6.2.6 Equation P_4

Equation P_4, multiplyed by its common divisor, is

$$f(x, y) \stackrel{\text{def}}{=} -2yy'' + y'^2 + 3y^4 + 8xy^3 + 4(x^2 - a)y^2 + 2b = 0, \tag{6.21}$$

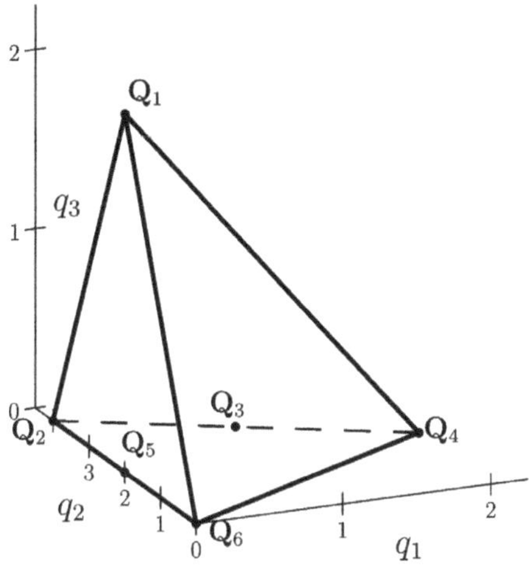

Figure 6.6. Support and polyhedron for equation (6.21).

where a and b are complex parameters. If $a, b \neq 0$, the support $\tilde{\mathbf{S}}$ has 6 points and the polyhedron $\mathbf{\Gamma}(f)$ is a tetrahedron, as shown in Figure 6.6. It has 4 faces $\mathbf{\Gamma}_j^{(2)}$ with normals $\mathbf{N}_1 = (0, 0, -1)$, $\mathbf{N}_2 = (-1, 0, 0)$, $\mathbf{N}_3 = (1, 1, 1)$, $\mathbf{N}_4 = (1, -1, 1)$. Only two of them ($\mathbf{N}_3 = (1, 1, 1)$ and $\mathbf{N}_4 = (1, -1, 1)$) have the external normal vector $\mathbf{N} = (n_1, n_2, n_3)$ with $n_1 + n_3 > 0$.

First consider the face $\mathbf{\Gamma}_3^{(2)}$. It corresponds to the truncated equation

$$\check{f}_3^{(2)}(x, y) \stackrel{\text{def}}{=} -2yy'' + y'^2 + 3y^4 + 8xy^3 + 4x^2y^2 = 0. \tag{6.22}$$

After power transformation (6.7) with $\alpha = 1$, $\beta = 2$, we obtain the truncated equation

$$\begin{aligned} \check{f}_3^{*(2)}(u, v) &\stackrel{\text{def}}{=} -2 \cdot 4u^{(1+2+1)/2}\ddot{v}v + 4u^2\dot{v}^2 + 3u^2v^4 + 8u^2v^3 + 4u^2v^2 \\ &= 4u^2[-2v\ddot{v} + \dot{v}^2 + (3/4)v^4 + 2v^3 + v^2] \stackrel{\text{def}}{=} 4u^2 g(v) = 0. \end{aligned} \tag{6.23}$$

The equation $g(v) = 0$ has the first integral $\dot{v}^2 = v^4/4 + v^3 + v^2 + C_0 v \stackrel{\text{def}}{=} P(v)$. The discriminant of the polynomial $P(v)$ is $\Delta = -C_0^3(27C_0 - 8)/16$. If it differs from zero, i.e., $C_0 \neq 0$ and $C_0 \neq 8/27$, then the last equation has a two–parameter family of elliptic solutions

$$v = \varphi(u) \stackrel{\text{def}}{=} 3C_0\left[12\wp\left(u + C_1; \frac{1}{12} - \frac{C_0}{4}, -\frac{C_0^2}{64} + \frac{C_0}{48} - \frac{1}{216}\right) - 1\right]^{-1}$$

(see [2, Sect. 13.5], [8]). Thus, initial equation P_4 has solutions with asymptotic forms $y \sim x\varphi(x^2)$.

Now consider the face $\mathbf{\Gamma}_4^{(2)}$ with the truncated equation

$$\check{f}_4^{(2)}(x, y) \stackrel{\text{def}}{=} -2yy'' + y'^2 + 4x^2y^2 + 2b = 0. \tag{6.24}$$

After power transformation (6.7) with $\alpha = -1$, $\beta = 2$, we obtain the truncated equation

$$\check{f}_4^{*(2)}(u,v) \stackrel{\text{def}}{=} -2 \cdot 4v\ddot{v} + 4\dot{v}^2 + 4v^2 + 2b \stackrel{\text{def}}{=} g(v) = 0.$$

It has the first integral

$$\dot{v}^2 = v^2 + C_0 v - b/2. \tag{6.25}$$

Equation (6.25) has periodic solutions $v = \varphi(u)$ with period $2\pi i$ and the equation P_4 has a two–parameter family of solutions with asymptotic forms $y = \varphi(x^2)/x$.

6.2.7 Equation P_6

Multiplied by its common divisor, equation P_6 has the support $\tilde{\mathbf{S}}(P_6)$ consisting of 34 points in the generic case. The polyhedron $\mathbf{\Gamma}$ for P_6 has 10 faces $\mathbf{\Gamma}_j^{(2)}$ with external normals $\mathbf{N}_j = (n_1, n_2, n_3)$: $(0,0,1)$, $(-1,0,1)$, $(-1,-1,1)$, $(0,-1,1)$, $(1,0,-1)$, $(1,1,-1)$, $(0,1,1)$, $(-2,-1,1)$, $(2,1,-3)$, $(0,0,-1)$. No normal $\mathbf{N}_j$ has $n_1 \neq 0$ and $n_1 + n_3 > 0$. Thus P_6 has nonperiodic and elliptic asymptotic forms of solutions [9]. The same is true for nongeneric cases as well [9].

Bibliography

[1] A. D. Bruno, Plane Power Geometry for one ODE and P_1–P_6, *Chapter 1, this volume.*

[2] H. Bateman and A. Erdĺyi, *Higher Transcendental Functions*, vol. 3, McGraw–Hill, NY, Toronto, London, 1955.

[3] P. Boutroux, Recherches sur les transcendantes de M. Painlevé et l'étude asymptotique des équations différentielles du second ordre, *Ann. Sci. Ecole Norm. Sup.*, **30**: 255–375, 1913; **31**: 99–159, 1914.

[4] N. A. Kudryashov, *Analytic Theory of Nonlinear Differential Equations,* Inst. Comp. Issl., Moscow-Izhevsk, 2004 (in Russian).

[5] A. D. Bruno and I. V. Goryuchkina, Boutroux asymptotic forms of solutions to Painlevé equations and Power Geometry, *Doklady Mathem.*, **78**(2): 681–685, 2008 (= *DAN,* **422**(2): 157–160, 2008).

[6] A. R. Its, A. A. Kapaev, V. Yu. Novokshenov, and A. S. Fokas, *Painlevé Transcendents: the Reimann–Hilbert approach,* AMS, Providence, 2006 (= R&C Dynamics, Moscow-Izhevsk, 2005, 727 p. (in Russian)).

[7] A. D. Bruno and I. V. Goryuchkina, Asymptotic forms of solutions to the third Painlevé equation, *Doklady Mathem.*, **78**(2): 765–768, 2008 (= *DAN,* **422**(6): 729–732, 2008).

[8] A. D. Bruno and I. V. Goryuchkina, Asymptotic forms of solutions to the fourth Painlevé equation, *Doklady Mathem.*, **78**(3): 868–873, 2008 (= *DAN,* **423**(4): 443–448, 2008).

[9] I. V. Goryuchkina, *Three-dimensional analysis of asymptotic forms of the solutions to the sixth Painlevé equation*, preprint nr. 56, Keldysh Inst. Appl. Math. Moscow, 2010, 24 p. (in Russian); available from the Keldysh Institute of Applied Mathematics of RAS: `http://library.keldysh.ru/preprints/Default.asp?lg=e`.

Author Information

Alexander D. Bruno, Keldysh Institute of Applied Mathematics (KIAM), Miusskaya Sq. 4, 125047, Moscow, Russia.
E-mail: abruno@keldysh.ru

Chapter 7

Elliptic and Periodic Asymptotic Forms of Solutions to P_5

Alexander D. Bruno and Anastasia V. Parusnikova

Abstract. We apply methods from Space Power Geometry to the fifth Painlevé equation. Near infinity we obtained 2 families of elliptic asymptotic forms and 4 families of periodic asymptotic forms of its solutions. All of these families are 2-parameter.

Keywords. Painlevé Equations. Space Power Geometry, Asymptotic Form, Elliptic Solution, Periodic Solution.

2010 Mathematics Subject Classification. 34M55, 34E05.

7.1 The Fifth Painlevé Equation

We consider the fifth Painlevé equation P_5

$$w'' = \left(\frac{1}{2w} + \frac{1}{w-1}\right)(w')^2 - \frac{w'}{z} + \frac{(w-1)^2}{z^2}\left(\alpha w + \frac{\beta}{w}\right) + \frac{\gamma w}{z} + \frac{\delta w(w+1)}{w-1}, \tag{7.1}$$

where $\alpha, \beta, \gamma, \delta$ are complex parameters, z is an independent and w a dependent complex variable, $w' = dw/dz$. The fifth Painlevé equation (7.1) has two singular points: $z = 0$ and $z = \infty$.

Let us represent the P_5 equation (7.1) in the form of a differential sum (a polynomial in z, w, w', w''), i.e., multiply it by its common divisor $z^2 w(w-1)$ and put all the terms of the equation to the right part of the equation P_5:

$$\begin{aligned} f(z,w) \stackrel{\text{def}}{=} & -z^2 w(w-1)w'' + z^2\left(\frac{3}{2}w - \frac{1}{2}\right)(w')^2 - zw(w-1)w' \\ & + (w-1)^3(\alpha w^2 + \beta) + \gamma z w^2(w-1) + \delta z^2 w^2(w+1) = 0. \end{aligned} \tag{7.2}$$

The P_5 equation (7.1) is invariant under the substitution

$$(z, w, \alpha, \beta, \gamma, \delta) = (\tilde{z}, \frac{1}{\tilde{w}}, -\tilde{\beta}, -\tilde{\alpha}, -\tilde{\gamma}, \tilde{\delta}). \tag{7.3}$$

The work was supported by RFBR, Grant Nr. 11-01-00023.

We now apply methods from Space Power Geometry [1] to equation P_5.

To each differential monomial $a(z, w)$ in the differential sum (7.2) we put in correspondence its *three-dimensional power exponent* $\mathbf{Q}(a(z,w)) = (q_1, q_2, q_3)$ according to the following rules:

$$\mathbf{Q}(cz^r w^s) = (r, s, 0);$$

$$\mathbf{Q}\left(\frac{d^l w}{dz^l}\right) = (0, 1, l);$$

$$\mathbf{Q}\big(a(z,w)b(z,w)\big) = \mathbf{Q}\big(a(z,w)\big) + \mathbf{Q}\big(b(z,w)\big).$$

The set of all vector power exponents of differential monomials in the differential sum $f(z, w)$ is called *the three-dimensional support of the differential sum* $f(z, w)$ and is denoted by $\tilde{\mathbf{S}}(f)$. The convex hull $\mathbf{\Gamma}(f)$ of a support $\tilde{\mathbf{S}}(f)$ is called *a polyhedron of the differential sum* $f(z, w)$, the boundary $\partial\mathbf{\Gamma}(f)$ of $\mathbf{\Gamma}(f)$ consists of vertices $\mathbf{\Gamma}_j^{(0)}$, edges $\mathbf{\Gamma}_j^{(1)}$ and faces $\mathbf{\Gamma}_j^{(2)}$.

7.2 The case $\delta \neq 0$

7.2.1 General Properties of the P_5 Equation

The support $\tilde{\mathbf{S}}$ of equation (7.2) consists of the following points:

$$\begin{gathered}\mathbf{Q}_1 = (2,2,2), \quad \mathbf{Q}_2 = (2,3,2), \quad \mathbf{Q}_3 = (1,2,1), \quad \mathbf{Q}_4 = (1,3,1),\\ \mathbf{Q}_5 = (0,0,0), \quad \mathbf{Q}_6 = (0,1,0), \quad \mathbf{Q}_7 = (0,2,0),\\ \mathbf{Q}_8 = (0,3,0), \quad \mathbf{Q}_9 = (0,4,0), \quad \mathbf{Q}_{10} = (0,5,0),\\ \mathbf{Q}_{11} = (1,2,0), \quad \mathbf{Q}_{12} = (1,3,0), \quad \mathbf{Q}_{13} = (2,2,0), \quad \mathbf{Q}_{14} = (2,3,0).\end{gathered} \tag{7.4}$$

The convex hull of the support of equation (7.2) is a pentahedron $\mathbf{\Gamma}(f)$ (see Figure 7.1). Its vertices are $\mathbf{Q}_1$, $\mathbf{Q}_2$, $\mathbf{Q}_5$, $\mathbf{Q}_{10}$, $\mathbf{Q}_{13}$, $\mathbf{Q}_{14}$. Its two-dimensional faces are

$$\begin{gathered}\mathbf{\Gamma}_1^{(2)} = \operatorname{conv}(\mathbf{Q}_1, \mathbf{Q}_2, \mathbf{Q}_{13}, \mathbf{Q}_{14}), \qquad \mathbf{\Gamma}_2^{(2)} = \operatorname{conv}(\mathbf{Q}_1, \mathbf{Q}_5, \mathbf{Q}_{13}),\\ \mathbf{\Gamma}_3^{(2)} = \operatorname{conv}(\mathbf{Q}_2, \mathbf{Q}_{10}, \mathbf{Q}_{14}), \qquad \mathbf{\Gamma}_4^{(2)} = \operatorname{conv}(\mathbf{Q}_1, \mathbf{Q}_2, \mathbf{Q}_5, \dots, \mathbf{Q}_{10}),\\ \mathbf{\Gamma}_5^{(2)} = \operatorname{conv}(\mathbf{Q}_5, \dots, \mathbf{Q}_{10}, \mathbf{Q}_{11}, \mathbf{Q}_{12}, \mathbf{Q}_{13}, \mathbf{Q}_{14}).\end{gathered}$$

According to [1], as we consider the case $z \to \infty$, we are interested only in such faces $\mathbf{\Gamma}_j^{(2)}$ of the polyhedron $\mathbf{\Gamma}(f)$, for which the first coordinate of the external normal $\mathbf{N}_j = (n_1, n_2, n_3)$ is greater than zero ($n_1 > 0$), assuming that $n_1 = 1$; according to Theorem 1.3 from [1], we consider only the faces for which the coordinates of external normals satisfy the condition $n_3 + 1 > 0$.

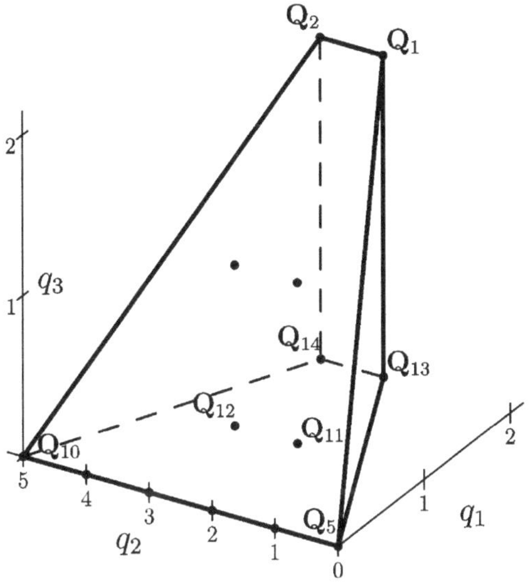

Figure 7.1. Support and polyhedron for equation (7.2).

We obtain only three following faces which satisfy the conditions described above:

1. $\mathbf{\Gamma}_1^{(2)}$, the equation of a plane containing the face is $q_1 - 2 = 0$, whereby the external normal for this face is $\mathbf{N}_1 = (1, 0, 0)$;
2. $\mathbf{\Gamma}_2^{(2)}$, the equation of a plane containing the face is $q_1 - q_2 = 0$, whereby he external normal to this face is $\mathbf{N}_2 = (1, -1, 0)$;
3. $\mathbf{\Gamma}_3^{(2)}$, the equation of a plane containing the face is $q_1 + q_2 - 5 = 0$, whereby the external normal to this face is $\mathbf{N}_3 = (1, 1, 0)$.

After the substitution (7.3) the face $\mathbf{\Gamma}_1^{(2)}$ transforms into itself and the faces $\mathbf{\Gamma}_2^{(2)}$ and $\mathbf{\Gamma}_3^{(2)}$ transform one into another.

The external normals to the faces $\mathbf{\Gamma}_4^{(2)}$ and $\mathbf{\Gamma}_5^{(2)}$ ($\mathbf{N}_4 = (-1, 0, 1)$ and $\mathbf{N}_5 = (0, 0, -1)$, respectively) do not satisfy the two conditions described above.

7.2.2 The First Family of Elliptic Asymptotic Forms

The face $\mathbf{\Gamma}_1^{(2)}$ corresponds to the following truncated equation

$$\check{f}_1^{(2)}(z, w) \stackrel{\text{def}}{=} -z^2 w(w-1)w'' + z^2\left(\frac{3}{2}w - \frac{1}{2}\right)(w')^2 + \delta z^2 w^2 (w+1) = 0. \quad (7.5)$$

We divide the equation (7.5) by z^2, make the substitution $(w')^2 = P$, after which $w'' = \frac{1}{2}\frac{dP}{dw}$, and obtain the linear equation

$$-w(w-1)\frac{dP}{dw} + (3w-1)\,P + 2\delta w^2(w+1) = 0, \tag{7.6}$$

which has the solution $P(w) = -2\delta w\left(C_0(w-1)^2 + w\right)$. Then we obtain the first integral of the equation (7.5)

$$(w')^2 = -2\delta w\big(C_0(w-1)^2 + w\big), \tag{7.7}$$

where C_0 is an arbitrary constant.

The discriminant Δ of the cubic polynomial $ax^3 + bx^2 + cx + d$ is

$$\Delta = b^2c^2 - 4ac^3 - 4b^3d + 18abcd - 27a^2d^2.$$

So, for the polynomial $w(C_0(w-1)^2 + w)$,

$$\Delta = C_0^2(1 - 4C_0).$$

The discriminant is equal to zero iff $C_0 = 0$ or $C_0 = 1/4$. If the discriminant is equal to zero, the polynomial $P(w)$ has less than 3 different roots. In other cases the polynomial under consideration has 3 different complex roots.

If $C_0 \neq 0$ and $C_0 \neq 1/4$, then the solutions to equation (7.7) are the elliptic functions $w = \varphi(z)$ [2].

Let us consider the following equation

$$\left(\frac{dx}{dz}\right)^2 = ax^3 + bx^2 + cx + d. \tag{7.8}$$

We assume that the discriminant of the cubic polynomial in the right side of equation (7.8) is not equal to zero. After the substitution

$$x = \frac{4}{a}y - \frac{b}{3a}$$

equation (7.8) is transformed to the Weierstrass normal form

$$\left(\frac{dy}{dz}\right)^2 = 4y^3 - g_2y - g_3, \tag{7.9}$$

where according to Section 13.5 in [2]

$$g_2 = \frac{b^2 - 3ac}{12}, \quad g_3 = \frac{-2b^3 + 9abc - 27a^2d}{432}. \tag{7.10}$$

We recall that the elliptic Weierstrass function $\wp(z; g_2, g_3)$ satisfies equation (7.9) with $y = \wp$, where g_2 and g_3 are constant.

According to the formulae (7.9), (7.10) we obtain that equation (7.7) (if $\Delta \neq 0$) has the following solution

$$w = -\frac{2}{\delta C_0}\wp\left(z + C_1; \frac{\delta^2}{3}\left(C_0^2 - 4C_0 + 1\right), \frac{\delta^3}{54}\left(C_0^2 + 8C_0 - 2\right)(2C_0 - 1)\right) + \frac{2C_0 - 1}{3C_0}, \tag{7.11}$$

where $\wp$ is the Weierstrass function, $C_0 \neq 0$, $C_0 \neq 1/4$, C_1 are arbitrary constants. This is a two-parameter (on C_0 and C_1) family of asymptotic forms.

If $C_0 = 0$, equation (7.6) has the solutions $w = \tilde{C}e^{\sqrt{-2\delta}z}$ and $w = \tilde{C}e^{-\sqrt{-2\delta}z}$. If $C_0 = 1/4$, equation (7.6) has the solutions $w = \tan^2(\sqrt{-\delta/8}z + \tilde{C})$, where C and $\tilde{C}$ are arbitrary constants.

7.2.3 The First Family of Periodic Asymptotic Forms

The face $\mathbf{\Gamma}_2^{(2)}$ corresponds to the following truncated equation:

$$\check{f}_2^{(2)}(z, w) \stackrel{\text{def}}{=} z^2 w w'' - \frac{z^2}{2}(w')^2 + \delta z^2 w^2 - \beta = 0. \tag{7.12}$$

According to Theorem 1.3 from [1], we make the transformation $w = v/z$ in equation (7.12). Then w' passes into v'/z and w'' passes into v''/z. After this substitution, equation (7.12) transforms to the following equation:

$$\check{f} \stackrel{\text{def}}{=} v v'' - \frac{(v')^2}{2} + \delta v^2 - \beta = 0, \ g(v) = \check{f}. \tag{7.13}$$

Now we pass to the solution of the equation $g(v) = 0$. After the transformation $(v')^2 = P$, we obtain the linear equation

$$v\frac{dP}{dv} - P + 2\delta v^2 - 2\beta = 0,$$

which has the following solution:

$$P = (v')^2 = -2\delta v^2 - 2C_0 v - 2\beta, \tag{7.14}$$

where C_0 is an arbitrary constant.

Let us consider an equation

$$(y')^2 = ay^2 + by + c, \tag{7.15}$$

where a, b, c are constant, $y' = dy/dx$. After differentiating equation (7.15) with respect to x, we obtain the following equation

$$2y'y'' = 2ayy' + by', \tag{7.16}$$

which has the solutions $y = \text{const}$ and

$$y = C_+ e^{\sqrt{a}x} + C_- e^{-\sqrt{a}x} - \frac{b}{2a}, \tag{7.17}$$

where C_+, C_- are arbitrary constants. The (7.17) expression is a solution of equation (7.15) iff

$$C_+C_- = \frac{b^2 - 4ac}{16a^2}. \tag{7.18}$$

Using the formulae (7.15), (7.17), and (7.18), we obtain that equation (7.14) has the solutions

$$v = C_+ e^{\sqrt{-2\delta}z} + C_- e^{-\sqrt{-2\delta}z} - \frac{C_0}{2\delta}, \tag{7.19}$$

where

$$C_+C_- = \frac{C_0^2}{16\delta^2} - \frac{\beta}{4\delta}. \tag{7.20}$$

Solution (7.19) is periodic in t with period $T = 2\pi i/\sqrt{-2\delta}$ if $\sqrt{-2\delta}z = it$, $t \in \mathbb{R}$.

7.2.4 The Second Family of Periodic Asymptotic Forms

The face $\mathbf{\Gamma}_3^{(2)}$ corresponds to the following truncated equation

$$\check{f}_3^{(2)}(z, w) \stackrel{\text{def}}{=} -z^2w^2w'' + \frac{3z^2}{2}(w')^2 + \delta z^2 w^3 + \alpha w^5 = 0. \tag{7.21}$$

According to Theorem 1.3 from [1], we make the transformation $w = vz$ in equation (7.12). Then w' passes into zv' and w'' passes into zv''. After this substitution, equation (7.21) transforms to the following equation:

$$\check{f} \stackrel{\text{def}}{=} -vv'' + \frac{3(v')^2}{2} + \delta v^3 + \alpha v^5 = 0,\ g(v) = \check{f}. \tag{7.22}$$

Now we pass to the solution of the equation $g(v) = 0$. After the transformation $(v')^2 = P$, we obtain the linear equation

$$-v^2\frac{dP}{dv} + 3vP + 2\delta v^3 + 2\alpha v^4 = 0,$$

which has the following solution:

$$P = (v')^2 = v^2(2\alpha v^2 + C_0 v - 2\delta), \tag{7.23}$$

where C_0 is an arbitrary constant. Substituting $\tilde{v} = 1/v$ to equation (7.23) we obtain the equation

$$(\tilde{v}')^2 = 2\alpha - 2C_0\tilde{v} - 2\delta\tilde{v}^2, \tag{7.24}$$

Using the formulae (7.15), (7.17), and (7.18), we obtain that equation (7.24) has the solutions

$$\tilde{v} = C_+ e^{\sqrt{-2\delta}z} + C_- e^{-\sqrt{-2\delta}z} - \frac{C_0}{2\delta},$$

where

$$C_+C_- = \frac{C_0^2}{16\delta^2} + \frac{\alpha}{4\delta}. \tag{7.25}$$

Thus,

$$w = z\big[C_+ e^{\sqrt{-2\delta}z} + C_- e^{-\sqrt{-2\delta}z} - C_0/(2\delta)\big]^{-1}. \tag{7.26}$$

Solutions (7.26) are periodic in t with period $T = 2\pi i/\sqrt{-2\delta}$ if $\sqrt{-2\delta}z = it$, $t \in \mathbb{R}$.

We can also obtain periodic asymptotic forms as solutions of the truncated equation corresponding to the face $\mathbf{\Gamma}_3^{(2)}$ from the asymptotic forms (7.19) found for the face $\mathbf{\Gamma}_2^{(2)}$ using the symmetry (7.3).

7.3 The Case $\delta = 0$, $\gamma \neq 0$

7.3.1 General Properties

The support of equation (7.2) in case $\delta = 0$ and $\gamma \neq 0$ consists of the points $\mathbf{Q}_1 - \mathbf{Q}_{12}$ (7.4).

In this case the convex hull of the support of equation (7.2) is a pentahedron $\tilde{\mathbf{\Gamma}}(f)$ (see Figure 7.2). Its vertices are $\mathbf{Q}_1, \mathbf{Q}_2, \mathbf{Q}_5, \mathbf{Q}_{10}, \mathbf{Q}_{11}, \mathbf{Q}_{12}$. Its two-dimensional faces are

$$\tilde{\mathbf{\Gamma}}_1^{(2)} = \operatorname{conv}(\mathbf{Q}_1, \mathbf{Q}_2, \mathbf{Q}_{11}, \mathbf{Q}_{12}), \qquad \tilde{\mathbf{\Gamma}}_2^{(2)} = \operatorname{conv}(\mathbf{Q}_1, \mathbf{Q}_5, \mathbf{Q}_{11}),$$
$$\tilde{\mathbf{\Gamma}}_3^{(2)} = \operatorname{conv}(\mathbf{Q}_2, \mathbf{Q}_{10}, \mathbf{Q}_{12}), \qquad \tilde{\mathbf{\Gamma}}_4^{(2)} = \operatorname{conv}(\mathbf{Q}_1, \mathbf{Q}_2, \mathbf{Q}_5, \dots, \mathbf{Q}_{10}),$$
$$\tilde{\mathbf{\Gamma}}_5^{(2)} = \operatorname{conv}(\mathbf{Q}_5, \dots, \mathbf{Q}_{10}, \mathbf{Q}_{11}, \mathbf{Q}_{12}).$$

According to [1], as we consider the case $z \to \infty$, we are interested only in such faces $\tilde{\mathbf{\Gamma}}_j^{(2)}$ of the polyhedron $\tilde{\mathbf{\Gamma}}(f)$, the first coordinate of the external normal $\tilde{\mathbf{N}}_j = (n_1, n_2, n_3)$ to which is greater than zero ($n_1 > 0$), we assume that $n_1 = 1$; according to Theorem 1.3 [1], we consider only the faces the coordinates of external normals to which satisfy the condition $n_3 + 1 > 0$.

We obtain only the three following edges which satisfy the conditions described above:

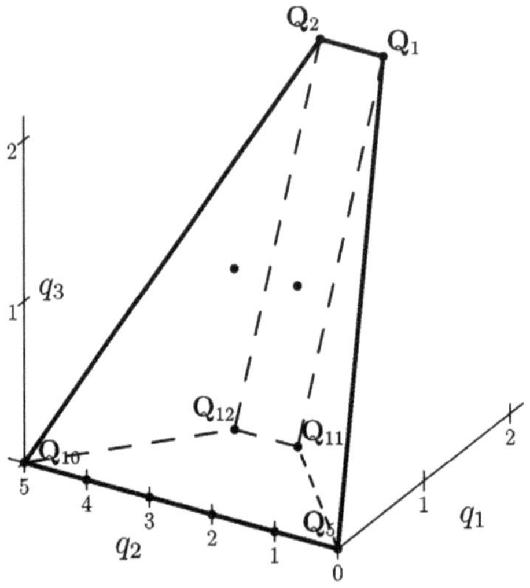

Figure 7.2. Support and polyhedron for equation (7.2) with $\delta = 0$.

1. $\tilde{\mathbf{\Gamma}}_1^{(2)}$, the equation of a plane containing the face is $2q_1 - q_3 - 2 = 0$, whereby the external normal to this face is $\tilde{\mathbf{N}}_1 = (1, 0, -1/2)$;

2. $\tilde{\mathbf{\Gamma}}_2^{(2)}$, the equation of a plane containing the face is $2q_1 - q_2 - q_3 = 0$, whereby the external normal to this face is $\tilde{\mathbf{N}}_2 = (1, -1/2, -1/2)$;

3. $\tilde{\mathbf{\Gamma}}_3^{(2)}$, the equation of a plane containing the face is $2q_1 + q_2 + q_3 - 5 = 0$, whereby the external normal to this face is $\tilde{\mathbf{N}}_3 = (1, 1/2, 1/2)$.

After the substitution (7.3), the face $\tilde{\mathbf{\Gamma}}_1^{(2)}$ transforms into itself and the faces $\tilde{\mathbf{\Gamma}}_2^{(2)}$ and $\tilde{\mathbf{\Gamma}}_3^{(2)}$ transform one into another.

The external normals to the faces $\tilde{\mathbf{\Gamma}}_4^{(2)}$ and $\tilde{\mathbf{\Gamma}}_5^{(2)}$ ($\tilde{\mathbf{N}}_4 = (-1, 0, 1)$ and $\tilde{\mathbf{N}}_5 = (0, 0, -1)$, respectively) do not satisfy the two conditions described above.

7.3.2 The Second Family of Elliptic Asymptotic Forms

The face $\tilde{\mathbf{\Gamma}}_1^{(2)}$ corresponds to the following truncated equation

$$\check{f}_1^{(2)}(z, w) \stackrel{\text{def}}{=} -z^2 w(w-1)w'' + z^2 \left(\frac{3}{2}w - \frac{u^2}{2}\right)(w')^2 + \gamma z w^2 (w-1) = 0. \quad (7.27)$$

According to Theorem 1.3 from [1], we make the transformation

$$u = \sqrt{z},$$

then w' passes into $\frac{1}{2u}\frac{dw}{du}$ and w'' passes into $\frac{1}{4u^2}\frac{d^2w}{du^2}$. After the substitution, equation (7.27) transforms to the following truncated equation

$$\check{f} = -w(w-1)\frac{u^2}{4}\frac{d^2w}{du^2} + \left(\frac{3}{2}w - \frac{1}{2}\right)\frac{u^2}{4}\left(\frac{dw}{du}\right)^2 + \gamma u^2 w^2 (w-1) = 0. \quad (7.28)$$

We obtain that

$$g(v) = 4u^{-2}\check{f} \stackrel{\text{def}}{=} -w(w-1)\frac{d^2w}{du^2} + \frac{3w-1}{2}\left(\frac{dw}{du}\right)^2 + 4\gamma w^2(w-1) = 0. \quad (7.29)$$

Now we pass to the solution of the equation $g(v) = 0$. After the transformation $(\frac{dw}{du})^2 = P(w)$, we obtain the linear equation

$$\frac{1}{2}w(w-1)\frac{dP}{dw} = \frac{3w-1}{2}P + 4\gamma w^2(w-1), \quad (7.30)$$

which has the following solution:

$$\left(\frac{dw}{du}\right)^2 = -8\gamma w(w-1)(1 + C_0(w-1)) \stackrel{\text{def}}{=} P(w), \quad (7.31)$$

where C_0 is an arbitrary constant.

The discriminant of the polynomial $w(w-1)(1 + C_0(w-1))$ is equal to $\Delta = (C_0 - 1)^2$. If $C_0 \neq 0$ and $\Delta \neq 0$, i.e., $C_0 \neq 1$, the solutions $w = \varphi(u)$ of the equation are the elliptic functions. This is why, if $\delta = 0$, $\gamma \neq 0$, the P_5 equation has a two-parameter family of elliptic asymptotic forms $w \sim \varphi(\sqrt{z})$ of the solutions. If $C_0 = 0$, then $P(w)$ is a polynomial of the second degree.

According to the formulae (7.9) and (7.10), we obtain that equation (7.31) (if $\Delta \neq 0$ and $C_0 \neq 0$) has the following solution:

$$w = -\frac{1}{2\gamma C_0}\wp\Bigg(\sqrt{z} + C_1; \frac{16\gamma^2}{3}(C_0^2 - C_0 + 1), \\ \frac{32\gamma^3}{27}(C_0 - 1)(4C_0 - 1)(5C_0 - 2)\Bigg) + \frac{2C_0 - 1}{3C_0}, \quad (7.32)$$

where $\wp$ is the Weierstrass function, $C_0 \neq 1$, $C_0 \neq 0$, C_3 are arbitrary constants. This is a two-parameter (on C_0 and C_3) family of asymptotic forms.

If $C_0 = 1$, equation (7.31) has the solution

$$w = \frac{1}{Ce^{\sqrt{8\gamma}u} - (16C)^{-1}e^{-\sqrt{8\gamma}u} + 1/2},$$

where C is an arbitrary constant.

If $C_0 = 0$, equation (7.31) has the following form:

$$\left(\frac{dw}{du}\right)^2 = -8\gamma w(w-1);$$

using the formulae (7.15), (7.17), and (7.18), we obtain that this equation has solutions

$$w = Ce^{\sqrt{-8\gamma}u} + \frac{C}{16}e^{-\sqrt{-8\gamma}u} + \frac{1}{2}, \tag{7.33}$$

where C is an arbitrary constant.

7.3.3 The Third Family of Periodic Asymptotic Forms

The face $\tilde{\mathbf{\Gamma}}_2^{(2)}$ corresponds to the following truncated equation:

$$z^2ww'' - \frac{z^2}{2}(w')^2 - \gamma zw^2 - \beta = 0. \tag{7.34}$$

According to Theorem 1.3 from [1], we make the transformation

$$w = \frac{v}{\sqrt{z}}, \ u = \sqrt{z},$$

in equation (7.34), and after this transformation w' passes into $\frac{1}{2u^2}\frac{dv}{du}$ and w'' passes into $\frac{1}{4u^3}\frac{d^2v}{du^2}$. After the substitution, equation (7.34) transforms to the equation

$$\begin{aligned} \check{f} &\stackrel{\text{def}}{=} \frac{1}{4}v\frac{d^2v}{du^2} - \frac{1}{8}\left(\frac{dv}{du}\right)^2 - \gamma v^2 - \beta = 0, \\ g(v) &= 4\check{f} = v\frac{d^2v}{du^2} - \frac{1}{2}\left(\frac{dv}{du}\right)^2 - 4\gamma v^2 - 4\beta. \end{aligned} \tag{7.35}$$

Now we pass to the solution of the equation $g(v) = 0$. After the transformation $\dot{v}^2 = P$, we obtain the linear equation

$$v\frac{dP}{dv} - P - 8\beta - 8\gamma v^2 = 0,$$

which has the following solution:

$$\dot{v}^2 = 8(\gamma v^2 + C_0 v - \beta) \stackrel{\text{def}}{=} P, \tag{7.36}$$

where $\dot{v} = dv/du$ and C_0 is an arbitrary constant.

Using the formulae (7.15), (7.17), and (7.18), we obtain that equation (7.36) has solutions

$$v = C_+e^{\sqrt{8\gamma z}} + C_-e^{-\sqrt{8\gamma z}} - \frac{C_0}{2\gamma}, \tag{7.37}$$

where

$$C_+C_- = \frac{\beta}{4\gamma} + \frac{C_0^2}{16\gamma^2}. \tag{7.38}$$

7.3.4 The Fourth Family of Periodic Asymptotic Forms

The face $\tilde{\mathbf{\Gamma}}_3^{(2)}$ corresponds to the following truncated equation:

$$-z^2w^2w'' + \frac{3z^2w}{2}(w')^2 + \gamma z w^3 + \alpha w^5 = 0. \tag{7.39}$$

According to Theorem 1.3 from [1], we make the transformation

$$w = \sqrt{z}v, \; u = \sqrt{z}, \tag{7.40}$$

in equation (7.39), and after this transformation w' passes into $\frac{1}{2}\frac{dv}{du}$ and w'' passes into $\frac{1}{4u}\frac{d^2v}{du^2}$. After the substitution, equation (7.39) transforms to the equation

$$\begin{gathered}\check{f} \overset{\text{def}}{=} -\frac{1}{4}v^2\frac{d^2v}{du^2} + \frac{3}{8}v\left(\frac{dv}{du}\right)^2 + \gamma v^3 + \alpha v^5 = 0,\\ g(v) = 8\check{f}.\end{gathered} \tag{7.41}$$

Now we pass to the solution of the equation $g(v) = 0$. After the transformation $\dot{v}^2 = P$, we obtain the linear equation

$$-v\frac{dP}{dv} + 3P + 8\alpha v^4 + 8\gamma v^2 = 0,$$

which has the following solution:

$$\left(\frac{dv}{du}\right)^2 = 8v^2(\alpha v^2 - C_0 v - \gamma) \overset{\text{def}}{=} P, \tag{7.42}$$

where C_0 is an arbitrary constant.

Substituting $\tilde{v} = 1/v$ to equation (7.42) we obtain the equation

$$\left(\frac{d\tilde{v}}{du}\right)^2 = 8(\alpha + C_0\tilde{v} - \gamma\tilde{v}^2), \tag{7.43}$$

Using the formulae (7.15), (7.17), and (7.18), we obtain that equation (7.43) has solutions

$$\tilde{v} = C_+e^{\sqrt{-8\gamma}u} + C_-e^{-\sqrt{-8\gamma}u} + \frac{C_0}{2\gamma},$$

where

$$C_+C_- = \frac{C_0^2}{16\gamma^2} + \frac{\alpha}{4\gamma}. \tag{7.44}$$

Thus,

$$w = \sqrt{z}\left[C_+e^{\sqrt{-8\gamma z}} + C_-e^{-\sqrt{-8\gamma z}} + C_0/(2\gamma)\right]^{-1}. \tag{7.45}$$

We can also obtain periodic asymptotic forms for the face $\tilde{\mathbf{\Gamma}}_3^{(2)}$ from the asymptotic forms (7.37) for the face $\tilde{\mathbf{\Gamma}}_2^{(2)}$ using the symmetry (7.3) .

7.4 The Results Obtained

Theorem 7.1. *If $\delta \neq 0$ as $z \to \infty$ asymptotic forms of solutions to the fifth Painlevé equation P_5 form the following* 3 *two-parameter families:*

1. *elliptic, determined by the formula* (7.11)*;*
2. *periodic, defined by $w = v/z$ and by the formulae* (7.19), (7.20) *with the period $T = 2\pi i/\sqrt{-2\delta}$;*
3. *periodic, defined by the formulae* (7.25), (7.26) *with the period $T = 2\pi i/\sqrt{-2\delta}$.*

Theorem 7.2. *If $\delta = 0$, $\gamma \neq 0$ as $z \to \infty$ asymptotic forms of solutions to the fifth Painlevé equation P_5 form the following* 3 *two-parameter families:*

1. *elliptic, determined by the formula* (7.32)*;*
2. *periodic, defined by $w = v/\sqrt{z}$ and by the formulae* (7.37), (7.38) *with the period $T = \pi i/\sqrt{2\gamma}$ for $\sqrt{z}$;*
3. *periodic defined by $w = \sqrt{z}v$ and by the formulae* (7.44), (7.45) *with the period $T = \pi i/\sqrt{-2\gamma}$ for $\sqrt{z}$.*

If $\delta = \gamma = 0$ the fifth Painlevé equation (7.2) can be solved directly [3, 4] and its solutions are neither elliptic nor periodic.

For the main results of this chapter, see [5] in Russian.

Bibliography

[1] A. D. Bruno. Space Power Geometry for one ODE and P_1–P_4, P_6, *Chapter 6, this volume.*

[2] H. Bateman and A. Erdélyi. *Higher transcendental functions.* vol. 3. McGraw-Hill B.C., NY, Toronto, London, 1955 (= M.: Nauka, 1967).

[3] A. D. Bruno and I. V. Goryuchkina, Asymptotic expansions of solutions to the sixth Painlevé equation, *Trans. Moscow Math. Soc.*, **71**: 1–105, 2010 (= *Tr. MMO,* **71**: 6–118, 2010).

[4] A. D. Bruno and A. V. Parusnikova, Local expansions of solutions to the fifth Painlevé equation, *Doklady Mathematics,* **83**(3): 348–352, 2011 (= *DAN,* **438**(4): 439–443, 2011).

[5] A. D. Bruno and A. V. Parusnikova, *Periodic and elliptic asymptotic forms of solutions to the fifth Painlevé equation,* preprint of KIAM Nr. 61, Moscow, 2011 (in Russian); available from the Keldysh Institute of Applied Mathematics of RAS: `http://library.keldysh.ru/preprints/Default.asp?lg=e`.

Author Information

Alexander D. Bruno, Keldysh Institute of Applied Mathematics (KIAM), Miusskaya Sq. 4, 125047, Moscow, Russia.
E-mail: abruno@keldysh.ru

Anastasia V. Parusnikova, Moscow State Institute of Electronics and Mathematics (Technical University), Bolshoy Trehsvyatitelskiy per. 3, 109028, Moscow, Russia.
E-mail: parus-a@mail.ru

Chapter 8

Regular Asymptotic Expansions of Solutions to One ODE and P_1–P_5

Alexander D. Bruno

Abstract. We consider an ordinary differential equation of a rather general form. We define a regular asymptotic expansion of its solutions. This includes expansions with periodic, elliptic, or hyper-elliptic coefficients. We also present a method for computing such expansions in the case when a truncated equation has the first integral. We applied this approach to Painlevé equations P_1–P_5 and found that for all eight families of elliptic asymptotic forms there exists the second elliptic term of expansion but, for all four periodic families, the second periodic term is absent and the regular expansions do not exist (with the exception of one special case for P_4).

Keywords. Asymptotic Expansion, Periodic Expansion, Elliptic Expansion.

2010 Mathematics Subject Classification. Primary: 34M55; Secondary: 34E05.

8.1 Introduction

We consider an ordinary differential equation

$$f(x, y) = 0, \tag{8.1}$$

where $f(x, y)$ is a differential sum [1], i.e., a finite sum of differential monomials, each of which is a product of an ordinary monomial and finite number of derivatives $d^l y/dx^l$.

Let $x \to \infty$. Asymptotic expansion of solutions to equation (8.1) gives us

$$y = x^\alpha \sum_{k=0}^{\infty} \varphi_k(u) u^{-k}, \tag{8.2}$$

where $u = x^\beta$, α and $\beta = \text{const} \in \mathbb{R}$, $\beta > 0$ and $\varphi_k(u)$, $k > 0$ are such complex functions of $\varphi_0(u)$ and its derivatives $d^l\varphi_0(u)/du^l$, which do not have a logarithmic branch, will be called *regular*. Functions $\varphi_k(u)$ of these expansions inherit such properties of the function $\varphi_0(u)$ as periodicity, ellipticity, or hyperellipticity.

This work was supported by RFBR, Grant Nr. 11-01-00023.

Our goal is to provide a way to compute these expansions and to find out their existence for the first five Painlevé equations P_1–P_5. The solution of the problem is done in two steps. The first step is to compute the initial term

$$y = x^{\alpha}\varphi_0(u) \tag{8.3}$$

of expansion (8.2), and then to compute its “tail”. The algorithm for solving the first problem is given in [2, Sect. 1]. Here it is briefly repeated in Section 8.2. The algorithm for computing the tail of expansion of the form (8.2) is given in Section 8.3. The initial terms (8.3) for Painlevé equations P_1–P_4 were found in [2, Sect. 2] and for P_5 in [3]. Here in Sections 8.4–8.8, the existence of regular expansions of solutions to equations P_1–P_5 in the generic case are investigated by this method .

8.2 Finding Asymptotic Forms [2]

Each differential monomial $a(x, y)$ is assigned the *three-dimensional power exponent* $\mathbf{Q}(a) = (q_1, q_2, q_3)$ by the following rules:

$$\mathbf{Q}(cx^r y^s) = (r, s, 0), \quad \mathbf{Q}(d^l y/dx^l) = (0, 1, l).$$

The power exponent of the product is the sum of the power exponents of the factors

$$\mathbf{Q}(ab) = \mathbf{Q}(a) + \mathbf{Q}(b).$$

The set $\tilde{\mathbf{S}}(f)$ of power exponents $\mathbf{Q}(a_i)$ of all the differential monomials of the sum $f(x, y) = \sum a_i(x, y)$ is called the *support of the sum* $f(x, y)$. The convex hull $\mathbf{\Gamma}(f)$ of the support $\tilde{\mathbf{S}}(f)$ is called the *polyhedron of the sum* $f(x, y)$. The boundary $\partial\mathbf{\Gamma}(f)$ of the polyhedron $\mathbf{\Gamma}(f)$ consists of vertices $\mathbf{\Gamma}_j^{(0)}$, edges $\mathbf{\Gamma}_j^{(1)}$ and faces $\mathbf{\Gamma}_j^{(2)}$. These are called *(generalized) faces* $\mathbf{\Gamma}_j^{(d)}$, where the upper index denotes the dimension of the face and lower index denotes its number. Each face $\mathbf{\Gamma}_j^{(d)}$ is assigned the *truncated sum*

$$\check{f}_j^{(d)}(x, y) = \sum a_i(x, y) \quad \text{over} \quad \mathbf{Q}(a_i) \in \tilde{\mathbf{S}}(f) \cap \mathbf{\Gamma}_j^{(d)}. \tag{8.4}$$

Let $\mathbf{N}_j = (n_1, n_2, n_3)$ be the outward normal of the two-dimensional face $\mathbf{\Gamma}_j^{(2)}$. If

$$n_1 \neq 0 \text{ and } n_1 + n_3 > 0. \tag{8.5}$$

Then, according to Theorem 1.3 from [2] the power transformation

$$y = x^{\alpha} v, \quad u = x^{\beta} \tag{8.6}$$

with $\alpha = n_2/|n_1|$ and $\beta = (n_1 + n_3)/|n_1|$ changes face $\mathbf{\Gamma}_j^{(2)}$ into $\mathbf{\Gamma}_j^{*(2)}$ with outward normal $\mathbf{N}_j^* = (1, 0, 0)$, and equation (8.1) is transformed into the form of equation (8.8) (see below).

Herewith the power transformation (8.6) induces the following formulas for derivatives:

$$
\begin{aligned}
y' &= x^\alpha v' + \alpha x^{\alpha-1} v,\\
y'' &= x^\alpha v'' + 2\alpha x^{\alpha-1} v' + \alpha(\alpha-1)x^{\alpha-2} v,\\
v' &= \dot v \beta x^{\beta-1},\\
v'' &= \ddot v \beta^2 x^{2(\beta-1)} + \dot v \beta(\beta-1)x^{\beta-2};\\
y' &= \beta x^{\alpha+\beta-1}\dot v + \alpha x^{\alpha-1} v,\\
y'' &= \beta^2 x^{2\beta-2+\alpha}\ddot v + \beta(\beta-1)x^{\beta+\alpha-2}\dot v + 2\alpha\beta x^{\alpha+\beta-2}\dot v + \alpha(\alpha-1)x^{\alpha-2} v\\
&= \beta^2 x^{2\beta+\alpha-2}\ddot v + \beta(2\alpha+\beta-1)x^{\beta+\alpha-2}\dot v + \alpha(\alpha-1)x^{\alpha-2} v.
\end{aligned}
\tag{8.7}
$$

Here and henceforth $\dot v = dv/du$.

8.3 Computation of Expansions (8.2)

Theorem 8.1. *Let an equation of order n*

$$g(v) + \sum_{k=1}^{m} h_k(v)u^{-k} = 0 \tag{8.8}$$

have a solution of the form

$$v = w + \sum_{k=1}^{\infty} b_k(w)u^{-k}, \tag{8.9}$$

where $w = w(u)$ *is the solution to the truncated equation*

$$g(w) = 0 \tag{8.10}$$

with the property

$$0 < \varepsilon < |w|, |\dot w|, \dots, |w^{(n)}| < \varepsilon^{-1} < \infty. \tag{8.11}$$

Then $b_k(w)$ *satisfies the linear equation*

$$\mathscr{L}(u)b_k(w) + \theta_k(w) = 0, \tag{8.12}$$

where

$$\mathscr{L}(u) = \left.\frac{\delta g}{\delta v}\right|_{v=w}, \tag{8.13}$$

$\theta_k(w)$ *is a polynomial on* $w^{(l)}$ *depending on* $g(w)$, $h_j(w)$, *and* $h_j^{(l)}(w)$ *for* $j < k$ *and* $l = 0, 1, 2, \dots, n$.

Now we consider the case when the truncated equation (8.10) has the first integral of the form

$$\dot{w}^2 = P(w) \stackrel{\text{def}}{=} \sum_{k=0}^{\varkappa} p_k w^k, \quad p_k = \text{const} \in \mathbb{C}. \tag{8.14}$$

Differentiating with respect to u and dividing by $2\dot{w}$, we obtain

$$\ddot{w} = \frac{1}{2} P'(w). \tag{8.15}$$

Here and henceforth the prime denotes the derivative with respect to w.

Using equations (8.14) and (8.15), any power series R of w and its derivatives $w^{(l)}$ can be written as the sum

$$R = R^*(w) + \dot{w} R^{**}(w),$$

where $R^*(w)$ and $R^{**}(w)$ are power series only of w.

Let

$$b_k(w) = F_k(w) + \dot{w} G_k(w), \tag{8.16}$$

where F_k and G_k are function only of w. Then, omitting the index k, by (8.14) and (8.15), we obtain

$$\begin{aligned} \dot{b} &= F'\dot{w} + PG' + \frac{1}{2}P'G, \\ \ddot{b} &= PF'' + \frac{1}{2}P'F' + \dot{w}\left(PG'' + \frac{3}{2}P'G' + \frac{1}{2}P''G\right). \end{aligned} \tag{8.17}$$

Further derivatives of b are not needed here, because we are considering only equation (8.8) of the second order. In our case

$$\mathcal{L} b = \mathcal{F}(w) F(w) + \dot{w} \mathcal{G}(w) G(w).$$

Thus, equation (8.12) splits in two

$$\begin{aligned} \mathcal{F}(w) F_k(w) + \theta_k^*(w) &= 0, \\ \mathcal{G}(w) G_k(w) + \theta_k^{**}(w) &= 0, \end{aligned} \tag{8.18}$$

where $\theta_k(w) = \theta_k^*(w) + \dot{w}\theta_k^{**}(w)$. Note that in equations (8.18) differential operators $\mathcal{F}(w)$ and $\mathcal{G}(w)$ are operators on w and do not depend on u. If the polynomial $P(w)$ in (8.14) does not have multiple roots and its degree $\varkappa$ is greater than one, i.e.,

$$\varkappa > 2 \quad \text{and} \quad \Delta(P) \neq 0, \tag{8.19}$$

where $\Delta(P)$ is discriminant of the polynomial $P(w)$, then the solution $w(u)$ to the truncated equation (8.10) is a periodic (if $\varkappa = 2$), elliptic (if $2 < \varkappa < 5$), or hyperelliptic (if $\varkappa \geqslant 5$) function.

If the solutions $F_k(w)$ and $G_k(w)$ to the system (8.18) have no more than second branch order and have it only at the roots of the polynomial $P(w)$, then they are also periodic or (hyper)elliptic functions. Finally, if for the sequence of equations (8.18) with $k = 1, 2, \dots$, there exists solutions $F_k(w)$ and $G_k(w)$, with only such a branch, the solutions to equation (8.8) have a regular asymptotic expansion of solutions (8.9).

Let operators $\mathscr{F}^{-1}(w)$ and $\mathscr{G}^{-1}(w)$ be inverse to operators $\mathscr{F}(w)$ and $\mathscr{G}(w)$, respectively. Then the solutions of equations (8.8) are of the forms

$$F_k(w) = -\mathscr{F}^{-1}(w)\theta_k^*(w), \quad G_k(w) = -\mathscr{G}^{-1}(w)\theta_k^{**}(w). \tag{8.20}$$

Theorem 8.2. *If functions $\theta_k^*(w)$ and $\theta_k^{**}(w)$ are regular, then the solutions to equations* (8.20) *can have a logarithmic branch only at infinity $w = \infty$ and at singular points of the operators $\mathscr{F}^{-1}(w)$ and $\mathscr{G}^{-1}(w)$.*

For the existence of a regular expansions (8.9) we need to prove the existence of a sequence of functions $F_k(w)$ and $G_k(w)$ that do not have a logarithmic branch. On the other hand, if it is shown that $F_k(w)$ or $G_k(w)$ have a logarithmic branch, then it proves the absence of regular expansion. Below, for each of the Painlevé equations, we write equation (8.8), functions $\mathscr{F}(w)$, $\mathscr{G}(w)$, and inverse ones $\mathscr{F}^{-1}(w)$, $\mathscr{G}^{-1}(w)$. We find their singular points, and we find the conditions on the parameters of the equation and of the solution $w(u)$ in which the functions $F_1(w)$ and $G_1(w)$ do not have a logarithmic branch, as well as the conditions under which at least one of these functions has such a branching. It is a wonder that for each Painlevé equation P_l the operators $\mathscr{F}$ and $\mathscr{G}$ are expressed in the same way in terms of polynomial P and different cases are distinguished only by this polynomial, polynomials θ_1^* and θ_1^{**} that are obtained from h_1. At the same time, for 12 cases of faces $\mathbf{\Gamma}_j^{(2)}$ of the five Painleve equations P_1–P_5, there are only four different pairs of operators $\mathscr{F}$ and $\mathscr{G}$. Singular points of the inverse operators $\mathscr{F}^{-1}(w)$ and $\mathscr{G}^{-1}(w)$ are points of double roots of the polynomial $P(w)$ as well, but we have excluded them by condition (8.19).

8.4 Equation P_1

$$f(x, y) \stackrel{\text{def}}{=} -y'' + 3y^2 + x. \tag{8.21}$$

According to Subsection 2.1 in [2] and (8.7) power transformation $y = x^{1/2}v$, $u = x^{5/4}$ changes equation (8.21) into

$$-\left(\frac{5}{4}\right)^2 \ddot{v} + 3v^2 + 1 - \left(\frac{5}{4}\right)^2 \frac{\dot{v}}{u} + \frac{1}{4}\frac{v}{u^2} = 0. \tag{8.22}$$

Here, in (8.8) we have $m = 2$,

$$g(v) = -\left(\frac{5}{4}\right)^2 \ddot{v} + 3v^2 + 1, \quad h_1 = -\left(\frac{5}{4}\right)^2 \dot{v}, \quad h_2 = \frac{1}{4}v. \tag{8.23}$$

In the first integral (8.14) we have

$$P(w) = \left(\frac{4}{5}\right)^2 \cdot 2\left(w^3 + w + C_0\right) \stackrel{\text{def}}{=} \left(\frac{4}{5}\right)^2 \tilde{P}(w), \tag{8.24}$$

where C_0 is an arbitrary constant. The discriminant of the polynomial $w^3 + w + C_0$ equals to $\Delta = -4 - 27C_0^2$. According to condition (8.19) we assume that $\Delta \neq 0$, i.e.,

$$4 + 27C_0^2 \neq 0. \tag{8.25}$$

Here, according to (8.13) and (8.23) we have

$$\mathscr{L}b = -\left(\frac{5}{4}\right)^2 \ddot{b} + \left(3w^2 + 1\right) b. \tag{8.26}$$

According to (8.17) and (8.24)

$$\begin{aligned} \mathscr{F}F &= -\left(\frac{5}{4}\right)^2 \left(\frac{1}{2}P'F' + PF''\right) + 6wF = \\ &= -\left(\frac{1}{2}\tilde{P}'F' + \tilde{P}F''\right) + \frac{1}{2}\tilde{P}''F = -\left[\tilde{P}^{3/2}\left(F/\tilde{P}^{1/2}\right)'\right]', \end{aligned} \tag{8.27}$$

$$\mathscr{G}G = -\left(\frac{3}{2}\tilde{P}'G' + \tilde{P}G''\right) = -\frac{1}{\tilde{P}^{1/2}}\left[\tilde{P}^{3/2}G'\right]'. \tag{8.28}$$

Inverting operators (8.27) and (8.28) we get solutions to equations (8.18) in the form

$$F_k = \tilde{P}^{1/2} \int \frac{1}{\tilde{P}^{3/2}} \int \theta_k^* \, dw \, dw, \quad G_k = \int \frac{1}{\tilde{P}^{3/2}} \int \tilde{P}^{1/2} \theta_k^{**} \, dw \, dw. \tag{8.29}$$

Due to condition (8.25) the branching is possible only at the point $w = \infty$. According to (8.23) for $k = 1$ we have $\theta_1^* \equiv 0$, $\theta_1^{**} = -(5/4)^2$. Therefore, $F_1 \equiv 0$ and

$$G_1 = -\left(\frac{5}{4}\right)^2 \int \frac{1}{\tilde{P}^{3/2}} \int \tilde{P}^{1/2} \, dw \, dw.$$

According to (8.24) and (8.28) near $w = \infty$ we have

$$\tilde{P}^{1/2} = 2^{1/2} w^{3/2} \left(1 + O(w^{-2})\right), \quad \tilde{P}^{-3/2} = 2^{-3/2} w^{-9/2} \left(1 + O(w^{-2})\right).$$

Thus,

$$\begin{aligned} G_1 &= -\left(\frac{5}{4}\right)^2 \int 2^{-3/2} w^{-9/2} \cdot 2^{1/2} \cdot \frac{2}{5} w^{5/2} \left(1 + O(w^{-2})\right) dw \\ &= \text{const}\, w^{-1} \left(1 + O(w^{-2})\right), \end{aligned}$$

i.e., $G_1(w)$ does not have logarithmic branching. It was shown in [5, Sect. 4] that in this case there is a regular expansion (8.2). For that, initial terms of the expansions $F_k(w)$ and $G_k(w)$, $k = 1, 2, \ldots, 6$ were explicitly computed.

8.5 Equation P_2

$$f(x,y) \stackrel{\text{def}}{=} -y'' + 2y^3 + xy + a = 0, \tag{8.30}$$

where a is a complex parameter.

8.5.1 Elliptic Asymptotic Forms, Face $\Gamma_3^{(2)}$

According to Section 2.2 in [2] and to (8.7), power transformation

$$y = x^{1/2} v, \quad u = x^{3/2} \tag{8.31}$$

reduces the equation (8.30) to

$$u\left[-\left(\frac{3}{2}\right)^2 \ddot{v} + 2v^3 + v\right] + a - \left(\frac{3}{2}\right)^2 \dot{v} + \frac{1}{4} v u^{-1} = 0. \tag{8.32}$$

Therefore, in the form (8.8) it has $m = 2$,

$$g(v) = -\left(\frac{3}{2}\right)^2 \ddot{v} + 2v^3 + v, \quad h_1 = a - \left(\frac{3}{2}\right)^2 \dot{v}, \quad h_2 = \frac{1}{4} v. \tag{8.33}$$

In the first integral (8.14)

$$P(w) = \left(\frac{2}{3}\right)^2 (w^4 + w^2 + C_0) \stackrel{\text{def}}{=} \left(\frac{2}{3}\right)^2 \tilde{P}(w), \tag{8.34}$$

C_0 is an arbitrary constant. Since the discriminant of the polynomial $\tilde{P} = w^4 + w^2 + C_0$ is $\Delta = C_0(C_0 - 1/4)^2$, then we assume that

$$C_0 \neq 0, \quad C_0 \neq \frac{1}{4}. \tag{8.35}$$

The condition (8.19) is fulfilled. Since

$$\frac{\delta g}{\delta v} = -\left(\frac{3}{2}\right)^2 \frac{d^2}{du^2} + 6v^2 + 1,$$

then

$$\mathscr{L}(w) b_k = -\left(\frac{3}{2}\right)^2 \ddot{b}_k + (6w^2 + 1) b_k.$$

According to (8.17) and (8.34),

$$\begin{aligned}\mathscr{F}F_k &= -\left(\frac{3}{2}\right)^2\left(\frac{1}{2}P'F_k' + PF_k''\right) + (6w^2+1)F_k\\ &= -\left(\frac{1}{2}\tilde{P}'F_k' + \tilde{P}F_k''\right) + \frac{1}{2}\tilde{P}''F_k = -\left[\tilde{P}^{3/2}\left(\frac{F}{\tilde{P}^{1/2}}\right)'\right]'\\ \mathscr{G}G_k &= -\left(\frac{3}{2}\right)^2\left(\frac{1}{2}P''G_k + \frac{3}{2}P'G_k' + PG_k''\right) + (6w^2+1)G_k\\ &= -\left(\frac{3}{2}\tilde{P}'G_k' + \tilde{P}G_k''\right) = -\frac{1}{\tilde{P}^{1/2}}\left[\tilde{P}^{3/2}G'\right]',\end{aligned} \tag{8.36}$$

because $-\tilde{P}''/2 + 6w^2 + 1 \equiv 0$.

Formulas (8.36) coincide with formulas (8.27), (8.28), so solutions to the equations (8.18) are given by formulas (8.29). In this case branching is also possible only at infinity $w = \infty$. According to (8.33), here $\theta_1^* = a$, $\theta_1^{**} = -(3/2)^2$. According to (8.34), for $w \to \infty$ we have

$$\tilde{P}^{1/2} = w^2\left(1 + O(w^{-2})\right), \quad \tilde{P}^{-3/2} = w^{-6}\left(1 + O(w^{-2})\right).$$

Thus, according to (8.29),

$$F_1 = aw^2\left(1 + O(w^{-2})\right)\int w^{-5}\left(1 + O(w^{-1})\right)dw = -\frac{a}{4}w^{-2}\left(1 + O(w^{-1})\right),$$

$$\begin{aligned}G_1 &= -\left(\frac{3}{2}\right)^2\int w^{-6}\left(1 + O(w^{-2})\right)\frac{1}{3}w^3\left(1 + O(w^{-1})\right)dw\\ &= \text{const}\cdot w^{-2}\left(1 + O(w^{-1})\right),\end{aligned}$$

i.e., F_1 and G_1 do not have a logarithmic branch. It was show in [5, Sect. 5] that there is a regular expansion (8.2) in this case. For that, the initial terms of the series F_k, G_k, $k = 1, 2, 3, 4$ were computed.

8.5.2 Periodic Asymptotic Forms, Face $\Gamma_4^{(2)}$

We apply transformation (8.6) to equation (8.21) with $\alpha = -1$, $\beta = 3/2$, i.e., $y = v/x$, $u = x^{3/2}$. According to (8.7), equation (8.30) changes into

$$f^*(u,v) \stackrel{\text{def}}{=} -\left(\frac{3}{2}\right)^2\ddot{v} + v + a + \left(\frac{3}{2}\right)^2\frac{\dot{v}}{u} + \frac{2v(v^2-1)}{u^2} = 0, \tag{8.37}$$

and the truncated equation changes into

$$\check{f}^{*(2)}(u,v) \stackrel{\text{def}}{=} -\left(\frac{3}{2}\right)^2\ddot{v} + v + a \stackrel{\text{def}}{=} g(v) = 0. \tag{8.38}$$

Equation (8.38) has the first integral

$$\dot v^2 = \left(\frac{2}{3}\right)^2 \left(v^2 + 2av + C_0\right) \stackrel{\text{def}}{=} P(v) \stackrel{\text{def}}{=} \left(\frac{2}{3}\right)^2 \tilde P(v). \tag{8.39}$$

Here, $\tilde P = w^2 + 2aw + C_0$. Formulas (8.29) are valid again here. According to (8.37), now $\theta_1^* = 0$, $\theta_2^{**} = (3/2)^2$, and, according to (8.39),

$$\tilde P^{1/2} = w\left(1 + O(w^{-1})\right), \quad \tilde P^{3/2} = w^{-3}\left(1 + O(w^{-1})\right).$$

Thus, from (8.29) we get that $F_1 \equiv 0$ and

$$G_1 = \left(\frac{3}{2}\right)^2 \int w^{-3}\left(1 + O(w^{-1})\right)\left(\frac{1}{2}w^2 + O(w)\right) dw = \left(\frac{3}{2}\right)^2 \left(\ln w + O(w^{-1})\right),$$

i.e., here is certainly the logarithmic branch and there is no regular expansion (8.2). For two other ways ti proof its absence, see [4, Sect. 4].

8.6 Equation P_3

$$f(x, y) \stackrel{\text{def}}{=} -xyy'' + xy' - yy' + ay^3 + by + cxy^4 + dx = 0, \tag{8.40}$$

where a, b, c, d are complex parameters.

8.6.1 Case $cd \neq 0$ [2, Sect. 2.3]

Here the power transformation (8.6) is identical.

Equation (8.40) with $cd \neq 0$ is of the form (8.8) with $m = 1$, where

$$g(w) \stackrel{\text{def}}{=} -w\ddot w + \dot w^2 + cw^4 + d = 0, \quad h_1 = -w\dot w + aw^3 + bw, \tag{8.41}$$

$$P(w) = cw^4 + C_0 w^2 - d, \quad \Delta(P) = -cd\left(C_0^2 + 4cd\right)^2/16 \neq 0. \tag{8.42}$$

Operators $\mathscr{F}$ and $\mathscr{G}$ are

$$\begin{aligned} \mathscr{F}F &= \left(\frac{1}{2}wP'' - P'\right)F + \left(2P - \frac{1}{2}wP'\right)F' - wPF'' \\ &= -w^3\left[\frac{P^{3/2}}{w^2}\left(\frac{F}{P^{1/2}}\right)'\right]', \\ \mathscr{G}G &= \left(2P - \frac{3}{2}wP'\right)G' - wPG'' = -\frac{w^3}{P^{1/2}}\left(\frac{P^{3/2}G'}{w^2}\right)'. \end{aligned} \tag{8.43}$$

Solutions to equations (8.18) are of the form

$$F_k = P^{1/2}\int \frac{w^2}{P^{3/2}} \int \frac{\theta_k^*}{w^3}\, dw\, dw, \quad G_k = \int \frac{w^2}{P^{3/2}} \int \frac{P^{1/2}\theta_k^{**}}{w^3}\, dw\, dw. \tag{8.44}$$

According to Theorem 8.2, F_k and G_k may have logarithmic branches only at two points, $w = 0$ and $w = \infty$. According to (8.41), here $\theta_1^* = aw^3 + bw$, $\theta_1^{**} = -w$.

For $w \to 0$ we have

$$P^{1/2} = (-d)^{1/2}\left(1 + O(w^2)\right), \quad P^{-3/2} = (-d)^{-3/2}\left(1 + O(w^2)\right),$$

and thus

$$\begin{aligned} F_1 &= P^{1/2} \int w^2(-d)^{-3/2}\left(1 + O(w^2)\right)\left(-\frac{b}{w} + C_1 + aw\right) dw \\ &= \text{const} \cdot P^{1/2} \int w\left(1 + O(w)\right) dw, \end{aligned}$$

where C_1 is an arbitrary constant,

$$G_1 = \text{const} \cdot \int w^2\left(1 + O(w^2)\right)\frac{1}{w}\left(1 + O(w)\right) dw = \text{const} \cdot \int w\left(1 + O(w)\right) dw,$$

i.e., they do not have branching.

For $w \to \infty$ we have

$$P^{1/2} = \sqrt{c}w^2\left(1 + O(w^{-2})\right), \quad P^{-3/2} = c^{-3/2}w^{-6}\left(1 + O(w^{-2})\right),$$

and thus

$$\begin{aligned} F_1 &= \text{const}\, P^{1/2} \int w^{-4}\left(1 + O(w^{-2})\right) w\left(1 + O(w^{-1})\right) dw, \\ G_1 &= \text{const} \int w^{-4}\left(1 + O(w^{-2})\right) w\left(1 + O(w^{-1})\right) dw, \end{aligned}$$

i.e., they do not have branching.

8.6.2 Case $c = 0$, $ad \neq 0$ [2, Sect. 2.4]

After the power transformation $y = x^{1/3}v$, $u = x^{2/3}$, the equation (8.40) with $c = 0$ takes the form (8.8) with $m = 1$, where

$$\begin{gathered} g(v) = -\left(\frac{2}{3}\right)^2 v\ddot{v} + \left(\frac{2}{3}\right)^2 \dot{v}^2 + av^3 + d, \quad h_1 = -\left(\frac{2}{3}\right)^2 v\dot{v} + bv, \\ P(v) = \left(\frac{3}{2}\right)^2\left(2av^3 + C_0v^2 - d\right), \quad \Delta(P) = \left(\frac{3}{2}\right)^8 4d\left(C_0^3 - 27a^2d\right) \neq 0. \end{gathered} \tag{8.45}$$

Formulas (8.43) and (8.44), with substitution w by v, are also valid here. According to (8.45), $\theta_1^* = bv$, $\theta_1^{**} = -(2/3)^2\, v$. For $v \to 0$ we get $P^{1/2} = \text{const} + \cdots$,

$P^{-3/2} = \text{const} + \cdots$. Thus, from (8.44) it turns out that

$$F_1 = \text{const}\, P^{1/2} \int [bv + \cdots]\, dv, \quad G_1 = \text{const} \int [v + \cdots]\, dv,$$

i.e., there is no branch.

For $v \to \infty$ we get $P^{1/2} = \text{const}\,[v^{3/2} + \cdots]$, $P^{3/2} = \text{const}\,[v^{-9/2} + \cdots]$. Thus, from (8.44) we get

$$F_1 = \text{const}\, P^{1/2} \int \left[v^{-7/2} + \cdots\right] dv, \quad G_1 = \text{const} \int \left[v^{-5/2} \cdot v^{1/2} + \cdots\right] dv,$$

i.e., there is no branching.

8.6.3 Case $c = d = 0$, $ab \neq 0$ [2, Sect. 2.5]

After the power transformation $y = v$, $u = x^{1/2}$, equation (8.40) with $c = d = 0$ takes the form (8.8) with $m = 1$, where

$$\begin{aligned} g &= -y\ddot{y} + \dot{y}^2 + 4ay^3 + 4by, & h_16 &= y\dot{y}, \\ P(y) &= 8\left(ay^3 + C_0y^2 - by\right), & \Delta(P) &= 2^{12}b^2\left(C_0^2 + 4ab\right) \neq 0. \end{aligned} \tag{8.46}$$

Formulas (8.43) and (8.44) with substitution w by y are also valid here. According to (8.46), $\theta_1^* = 0$, $\theta_1^{**} = y$. Thus, $F_1 \equiv 0$.

For $y \to 0$ we get $P^{1/2} \sim \text{const}\, y^{1/2}$, $P^{-3/2} \sim \text{const}\, y^{-3/2}$. Therefore, from (8.44) we get

$$G_1 \sim \int [y^{1/2-1/2}]\, dy,$$

i.e., there is no branching.

For $y \to \infty$ we get $P^{1/2} \sim y^{3/2}$, $P^{-3/2} \sim y^{-9/2}$, so, from (8.44) we get

$$G_1 \sim \int y^{-5/2+1/2}\, dy,$$

i.e., there is no branching.

8.7 Equation P_4 [2, Sect. 2.6]

$$f(x, y) \stackrel{\text{def}}{=} -2yy'' + y'^2 + 3y^4 + 8xy^3 + 4(x^2 - a)y^2 + 2b = 0. \tag{8.47}$$

8.7.1 Elliptic Asymptotic Forms, Face $\Gamma_3^{(2)}$

After power transformation $y = xv$, $u = x^2$, we get equation (8.8) with $m = 2$,

$$g(v) = -2v\ddot{v} + \dot{v}^2 + (3/4)v^4 + 2v^3 + v^2,$$

$$h_1 = -av^2 + 2v\dot{v}, h_2 = \frac{b}{2} + \frac{1}{4}v^2, \tag{8.48}$$

$$P(v) = \frac{1}{4}v^4 + v^3 + v^2 + C_0 v, \quad \Delta = -C_0^3(27C_0 - 8)/2^{12} \neq 0. \tag{8.49}$$

Now

$$\mathscr{F}F = (vP'' - P')F + (2P - P'v)F' - 2vPF'' = -2v^2\left[\frac{P^{3/2}}{v}\left(\frac{F}{P^{1/2}}\right)'\right]',$$

$$\mathscr{G}G = (2P - 3P'v)G' - 2vPG'' = -\frac{2v^2}{P^{1/2}}\left(\frac{P^{3/2}G'}{v}\right)'.$$

Therefore, the solutions to equations (8.18) are

$$F_k = \frac{1}{2}P^{1/2}\int\frac{v}{P^{3/2}}\int\frac{\theta_k^*}{v^2}\,dv\,dv, \quad G_k = \frac{1}{2}\int\frac{v}{P^{3/2}}\int\frac{P^{1/2}\theta_k^{**}}{v^2}\,dv\,dv. \tag{8.50}$$

Here, singular points are $v = 0$ and $v = \infty$. According to (8.48) we get $\theta_1^* = -av^2$, $\theta_1^{**} = 2v$.

For $v \to 0$, we have $P^{1/2} = \sqrt{C_0}v^{1/2}(1 + O(v))$, $P^{-3/2} = C_0^{-3/2}v^{-3/2}(1 + O(v))$. So,

$$F_1 = \text{const}\,P^{1/2}\int v^{-1/2}\cdot v\big(1 + O(v)\big)\,dv,$$

$$G_1 = \text{const}\int v^{-1/2}\cdot v^{1/2}\big(1 + O(v)\big)\,dv,$$

i.e., they do not have branching.

For $v \to \infty$ we get $P^{1/2} = (1/2)v^2(1 + O(v^{-1}))$, $P^{-3/2} = 8v^{-6}(1 + O(v^{-1}))$. Thus,

$$F_1 = \text{const}\,P^{1/2}\int v^{-4}\big(1 + O(v^{-1})\big)\,dv,$$

$$G_1 = \text{const}\int v^{-5}v^2\big(1 + O(v^{-1})\big)\,dv,$$

i.e., they do not have branching.

8.7.2 Periodic Asymptotic Forms, Face $\Gamma_4^{(2)}$ [2, Sect. 2.6]

After the change (8.6) with $\alpha = -1$, $\beta = 2$, equations (8.47) becomes

$$f^*(u, v) \stackrel{\text{def}}{=} -8v\ddot{v} + 4\dot{v}^2 + 4v^2 + 2b + 8\frac{v^3}{u} - 4a\frac{v^2}{u} + \frac{3(v^4 - v^2)}{u^2} = 0, \tag{8.51}$$

and the truncated equation becomes

$$f_4^{*(2)}(u,v) \stackrel{\text{def}}{=} -8v\ddot{v} + 4\dot{v}^2 + 4v^2 + 2b \stackrel{\text{def}}{=} g(v) = 0. \tag{8.52}$$

It has the first integral

$$\dot{v}^2 = v^2 + C_0 v - \frac{b}{2} \stackrel{\text{def}}{=} P(v). \tag{8.53}$$

Here formulas (8.50) are again valid, $\theta_1^* = 8v^3 - 4av^2$, $\theta_1^{**} = 0$. So, $G_1 \equiv 0$.

For $v \to 0$ we get $P^{-3/2} = (-b/2)^{-3/2}(1 + O(v))$, $F_1 = \text{const}\, P^{1/2} \int v(1 + O(v))(C_1 - 4av + 4v^2)\, dv$, i.e., there is no branch.

For $v \to \infty$ we get $P^{-3/2} = v^{-3}\left(1 - (3C_0/2)v^{-1} + O(v^{-2})\right)$,

$$F_1 = \text{const}\, P^{1/2} \int v^{-2}\left(1 - (3C_0/2)v^{-1} + O(v^{-2}\right)\left(C_1 - 4av + 4v^2\right) dv =$$
$$= \text{const}\, P^{1/2} \int \left[4 - (4a + 6C_0)v^{-1} + O(v^{-2})\right] dv.$$

Here there is logarithmic branching for $2a + 3C_0 \neq 0$. This problem was considered in [4, Sect. 5] and [6], but without such an accurate proof.

8.8 Equation P_5 [3]

$$f(x,y) \stackrel{\text{def}}{=} -x^2 y(y-1)y'' + x^2 \frac{3y-1}{2} y'^2 - xy(y-1)y'$$
$$+ (y-1)^3(ay^2 + b) + cxy^2(y-1) + dx^2y^2(y+1) = 0, \tag{8.54}$$

where a, b, c, d are complex parameters.

8.8.1 Case $d \neq 0$, Elliptic Asymptotic Forms, Face $\Gamma_1^{(2)}$ [3, Sect. 2.2]

Here transformation (8.6) is identical $y = v$, $x = u$. Thus, in equation (8.8) $m = 2$,

$$\begin{aligned} g(v) &= -y(y-1)y'' + (3y-1)y'^2/2 + dy^2(y+1), \\ h_1 &= -y(y-1)y' + cy^2(y-1), \qquad h_2 = (y-1)^3\left(ay^2 + b\right). \end{aligned} \tag{8.55}$$

Here

$$P = -2dy\left[C_0(y-1)^2 + y\right], \qquad \Delta(P) = (2d)^4 C_0^2(1 - 4C_0) \neq 0. \tag{8.56}$$

Now

$$\begin{aligned}
\mathscr{F}F &= \left[\frac{1}{2}y(y-1)P'' - \frac{3y-1}{2}P'\right]F + \left[(3y-1)P - \frac{1}{2}y(y-1)P'\right]F' \\
&\quad - y(y-1)PF'' = -y^2(y-1)^3\left[\frac{P^{3/2}}{y(y-1)^2}\left(\frac{F}{P^{1/2}}\right)'\right]', \\
\mathscr{G}G &= \left[(3y-1)P - \frac{3}{2}y(y-1)P'\right]G' - y(y-1)PG'' \\
&\quad - \frac{y^2(y-1)^3}{P^{1/2}}\left[\frac{P^{3/2}G'}{y(y-1)^2}\right]'.
\end{aligned} \tag{8.57}$$

Therefore, the solutions to equations (8.18) are

$$\begin{aligned}
F_k &= P^{1/2}\int \frac{y(y-1)^2}{P^{3/2}} \int \frac{\theta_k^*}{y^2(y-1)^3}\, dy\, dy, \\
G_k &= \int \frac{y(y-1)^2}{P^{3/2}} \int \frac{P^{1/2}\theta_k^{**}}{y^2(y-1)^3}\, dy\, dy.
\end{aligned} \tag{8.58}$$

Here we have three singular points: $y = 0$, $y = 1$, and $y = \infty$. According to (8.55) we get $\theta_1^* = cy^2(y-1)$, $\theta_1^{**} = -y(y-1)$,

$$\int \frac{\theta_1^*}{y^2(y-1)^3}\, dy = c\int \frac{dy}{(y-1)^2} = -\frac{c}{y-1} + C_1.$$

For $y \to 0$ we get $P^{1/2} = \text{const} \cdot y^{1/2} + \cdots$, $P^{-3/2} = \text{const} \cdot y^{-3/2} + \cdots$, $y - 1 = -1 + \cdots$. Thus,

$$\begin{aligned}
F_1 &= \text{const}\, P^{1/2}\int y^{-1/2}(c + C_1 + \ldots)\, dy, \\
G_1 &= \text{const}\int y^{-1/2}y^{1/2}(1 + O(y))\, dy,
\end{aligned}$$

i.e., there is no branching.

For $y \to 1$ we get $P^{1/2} = \text{const} + \cdots$, $P^{-3/2} = \text{const} + \cdots$, so,

$$\begin{aligned}
F_1 &= \text{const}\, P^{1/2}\int \left[-y(y-1)c + C_1 y(y-1)^2 + \cdots\right] dy, \\
G_1 &= \text{const}\int [(y-1) + \ldots]\, dy,
\end{aligned}$$

i.e., there is no branch and no singularity.

For $y \to \infty$ we get $P^{1/2} = \text{const} \cdot y^{3/2} + \cdots$, $P^{-3/2} = \text{const} \cdot y^{-9/2} + \cdots$, so,

$$\begin{aligned}
F_1 &= \text{const}\, P^{1/2}\int \left[y^{-3/2} + \cdots\right] dy, \\
G_1 &= \text{const}\int \left(y^{-3/2} + \cdots\right)\left(y^{-1/2} + \cdots\right) dy,
\end{aligned}$$

i.e., there is no branch.

8.8.2 Case $d \neq 0$, Periodic Asymptotic Forms, Face $\Gamma_2^{(2)}$ [3, Sect. 2.3]

After the change $y = v/x$, equation (8.54) takes the form (8.8), where $m = 5$,

$$g(v) = v\ddot{v} - \frac{1}{2}\dot{v}^2 + dv^2 - b, \quad h_1 = dv^3 - cv^2 + 3bv - v\dot{v}. \tag{8.59}$$

Here $P = -2dv^2 - 2C_0 v - 2b$. Solutions to the equations (8.8) have the form (8.58) again with y replaced by v. According to (8.59), $\theta_1^* = dv^3 - cv^2 + 3bv$, $\theta_1^{**} = -v$.

For $x \to \infty$ we get $P^{1/2} = \text{const}\, v + \cdots$, $P^{-3/2} = \text{const}\, v^{-3} + \cdots$ According to (8.58)

$$F_1 = \text{const}\, P^{1/2} \int \left(\frac{1}{v} + \cdots\right) dy, \qquad G_1 = \text{const} \int (1 + \cdots)\left(\frac{1}{v^2} + \cdots\right) dy,$$

i.e., F_1 always has logarithmic branching, and the regular expansion is absent.

8.8.3 Case $d = 0$, $c \neq 0$, Elliptic Asymptotic Forms, Face $\tilde{\Gamma}_1^{(2)}$ [3, Sect. 3.2]

After the change $y = v$, $u = x^{1/2}$, equation (8.54) takes the form (8.8) with $m = 2$, where

$$\begin{aligned} g(y) &= -y(y-1)\ddot{y} + \frac{3y-1}{2}\dot{y}^2 + 4cy^2(y-1), \\ h_1 &= -y(y-1)\dot{y}, \quad h_2 = 4(y-1)^3(ay^2 + b). \end{aligned} \tag{8.60}$$

Here $P = -8cy(y-1)[C_0(y-1)+1]$, $\Delta(P) = (C_0 - 1)^2 \neq 0$, and $C_0 \neq 0$. Formulas (8.58) are again valid. According to (8.60), $\theta_1^* = 0$, $\theta_1^{**} = -y(y-1)$. So,

$$F_1 \equiv 0, \quad G_1 = -\int \frac{y(y-1)^2}{P^{3/2}} \int \frac{P^{1/2}}{y(y-1)^2}\, dy\, dy.$$

For $y \to 0$ we get $P^{1/2} = \text{const}\, y^{1/2} + \cdots$, $P^{-3/2} = \text{const}\, y^{-3/2} + \cdots$, $G_1 = \text{const} \int y^{-1/2} \cdot y^{1/2}(1 + \cdots)\, dy$, i.e., there is no branching.

For $y \to 1$ we get $P^{1/2} \sim \text{const}\,(y-1)^{1/2}$, $P^{-3/2} \sim \text{const}\,(y^{-1})^{-3/2}$, $G_1 = \text{const} \int (y-1)^{1/2}(y-1)^{-1/2}(1 + \cdots)\, dy$, i.e., there is no branching.

For $y \to \infty$ we get $P^{1/2} = \text{const}\, y^{3/2}(1 + O(y^{-1}))$, $P^{-3/2} = \text{const}\, y^{-9/2}(1 + O(y^{-1}))$. So $G_1 = \text{const} \int y^{-3/2} \cdot y^{-1/2}(1 + \cdots)\, dy$, i.e., there is no branching.

8.8.4 Case $d = 0$, $c \neq 0$, Periodic Asymptotic Forms, Face $\tilde{\Gamma}_2^{(2)}$ [3, Sect. 3.3]

After the change $y = x^{-1/2}v$, $u = x^{1/2}$, equation (8.54) takes the form (8.8), where $m = 5$,

$$g(v) = v\ddot{v} - \frac{1}{2}\dot{v}^2 - 4cv^2 - 4b, \qquad h_1(v) = -\frac{1}{2}v^2 P' + \frac{3}{2}P + bv + cv^2 - \frac{7}{2}v\dot{v}.$$

Here $P = 8\left(cv^2 + C_0 v - b\right)$. Thus, for $x \to \infty$ we get $\theta_1^* = \operatorname{const} v^3 + \cdots$, $P^{1/2} \sim v$, $P^{-3/2} \sim v^{-3}$, and according to (8.58) $F_1 = \operatorname{const} P^{1/2} \int \left(v^{-1} + \cdots\right) dv$, i.e., F_1 always has branching, and a regular expansion is absent.

Bibliography

[1] A. D. Bruno, Plane Power Geometry for one ODE and P_1–P_6, *Chapter 1, this volume.*

[2] A. D. Bruno, Space Power Geometry for one ODE and P_1–P_4, P_6, *Chapter 6, this volume.*

[3] A. D. Bruno, A. V. Parusnikova, Elliptic and periodic asymptotic forms of solutions to P_5, *Chapter 7, this volume.*

[4] A. D. Bruno, *Power-exponential expansions of solutions to an ODE*, preprint Nr. 54 of KIAM, Moscow, 2011 (in Russian); available from the Keldysh Institute of Applied Mathematics of RAS: `http://library.keldysh.ru/preprints/Default.asp?lg=e`.

[5] A. D. Bruno, *Power-elliptic expansions of solutions to an ODE*, preprint Nr. 60 of KIAM, Moscow, 2011 (in Russian); available from the Keldysh Institute of Applied Mathematics of RAS: `http://library.keldysh.ru/preprints/Default.asp?lg=e`.

[6] A. D. Bruno, Power-exponential expansions of soltions to an ordinary differential equation, *Doklady Mathematics*, **85**(3), 2012 (= *DAN,* **444**(2), 2012).

Author Information

Alexander D. Bruno, Keldysh Institute of Applied Mathematics (KIAM), Miusskaya Sq. 4, 125047, Moscow, Russia.
E-mail: `abruno@keldysh.ru`

Part III

Isomondromy Deformations

Chapter 9

Isomonodromic Deformations on Riemann Surfaces

Dmitry V. Artamonov

Abstract. Some representations of a pair "a bundle with a connection" on a Riemann surface, based on a representation of a surface as a factor of an exterior of a unit disk, is introduced. In this representation isomonodromic deformations of bundles with logarithmic connections are described by some modification of the Schlesinger system on a Riemann sphere (typically this modification is just the ordinary Schlesinger system) plus some linear system.

Keywords. Isomonodormic Deformations, The Schlesinger System, Bundles on Riemann Surfaces.

2010 Mathematics Subject Classification. Primary 34G56; Secondary 32G08, 32G34.

9.1 Introduction

Consider a linear fuchsian system on a Riemann sphere: $\frac{dy}{dz} = \sum_{i=1}^{n} \frac{B_i}{z-a_i} y$. Let us change positions of singularities a_i in such a way that the monodromy is preserved and the singularities do not confluence. Then the residues B_i become functions of a_i. If one considers only the so-called Schlesinger deformations (for typical fuchsian systems all the deformations are Schlesinger deformations), these functions are exactly solutions of the Schlesinger system

$$dB_i = -\sum_{\substack{j=1 \\ i \neq j}}^{n} \frac{[B_i, B_j]}{a_i - a_j} d(a_i - a_j).$$

Take a Riemann surfaces of positive genus instead of the Riemann sphere. In this case it is natural to consider the deformations of bundles with connections. Also it is natural to allow changing the complex structure on the surface.

The isomonodromic deformations on Riemann surfaces were considered e.g. in [1–3]. The general case was considered by Krichever [2]. His approach is based on the fact that all holomorphic vector bundles are meromorphically trivial. But the equations of isomonodromic deformations from [2] differ greatly from the Schlesinger system.

This work was completed with the support of the grant MK-4270.2011.1.

In the case of genus 1, another approach is known: the elliptic Schlesinger system (see [1,3]). This system of equations describes isomonodromic deformations on a torus and looks like an ordinary Schlesinger system.

The problem of generalizing the Schlesinger system to higher genus surfaces was posed in [1]. It will be proved that in the case of Riemann surfaces the isomonodromic deformations can be described by the Schlesinger system plus some collection of linear equations.

9.2 The Space of Parameters $\tilde{T}$

Let M be a Riemann surface of genus $g > 1$. Let us fix some initial point x_0 in M.

Definition 9.1. The space $\tilde{T}$ of parameters of deformations is defined as follows. Let T be the Teichmuller space with n marked points $a_1, \dots, a_n$, where $a_i \neq a_j$ for $i \neq j$. Then $\tilde{T}$ is the universal covering of T.

Definition 9.2. The space $\tilde{T}_1$, on which the isomonodromic families of pairs "a bundle with a connection" are defined, is constructed as follows. Let T_1 be the Teichmuller space with $(n+1)$ marked points $z, a_1, \dots, a_n$, where $a_i \neq a_j$ for $i \neq j$. Then $\tilde{T}_1$ is the universal covering of T_1 taken by variables a_i.

There exists a mapping $\tilde{T}_1 \to \tilde{T}$ which forgets the marked point z. Let $\tau \in \tilde{T}$. Denote as $\tilde{T}_1 \mid_\tau$ the preimage of τ under the mapping $\tilde{T}_1 \to \tilde{T}$.

Only the local equations of isomonodromic deformations will be written. This is why we are allowed not to take the universal covering.

The deformed objects are pairs "a bundle with a connection" (not a form = a system of linear equations = a connection in a trivial bundle as in the case of genus 0).

9.3 The Description of Bundles with Connections on a Riemann Surface

The surface is represented as a factor of an exterior of a unit disk by an action of a fuchsian group. A fundamental polygon U with $4g$ vertices (g is a genus of a surface) is chosen in such a canonical way that one of its vertices is identically ∞. The singularities become points in this polygon.

The deformed objects, bundles with connections (E, ∇) on the Riemann surface, are described by the following data:

(i) a form ω on a Riemann Sphere. This form is of type $\omega = (\frac{C_k}{z^k} + \dots \frac{C_1}{z} + \sum_i \frac{B_i}{z-a_i})dz$. All its singularities, except 0, belong to the fundamental polygon U. The 0 is a regular singularity;

(ii) a collection of nondegenerate matrices $S_{x_0^1,x_0^i}$, $i = 2,\dots,4g$, where the symbols x_0^i denote the vertices of the fundamental polygon.

The form ω is constructed as follows. Take an inverse image of a pair (E,∇) on a surface under factorization $U \to M$ and get some pair (E_U,∇_U) on a fundamental polygon U. Continue this pair to a bundle with a connection on the Riemann sphere. Take a meromorphic trivialization of this bundle, which is holomorphic in $\overline{\mathbb{C}} \setminus \{0\}$. The connection in this bundle is defined by a form with a regular singularity in zero. This is the form ω. For typical monodromy and typical positions of singularities, the trivialization can be chosen in such a way that the form is written just as $\omega = (\frac{C_1}{z} + \sum_{i=1}^n \frac{B_i}{z-a_i})dz$.

Define matrices $S_{x_0^1 x_0^i}$. All vertices $x_0^1,\dots,x_0^{4g}$ of the polygon are glued to x_0 under factorization $U \to M$. A bundle with a connection (E_U,∇_U) is an inverse image of the bundle with connection (E,∇). There exist operators that identify stalks $E_{Ux_0^1}$, $E_{Ux_0^i}$ of the bundle E_U over the vertices x_0^1 and x_0^i. The matrix $S_{x_0^1,x_0^i}$ is the matrix of this operator.

The procedure of construction of the form and the matrices is noncanonical. Different forms and matrices can give equivalent bundles with connections on a surface. We shall in fact not consider bundles with connections, but rather the data described above.

9.4 Isomonodromic Deformations

In this section (E,∇) is a logarithmic connection. Let (E^1,∇^1) be a pair on $\tilde{T}_1$. For $\tau \in \tilde{T}$ denote as $(E^1,\nabla^1)\mid_\tau$ the restriction of a bundle and a connection (E^1,∇^1) to the subspace $\tilde{T}_1\mid_\tau$.

Definition 9.3. An isomonodromic family is a pair (E^1,∇^1) on $\tilde{T}_1$, such that the following holds:

(i) the pair (E^1,∇^1) has singularities on hypersurfaces $z = a_i$ (i.e., hypersurfaces in $\tilde{T}_1$, which are preimages of hypersurfaces $z = a_i$ in T_1);

(ii) for all $\tau \in \tilde{T}$ the pairs $(E^1,\nabla^1)\mid_\tau$ have the same monodromy.

Such a family describes a deformation of a pair (E,∇) on the marked Riemann surface corresponding to τ_0, with singularities corresponding to marked points of τ_0, if $(E^1,\nabla^1)\mid_{\tau_0} = (E,\nabla)$.

Definition 9.4. A family (E^1,∇^1) is a Schlesinger family if the following holds: when the point in the Teichmuller space is fixed in some neighborhood of a hypersurface $z = a_i$, the connection ∇^1 is given in local coordinates by a form of type $\frac{B_i}{\zeta-a_i}d(\zeta - a_i) + h(\zeta,a_i)$, here $h(\zeta,a_i)$ is some holomorphic form, B_i are holomorphic functions of a_i.

Proposition 9.5. *For every logarithmic initial pair (E, ∇) at $t_0 \in \tilde{T}$ there exists its unique continuation to a Schlesinger isomonodromic family (E^1, ∇^1) on $\tilde{T}_1$.*

Now let us write the equations of Schlesinger isomonodromic deformations. For each $\tau \in T$ close to τ_0, take a pair $(E^1, \nabla^1) |_\tau$ and consider the corresponding form and matrices. Let us write the equations describing the resulting family of forms and matrices.

Proposition 9.6. *The equations of the isomonodromic deformations of ω are:*

$$dB_i = -\sum_{j=1, i\neq j}^{n} \frac{[B_i, B_j]}{a_i - a_j} d(a_i - a_j) + \frac{\partial C_1}{\partial a_i} da_i, \qquad i = 1, \dots, n,$$

$$\frac{\partial C_2}{\partial a_i} - \frac{\partial C_1}{\partial a_i} a_i = -[B_i, C_1], \dots, \qquad \frac{\partial C_{l+1}}{\partial a_i} - \frac{\partial C_l}{\partial a_i} a_i = -[B_i, C_l], \dots,$$

$$-\frac{\partial C_k}{\partial a_i} a_i = -[B_i, C_k].$$

In the typical case, when $C_l = 0$ for $l > 1$, in the notations $a_0 = 0$, $B_0 = C_1 = -\sum_{i=1}^{n} B_i$ this system turns into an ordinary Schlesinger system

$$dB_i = -\sum_{j=0,\, i\neq j}^{n} \frac{[B_i, B_j]}{a_i - a_j} d(a_i - a_j).$$

Adding the equation describing how the matrices $S_{x_0^1, x_0^i}$ are changing, we get the main result.

Theorem 9.7. *In terms of the data presented in Section 9.3, the Schlesinger isomodromic deformation locally are described as follows:*

(i) *The vertices x_0^i of the fundamental polygon change according to the change of a point in the Teichmuller space.*

(ii) *The coefficients of the form*

$$\omega = \left(\frac{C_k}{z^k} + \dots + \frac{C_1}{z} + \sum_i \frac{B_i}{z - a_i} \right) dz$$

satisfy the equations listed in the Proposition 9.6. In the typical case $C_l = 0$, $l > 0$ in notations $a_0 = 0$, $B_0 = -\sum_{i=1}^{n} B_i$ this is just the Schlesinger system

$$dB_i = -\sum_{j=0,\, i\neq j}^{n} \frac{[B_i, B_j]}{a_i - a_j} d(a_i - a_j).$$

(iii) *The matrices $S_{x_0^1, x_0^i}$ are isomonodromic families of solutions of the equation $dS_{x_0^1 z} = \omega S_{x_0^1 z}$ at the point $z = x_0^i$.*

Bibliography

[1] D. Korotkin, N. Manojlović, and H. Samtleben, Schlesinger transformations for elliptic isomonodromic deformations, *J. Math. Phys.*, **41**: 3125, 2000.

[2] I. M. Krichever, Isomonodromy equations on algebraic curves, canonical transformations and Whitham equations, *Mosc. Math. J.*, **2**(4): 717–752, 2002.

[3] Yu. Chernyakov, A. M. Levin, M. Olshanetsky, and A. Zotov, Elliptic Schlesinger system and Painleve VI, *J. Phys. A: Math. Gen.*, **39**: 12083, 2006, 12083.

Author Information

Dmitry V. Artamonov, 119421, Novatorov 4-3-45, Moscow, Russia.
E-mail: `artamonov.dmitri@gmail.com`

Chapter 10

On Birational Darboux Coordinates of Isomonodromic Deformation Equations Phase Space

Mikhail V. Babich

Abstract. I propose the transparent geometric model that makes it possible to present the set of the rational Darboux coordinates on the phase space of the Isomonodromic Deformation equation. At the base of this construction lies an observation that for the matrix from the orbit the projections of the kernel and image to the corresponding complementary coordinate subspaces are conjugated with each other with respect to the canonical structure of the orbit.

Keywords. Standard Jordan Form, Momentum Map, Lie–Poisson–Kirillov–Kostant Form, Rational Symplectic Coordinates.

2010 Mathematics Subject Classification. Primary: 34M55.

Let us consider the deformation of the Fuchs equation

$$\frac{d}{dz}\Psi = \sum_{k=1}^{M} \frac{A^k}{z - z^k}\Psi; \; A^k \in \mathrm{sl}(N, \mathbb{C}); \; z, z^k \in \mathbb{C}. \tag{1}$$

It is known that the isomonodromic deformation of this equation may be associated with some Hamiltonian system defined on the space that we denote by $\mathcal{O}_{J^1} \times \mathcal{O}_{J^2} \times \cdots \times \mathcal{O}_{J^M} /\!/ \mathrm{SL}(N, \mathbb{C})$. This space is the quotient of the product of the several coadjoint orbits $\mathcal{O}_{J^k} := \cup_{g \in SL(N,\mathbb{C})} g J^k g^{-1} \ni A^k$ over the diagonal action of $\mathrm{SL}(N, \mathbb{C})$ intersected by the subspace $\sum_{k=1}^{M} A^k = 0$.

It is known that both the orbit $\mathcal{O}_J$ and $\mathcal{O}_{J^1} \times \mathcal{O}_{J^2} \times \cdots \times \mathcal{O}_{J^M} /\!/ \mathrm{SL}(N, \mathbb{C})$ are the symplectic manifolds equipped with the canonical symplectic forms ω_J and $\omega = \sum_k \omega_{J^k}$. First, we will introduce the birational Darboux parametrization of $\mathcal{O}_J$. The birationality means that the transformations (in both directions) between the set of the matrix elements of $A \in \mathcal{O}_J$ and the set of the canonical coordinate functions in question are given by the rational functions.

Let us build a set of the canonical coordinates on the orbits $\mathcal{O}_{J^k}$. It is the heart of the construction.

We get the coordinates as a result of the following iterative process. Let us put in any order the eigenvalues of J: $\lambda_1, \lambda_2, \ldots, \lambda_{n_{\max}}$, where the eigenvalues corresponding to

Jordan blocks are written several times, in accordance with the maximal size of the blocks corresponding to the eigenvalue. Let us define the iteration $A_{j-1} \to A_j$:

$$A_{j-1} = \begin{pmatrix} I & 0 \\ p_j & I \end{pmatrix} \begin{pmatrix} \lambda_j & q_j \\ 0 & A_j \end{pmatrix} \begin{pmatrix} I & 0 \\ p_j & I \end{pmatrix}^{-1}, \quad A_0 := A,$$

where the columns of $(I p_j)^T$ form the basis of the eigenspace $\ker(A_{j-1} - \lambda_j I)$. It can be proved that

the process is correctly defined on the Zariski-open subset of the orbit and define the rectangular matrices p_j, q_j, which matrix elements $(p_j)_{st}, (q_j)_{ts}$ are canonically conjugated with respect to the Lie–Poisson–Kirillov–Kostant form on the orbit $\mathcal{O}_J$.

The coordinates presented are similar to the cylindrical coordinates introduced by Archimedes on a sphere (it is an orbit of O(3)) for the calculation of its area. Namely, the family of p-coordinates has a group nature; they parameterize some Abelian subgroups of triangular matrices, an analog of the longitude. The conjugated q-coordinates form another family. These coordinates are invariant with respect to the action of the corresponding Abelian subgroups on $\mathcal{O}_J$. They are an analog of the projection of the sphere on the diameter connecting the north and the south poles. It is important for us that the matrix elements depend on the coordinates from the q-family linearly.

Let us consider the isomonodromic deformations of the Schlesinger type:

$$dA^k + [A^k, dRR^{-1} + \sum_{i \neq k} A^i d \log(z^k - z^i)] = 0. \tag{2}$$

The matrix function $R = R(z^1, \dots, z^M)$ is the normalization matrix. It is the value of the isomonodromic fundamental system of the solutions (1) in the infinity $R := \Psi(z|_{z=\infty}, z^1, \dots z^M)$.

The fixation of the normalization R is equivalent to the fixation of the section of $\mathcal{O}_{J1} \times \dots \times \mathcal{O}_{JM} \to \mathcal{O}_{J1} \times \dots \times \mathcal{O}_{JM} /\!/ \mathrm{SL}(N, \mathbb{C})$. We will fix this section explicitly in terms of the constructed coordinates on $\mathcal{O}_{J1} \times \dots \times \mathcal{O}_{JM}$.

If the dimension of $\mathcal{O}_{J1} \times \mathcal{O}_{J2} \times \dots \times \mathcal{O}_{JM} /\!/ \mathrm{SL}(N, \mathbb{C})$ is high enough we can normalize (2) by setting equal to the constant values the coordinates from the special subset $p_1^{norm}, p_2^{norm}, \dots, p_{N^2-1}^{norm}$ from the set of all p_i^k, and the special choice of the values of the coordinates q_i^{Σ} conjugated p_i^{norm} provides the necessary restriction $\sum_{k=1}^M A^k = 0$.

The dependence A^k on the coordinates q_i^k is linear. Consequently *the solution of the linear equations $\sum_{k=1}^M A^k = 0$ is given by the rational functions $q_i^{\Sigma} = q_i^{\Sigma}(\hat{p}, \hat{q})$*, where by $\hat{p}, \hat{q}$ we denote the sets of all the coordinates without p_i^{norm} and q_i^{Σ}. They are removed from the sets.

The coordinates p_i^{norm} conjugated to q_i^{Σ} are equal to constants, and consequently the symplectic form ω is equal to $\sum_i d\hat{p}_i \wedge d\hat{q}_i$.

Functions $\hat{p}, \hat{q}$ form the set of the birational canonical coordinates.

Example. Let two matrices residues in (1), say A^M and A^{M-1}, have one-dimensional eigenspaces only, and at least one eigenspacc of A^{M-2}, say, $\ker(A^{M-2} - \lambda_N^{M-2} I)$ is also one-dimensional. We present the matrices A^M, A^{M-1}, A^{M-2} for this case:

$$A^M = \begin{pmatrix} \lambda_1^M & q_1^\Sigma & \dots & q_{N-1}^\Sigma \\ 0 & \lambda_2^M & \dots & q_{N-1+N-2}^\Sigma \\ \vdots & \vdots & \ddots & \vdots \\ 0 & 0 & \dots & \lambda_N^M \end{pmatrix},$$

$$A^{M-1} = \begin{pmatrix} \lambda_1^{M-1} & 0 & \dots & 0 \\ q_{N(N-1)/2+1}^\Sigma & \lambda_2^{M-1} & \dots & 0 \\ \vdots & \vdots & \ddots & \vdots \\ q_{N(N-1)/2+N-1}^\Sigma & q_{N(N-1)/2+N-1+N-2}^\Sigma & \dots & \lambda_N^{M-1} \end{pmatrix},$$

$$A^{M-2} = \begin{pmatrix} (A'^{N-2})_{11} + q_{N(N-1)+1}^\Sigma & (A^{N-2})_{12} & \dots & (A^{N-2})_{1N} \\ (A^{N-2})_{21} & (A'^{N-2})_{22} + q_{N(N-1)+2}^\Sigma & \dots & (A^{N-2})_{2N} \\ \vdots & \vdots & \ddots & \vdots \\ (A^{N-2})_{N1} & (A^{N-2})_{N2} & \dots & (A^{N-2})_{NN} \end{pmatrix}.$$

It is important that the values $(A'^{N-2})_{ii}, i = 1, \dots, N-1$, and $(A^{N-2})_{NN}$ from the diagonal *do not* depend on the variables q_j^Σ, and all the matrix elements of A^{N-2} *do not* depend on q_j^Σ, $4j \le N(N-1)$.

We can see that the equation $\sum A^k = 0$ just fix the values of all q_i^Σ: to get $(\sum A^k)_{jj} = 0, j < N$ we fix $q_i^\Sigma, i > N(N-1)$; after that we fix $q_i^\Sigma, i \le N(N-1)$ to vanish off-diagonal entries of the equality $\sum A^k = 0$ using the freedom to set upper and lower triangular parts of A^M and A^{M-1}. The last term $(\sum A^k)_{NN}$ is automatically equal to zero because all the matrices are traceless.

We note that it is a case of a general position for J^M, J^{M-1}, J^{M-2}, and we do not put any restrictions on the structure of all other matrices $J^k, k < M-2$.

Bibliography

[1] M. V. Babich, About a canonical parametrization of the phase spaces of equations of isomonodromic deformations of Fuchs systems of dimension 2×2, *Russian Math. Surveys*, **64**:1 (2009), 45–127.

[2] M. V. Babich, Rational symplectic coordinates on the space of Fuchs equations, $m \times m$-case, *Lett. Math. Phys.*, **86** (2008), 63–77.

[3] M. V. Babich and S. E. Derkachov, On rational symplectic parametrization of coajoint orbit of $GL(N)$. Diagonalizable case, *Algebra and Analiz*, **22** 3 (2010), 16–30, (in Russian).

[4] M. V. Babich, *Rational version of Archimedes symplectomorphysm and birational Darboux coordinates on coadjoint orbit of GL(N,C)*, Preprint MPIM Bonn, MPIM2010-59, (2010), 1–36, `http://www.mpim-bonn.mpg.de/Research/MPIM+Preprint+Series/`.

Author Information

Mikhail V. Babich, Petersburg Department of Mathematical Institute, Fontanka 27, St. Petersburg, Russia.
E-mail: mbabich@pdmi.ras.ru, misha.babich@gmail.com

Chapter 11

On the Malgrange Isomonodromic Deformations of Nonresonant Irregular Systems

Yuliya P. Bibilo and Renat R. Gontsov

Abstract. Movable singularities of the equations governing the Malgrange isomonodromic deformation of a nonresonant rank 2 meromorphic connection are studied: we describe the theta-divisor of the deformation and estimate orders of movable poles of the equations in the case of irreducible monodromy.

Keywords. Holomorphic Vector Bundle, Meromorphic Connection, Irregular Singularity, Isomonodromic Deformation, Theta-Divisor, Painlevé Property.

2010 Mathematics Subject Classification. Primary: 34M56; Secondary: 34M40.

11.1 Introduction

Consider a *meromorphic linear system* on the Riemann sphere $\overline{\mathbb{C}}$, i.e., a system of p linear ordinary differential equations with singularities $a_1^0, \dots, a_n^0 \in \mathbb{C}$ and possibly ∞. By a conformal mapping one can always arrange that all the singularities are in the complex plane only. This means that one can reduce the system to the form

$$\frac{dy}{dz} = B(z)\,y, \qquad B(z) = \sum_{i=1}^{n}\sum_{j=1}^{r_i+1} \frac{B_{ij}^0}{(z-a_i^0)^j}, \tag{11.1}$$

where $y(z) \in \mathbb{C}^p$, B_{ij}^0 are $(p \times p)$-matrices and $\sum_{i=1}^n B_{i1}^0 = 0$, to ensure that ∞ is not a singular point.

The nonnegative integers $r_1, \dots, r_n$ are called the *Poincaré ranks* of the singularities $a_1^0, \dots, a_n^0$ respectively. One can assume that the Poincaré ranks $r_1, \dots, r_m$ are positive and $r_{m+1} = \dots = r_n = 0$ (that is, the singular points $a_{m+1}^0, \dots, a_n^0$ are *Fuchsian*) for some $0 \leqslant m \leqslant n$.

We consider the *nonresonant* case. This means that the leading term B_{i,r_i+1}^0 of each non-Fuchsian singularity a_i^0, $i = 1, \dots, m$, has p distinct eigenvalues.

This work is supported by the Russian Foundation for Basic Research (RFBR-11-01-00339) and Russian Federation presidential program grants NSh-8508.2010.1 and MK-4270.2011.1.

The system (11.1) can be thought of as a meromorphic connection ∇^0 (more precisely, as an equation for horizontal sections with respect to this connection) on a holomorphically trivial vector bundle E^0 of rank p over $\overline{\mathbb{C}}$. In a neighborhood of each (nonresonant) irregular singularity a_i^0 the local connection form $\omega^0 = -B(z)dz$ of ∇^0 is formally equivalent to the 1-form

$$\omega_{\Lambda_i^0} = \sum_{j=1}^{r_i+1} \frac{\Lambda_{ij}^0}{\left(z - a_i^0\right)^j}\, dz,$$

where $\Lambda_{i1}^0, \ldots, \Lambda_{i,r_i+1}^0$ are diagonal matrices and the leading term Λ_{i,r_i+1}^0 are conjugated to B_{i,r_i+1}^0.

One should note that formally equivalent systems in a neighborhood $O_{a_i^0}$ of an irregular singularity a_i^0 are not necessarily holomorphically or meromorphically equivalent. The system (11.1) has in $O_{a_i^0}$ a formal fundamental matrix of the form

$$\widehat{Y}(z) = \widehat{F}(z)\left(z - a_i^0\right)^{-\Lambda_{i1}^0} e^{Q(z)}, \qquad Q(z) = \sum_{j=1}^{r_i} \frac{\Lambda_{ij}^0}{j}\left(z - a_i^0\right)^{-j},$$

where $\widehat{F}(z)$ is an invertible formal Taylor series in $(z - a_i^0)$. One can cover $O_{a_i^0}$ by a set of sufficiently small sectors $S_1, \ldots, S_N$ such that in each S_k there exists a unique fundamental matrix $Y_k(z) = F_k(z)(z-a_i^0)^{-\Lambda_{i1}^0} e^{Q(z)}$ of the system with $F_k(z)$ having $\widehat{F}(z)$ as asymptotic series in S_k. In every intersection $S_k \cap S_{k+1}$ the fundamental matrices $Y_k(z)$, $Y_{k+1}(z)$ are connected by a constant matrix C_k: $Y_{k+1}(z) = Y_k(z)C_k$, which is called a *Stokes matrix*. If a_i^0 is a nonresonant singularity, then two formally equivalent systems are holomorphically equivalent in $O_{a_i^0}$ if and only if they have the same sets of Stokes matrices.

Further we will focus on the deformations of the system (11.1) (of the pair (E^0, ∇^0)) that allow the local formal equivalence class

$$\omega_{\Lambda_i} = \sum_{j=2}^{r_i+1} \frac{\Lambda_{ij}}{(z - a_i)^j}\, dz + \frac{\Lambda_{i1}^0}{z - a_i}\, dz, \qquad i = 1, \ldots, m,$$

to vary in the sense that the diagonal matrices $\Lambda_{i2}, \ldots, \Lambda_{i,r_i+1}$ vary in a neighborhood of $\Lambda_{i2}^0, \ldots, \Lambda_{i,r_i+1}^0$ with Λ_{i1}^0 held fixed. Thus, for the set $\Lambda_i = \{\Lambda_{i2}, \ldots, \Lambda_{i,r_i+1}\}$ of r_i diagonal matrices we denote by ∇_{Λ_i} the meromorphic connection on the holomorphically trivial vector bundle of rank p over O_{a_i} whose 1-form is ω_{Λ_i}.

11.2 The Malgrange Isomonodromic Deformation of the Pair (E^0, ∇^0)

First we describe in more detail the deformation space. For $k \in \mathbb{N}$ let us denote by Z^k the subset of the space $\mathbb{C}^k$ whose points have pairwise distinct coordinates. Then Z^n

will be the space of pole locations and

$$\mathcal{C}_i = \underbrace{\mathbb{C}^p \times \ldots \times \mathbb{C}^p}_{r_i-1} \times Z^p, \qquad i = 1, \ldots, m,$$

will be the space of local formal equivalence classes at the pole a_i. Define the deformation space $\mathcal{D}$ as the universal cover

$$\mathcal{D} = \widetilde{Z^n} \times \widetilde{\mathcal{C}}_1 \times \cdots \times \widetilde{\mathcal{C}}_m$$

of the Cartesian product $Z^n \times \mathcal{C}_1 \times \cdots \times \mathcal{C}_m$.

One has the standard projections

$$\begin{aligned} a = (a_1, \ldots, a_n) : \mathcal{D} &\to Z^n, \\ \Lambda_i = (\Lambda_{i2}, \ldots, \Lambda_{i,r_i+1}) : \mathcal{D} &\to \mathcal{C}_i, \qquad i = 1, \ldots, m. \end{aligned}$$

For $t \in \mathcal{D}$ we denote by $a_i(t)$ the i-th coordinate of the image of t under the first projection and by $\Lambda_i(t)$ the image of t under the second one. Denote then by t^0 the base point of the deformation space $\mathcal{D}$ corresponding to the system (11.1) (to the initial connection ∇^0), i.e., $a(t^0) = (a_1^0, \ldots, a_n^0)$, $\Lambda_i(t^0) = (\Lambda_{i2}^0, \ldots, \Lambda_{i,r_i+1}^0)$. Consider also the singular hypersurfaces

$$Y_i = \{(z,t) \in \overline{\mathbb{C}} \times \mathcal{D} \mid z = a_i(t)\} \subset \overline{\mathbb{C}} \times \mathcal{D}, \qquad i = 1, \ldots, n.$$

Now consider the fiber bundle $\mathcal{M}_i \to \mathcal{C}_i$, whose fiber over each point $\Lambda_i \in \mathcal{C}_i$ is the moduli space of local holomorphic equivalence classes of connections that are all formally equivalent to the connection ∇_{Λ_i}. A point of this fiber (a holomorphic equivalence class of connections) is determined by a corresponding set of Stokes matrices. Let $\sigma_i^0 \in \mathcal{M}_i$ denote the holomorphic equivalence class of the connection $\nabla^0|_{O_{a_i^0}} \sim \nabla_{\Lambda_i^0}$ and let σ_i denote the unique (local) horizontal section of the fibre bundle $\mathcal{M}_i \to \mathcal{C}_i$ such that $\sigma_i(\Lambda_i^0) = \sigma_i^0$.

There exists ([1, Th. 3.1]; see also [2, Th. 2.9]) the Malgrange isomonodromic deformation (E, ∇) of the pair (E^0, ∇^0), i.e., the rank p holomorphic vector bundle E over $\overline{\mathbb{C}} \times \mathcal{D}$ and the integrable meromorphic connection ∇ on E with a simple type r_i singularity over Y_i, $i = 1, \ldots, n$, satisfying the following properties:

(i) the restriction of (E, ∇) to $\overline{\mathbb{C}} \times \{t^0\}$ is equivalent to (E^0, ∇^0);

(ii) the restriction of ∇ to $\overline{\mathbb{C}} \times \{t\}$ is formally equivalent to the local connection $\nabla_{\Lambda_i(t)}$ near $z = a_i(t)$;

(iii) the latter restriction belongs to the local holomorphic equivalence class $\sigma_i(\Lambda_i(t)) \in \mathcal{M}_i$.

According to the Malgrange–Helminck–Palmer theorem (see [2, §3]) the set

$$\Theta = \{t \in \mathcal{D} \mid E|_{\overline{\mathbb{C}} \times \{t\}} \text{ is nontrivial }\}$$

is either empty or $\Theta \subset \mathcal{D}$ is an analytic subset of codimension one. Thus, the Malgrange isomonodromic deformation of the pair (E^0, ∇^0) determines an isomonodromic deformation

$$\frac{dy}{dz} = \left(\sum_{i=1}^{n} \sum_{j=1}^{r_i+1} \frac{B_{ij}(t)}{\left(z - a_i(t)\right)^j} \right) y, \qquad B_{ij}(t^0) = B_{ij}^0,$$

of the system (11.1) for $t \in D(t^0)$, where $D(t^0)$ is a neighborhood of the point t^0 in the space $\mathcal{D}$. The matrix functions $B_{ij}(t)$, holomorphic in $D(t^0)$, can be extended meromorphically to the whole space $\mathcal{D}$ having Θ as a polar locus.

11.3 Specificity of Meromorphic 2×2-Connections

The polar locus Θ (which is called the *Malgrange Θ-divisor*) possesses a local description analogous to that of the Θ-divisor for the Schlesinger equation governing isomonodromic deformations of Fuchsian systems (see [3]).

Now let us consider the two-dimensional case ($p = 2$). Suppose that the initial system has at most $m = 2$ irregular (nonresonant) singularities and they are of Poincaré rank 1. Thus we consider the system of the form

$$\frac{dy}{dz} = \left(\frac{B_{12}^0}{(z - a_1^0)^2} + \frac{B_{22}^0}{(z - a_2^0)^2} + \sum_{i=1}^{n} \frac{B_{i1}^0}{z - a_i^0} \right) y. \qquad (11.2)$$

Theorem 11.1. *Let the monodromy of the system* (11.2) *be* irreducible *and let* (E, ∇) *be the Malgrange isomonodromic deformation of the system* (11.2). *Consider any point* $t^* \in \Theta$ *such that* $E|_{\overline{\mathbb{C}} \times \{t^*\}} \cong \mathcal{O}(1) \oplus \mathcal{O}(-1)$.

Then in a neighborhood $D(t^*)$ *of* t^* *the Θ-divisor is given as a zero set of an* irreducible *holomorphic function* τ, *and the matrix functions* $B_{ij}(t)$ *have poles of at most* second *order along* $D(t^*) \cap \Theta$.

The latter means that $\tau^2(t) B_{ij}(t)$ are holomorphic matrix functions in $D(t^*)$.

Remark 11.2. For example, the Painlevé P_3 and P_5 equations can be described in terms of isomonodromic deformations satisfying the above theorem (see details in [4, Chap. 5, Sects. 4,5]): for P_3 one has $m = n = 2$ and for P_5 one has $m = 1, n = 3$. If $t^* \in \Theta$ and $E|_{\overline{\mathbb{C}} \times \{t^*\}} \cong \mathcal{O}(k) \oplus \mathcal{O}(-k)$, then the estimate $2k \leqslant m + n - 2$ holds [5] when the monodromy of a connection is irreducible. Thus $2k \leqslant 2$ and hence $k = 1$ in both cases.

Bibliography

[1] B. Malgrange, Sur les déformations isomonodromiques. II. Singularités irrégulières, *Progr. Math.* **37** (1983), 427–438.

[2] J. Palmer, Zeros of the Jimbo, Miwa, Ueno tau function, *J. Math. Phys.* **40**(12) (1999), 6638–6681.

[3] A. A. Bolibrukh, On the tau-function for the Schlesinger equation of isomonodromic deformations, *Math. Notes* **74**(2) (2003), 177–184.

[4] A. R. Its, A. A. Kapaev, V. Yu. Novokshenov, A. S. Fokas, *Painlevé transcendents: the Riemann–Hilbert approach,* AMS, Providence (= Moscow-Izhevsk, R&C Dynamics, 2005 (in Russian)).

[5] A. A. Bolibruch, S. Malek, C. Mitschi, On the generalized Riemann–Hilbert problem with irregular singularities, *Expo. Math.* **24**(3) (2006), 235–272.

Author Information

Yuliya P. Bibilo, Higher School of Economics (National Research University), Myasnitskaya st., 20, Moscow, Russia.
E-mail: `y.bibilo@gmail.com`

Renat R. Gontsov, Inst. for Information Transmission Problems, Bolshoy Karetny per. 19, 127994, Moscow, Russia.
E-mail: `gontsovrr@gmail.com`

Chapter 12

Critical behavior of P_6 Functions from the Isomonodromy Deformations Approach

Davide Guzzetti

Abstract. The critical behaviors of P_6 functions, obtained in the framework of the monodromy-preserving deformation method, are synthetically reviewed.

Keywords. Isomonodromy Deformation, Painlevé Equations.

2010 Mathematics Subject Classification. 34M55.

12.1 Introduction

We consider the sixth Painlevé equation, with independent variable x and dependent variable $y(x)$, in the standard form [8], with parameters α, β, γ, and $\delta \in \mathbf{C}$. *Solving* P_6 means that:

(i) we determine the critical behavior (or asymptotic expansion) of $y(x)$ by an explicit formula in terms of two integration constants;

(ii) we solve the *connection problem*, namely, we find the *explicit* relations among couples of integration constants at different critical points.

We review here the results of the isomonodromy deformation method, which answers (i) and (ii).

PVI is the isomonodromy deformation equation [9] of a 2×2-Fuchsian system

$$\begin{aligned} \frac{d\Psi}{d\lambda} &= \left[\frac{A_0(x)}{\lambda} + \frac{A_x(x)}{\lambda - x} + \frac{A_1(x)}{\lambda - 1}\right]\Psi, \\ y(x) &= \frac{x(A_0)_{12}}{x[(A_0)_{12} + (A_1)_{12}] - (A_1)_{12}}. \end{aligned} \tag{12.1}$$

The eigenvalues of $A_i(x)$ are algebraic functions of α, β, γ and δ. Let M_0, M_x, M_1 be the x-independent monodromy matrices of Ψ w.r.t. a given basis of loops. There is a one-to-one correspondence between a point in the space of monodromy data and a branch $y(x)$ (except for one M_i or $M_1 M_x M_0 = I$; see [5, 6]). As a consequence, the integration constants of $y(x)$ can be parameterized in terms of the monodromy data. Jimbo [8] provided the critical behavior and its parameterization in terms of monodromy data for a wide class of solutions. We review here the matching procedure of [4], which gives the results of [8] plus the other critical behaviors and their parameterization.

12.2 Behavior of $y(x)$

We consider $x \to 0$ ($x \to 1$ and $x \to \infty$ can be obtained from the results at $x = 0$ by symmetries). Let us divide the λ-plane into two domains. The "out" domain is $|\lambda| \geq |x|^{\delta_{\text{OUT}}}$, $\delta_{\text{OUT}} > 0$. (12.1) can be approximated by:

$$\frac{d\Psi_{\text{OUT}}}{d\lambda} = \left[\frac{A_0 + A_x}{\lambda} + \frac{xA_x}{\lambda^2} \sum_{n=0}^{N_{\text{OUT}}} \left(\frac{x}{\lambda}\right)^n + \frac{A_1}{\lambda - 1}\right] \Psi_{\text{OUT}}, \tag{12.2}$$

The "in" domain is $|\lambda| \leq |x|^{\delta_{\text{IN}}}$, $0 < \delta_{\text{IN}} \leq \delta_{\text{OUT}}$. (12.1) can be approximated by:

$$\frac{d\Psi_{\text{IN}}}{d\lambda} = \left[\frac{A_0}{\lambda} + \frac{A_x}{\lambda - x} - A_1 \sum_{n=0}^{N_{\text{IN}}} \lambda^n\right] \Psi_{\text{IN}}, \tag{12.3}$$

where N_{IN}, N_{OUT} are suitable integers. We impose that

$$\Psi_{\text{OUT}}(\lambda, x) \sim \Psi_{\text{IN}}(\lambda, x), \qquad |x|^{\delta_{\text{OUT}}} \leq |\lambda| \leq |x|^{\delta_{\text{IN}}}, \qquad x \to 0 \tag{12.4}$$

This *matching* produces the leading terms of $A_0(x)$, $A_1(x)$, and $A_x(x)$ as $x \to 0$, and therefore it produces the leading term(s) of $y(x)$ in (12.1). The results are:

Case 1. Two-complex parameter ($c_{11}, \sigma \in \mathbb{C}, 0 \leq \operatorname{Re} \sigma < 1$) solutions [3,4,6,8]

$$y(x) = \sum_{n=1}^{\infty} x^n \sum_{m=-n}^{n} c_{nm} x^{m\sigma}, \quad c_{11} \neq 0, \quad c_{nm} = c_{nm}(c_{11}, \sigma)$$

convergent. In particular, for $\sigma = 2i\nu$, $\nu \in \mathbb{R}$ the above becomes a three-real parameter solution

$$y(x) = x\Big[A \sin(\nu \ln x + \phi) + B + \sum_{n=1}^{\infty} x^n \sum_{m=-n-1}^{n+1} c_{nm} x^{m\sigma}\Big], \quad \phi = \phi(c_{11}) \tag{12.5}$$

Case 2 Three-real parameters ($\nu \in \mathbf{R}$, $d \in \mathbf{C}$) solutions [6]

$$\begin{aligned} y(x) &= \Big[\sum_{n=0}^{\infty} x^n \sum_{m=-n-1}^{n+1} c_{nm}\, e^{2imd} x^{2im\nu}\Big]^{-1} \quad \text{convergent} \\ &= \Big[A \sin(\nu \ln x + \phi(d)) + B + \sum_{n=1}^{\infty} x^n \sum_{m=-n-1}^{n+1} c_{nm} e^{2imd} x^{2im\nu}\Big]^{-1} \end{aligned} \tag{12.6}$$

All c_{nm}, A, B are determined by d and ν. (12.6) is obtained form (12.5) by the symmetry $y(x) \mapsto x/y(x)$. If $\arg x$ and $\arg(1-x)$ are bounded (namely, when we considered $y(x)$ as a branch), then (12.6) admits two infinite sequences of *movable poles* in the neighborhood of $x = 0$ (see [7]).

Case 3. If $\sqrt{-2\beta} \pm \sqrt{1-2\delta} \notin \mathbf{Z}$, there are one-complex parameter ($c_{11} \in \mathbf{C}$) solutions [4,6]

$$y(x) = \sum_{n=1}^{\infty} x^n \sum_{m=0}^{n} c_{nm} x^{m\sigma} = \sum_{k=0}^{\infty} y_k(x)\,(c_{11}\,x^{\sigma})^k, \quad \sigma \in \{\sigma_{\beta\delta}^+, \sigma_{\beta\delta}^-\}, \quad \text{convergent}$$

where $\sigma_{\beta\delta}^{\pm} := (\sqrt{-2\beta} \pm \sqrt{1-2\delta})\mathrm{sgn}(\mathrm{Re}(\sqrt{-2\beta} \pm \sqrt{1-2\delta}))$, and $y_k(x)$ are Taylor expansions of the form $y_1(x) = O(x)$, $y_k(x) = O(x^k)$. c_{nm} are determined by c_{11}. For $c_{11} = 0$, the above becomes a Taylor series with no free parameter. If $|\sqrt{-2\beta} \pm \sqrt{1-2\delta}| < 1$, the above solutions are defined also for $\sigma \in \{\pm(\sqrt{-2\beta} + \sqrt{1-2\delta}), \pm(\sqrt{-2\beta} - \sqrt{1-2\delta})\}$.

Case 4. If $\sqrt{-2\beta} + \sqrt{1-2\delta} = N$ or $\sqrt{-2\beta} - \sqrt{1-2\delta} = N$, $N \in \mathbf{Z}$, there are one-complex parameter ($a \in \mathbf{C}$) solutions [4,5]

$$y(x) = \sum_{n=1}^{|N|} b_n x^n + \left(a + b_{|N|+1} \ln x\right) x^{|N|+1} + \sum_{n=|N|+2}^{\infty} P_n(\ln x; a) x^n, \quad a \in \mathbf{C} \tag{12.7}$$

where b_n are functions of α, β, γ and δ, a is a free parameter, $P_n(\ln x; a)$ are polynomials of $\ln x$ of degree $n - |N|$, with coefficients determined by a, α, β, γ and δ. In particular, the above reduces to a Taylor series [4, 10]

$$y(x) = \sum_{n=1}^{|N|} b_n x^n + a x^{|N|+1} + \sum_{n=|N|+2}^{\infty} b_n(a) x^n \quad \text{convergent}$$

if either $\{\sqrt{2\alpha} + \sqrt{2\gamma}, \sqrt{2\alpha} - \sqrt{2\gamma}\} \cap \{|N| - 1, 1 - |N|\} \neq \emptyset$ or if $2\beta = -N^2$.

Case 5. If $2\beta \neq 2\delta - 1$, besides log solutions (12.7), there are one-complex parameter ($a \in \mathbf{C}$) logarithmic solutions [4,5,8]

$$y(x) = \left[\frac{2\beta + 1 - 2\delta}{4}(a + \ln x)^2 + \frac{2\beta}{2\beta + 1 - 2\delta}\right] x + \sum_{n \geq 2}^{\infty} P_n(\ln x; a) x^n, \quad a \in \mathbf{C}.$$

Case 3.1. The symmetry $y(x) \mapsto x/y(x)$ applied to Case 3 gives one-parameter ($a \in \mathbf{C}$) solutions

$$y(x) = \sum_{k=0}^{\infty} y_k(x)(a x^{\omega})^k, \quad \omega \in \{\omega_{\alpha\gamma}^+, \omega_{\alpha\gamma}^-\}, \quad y_0(x) = O(1), \quad y_k(x) = O(x^{k-1}),$$

where $\omega_{\alpha\gamma}^{\pm} := (\sqrt{2\alpha} \pm \sqrt{2\gamma})\mathrm{sgn}(\mathrm{Re}(\sqrt{2\alpha} \pm \sqrt{2\gamma})) \notin \mathbf{Z}$, and $y_k(x)$ are Taylor series.

Case 4.1. The symmetry $y(x) \mapsto x/y(x)$ applied to Case 4 with $N \neq 0$, gives solutions existing when $\sqrt{2\alpha} + \sqrt{2\gamma} = N$ or $\sqrt{2\alpha} - \sqrt{2\gamma} = N$:

$$y(x) = \sum_{n=0}^{|N|-1} b_n x^n + \left(a + b_N \ln x\right)x^{|N|} + \sum_{n=|N|+1}^{\infty} P_n(\ln x; a)x^n, \qquad N \neq 0$$

where b_n are functions of α, β, γ and δ, a is a free parameter, $P_n(\ln x; a)$ are polynomials of $\ln x$ of degree $n - |N| + 1$, with coefficients determined by a, α, β, γ and δ. If $\{\sqrt{-2\beta} + \sqrt{1-2\delta}, \sqrt{-2\beta} - \sqrt{1-2\delta}\} \cap \{|N| - 1, 1 - |N|\} \neq \emptyset$ or if $2\alpha = N^2$, the solutions become convergent Taylor expansions. The symmetry $y(x) \mapsto x/y(x)$ applied to Case 4 for $N = 0$ gives the solutions

$$y(x) = \frac{1}{(a \pm \sqrt{2\alpha}\ln x) + \sum_{n=1}^{\infty} P_n(\ln x; a)} \sim \pm\frac{1}{\sqrt{2\alpha}\ln x}, \qquad \alpha = \gamma.$$

Case 5.1. Symmetry $y(x) \mapsto x/y(x)$ applied to Case 6 gives

$$y(x) = \frac{2}{(\gamma - \alpha)\ln^2 x}\left[1 + \frac{1}{\ln x} + O\left(\frac{1}{\ln^2 x}\right)\right], \qquad \alpha \neq \gamma.$$

⋄ Cases 1 and 3 for $|\operatorname{Re}\sigma| < 1$, Case 4 for $N = 0$, and Case 5 are obtained by matching solutions of OUT- and IN-systems with $N_{\mathrm{OUT}} = N_{\mathrm{IN}} = 0$. Case 2 is obtained from (12.5) via symmetry $y(x) \mapsto x/y(x)$. Cases 3 and 4 with $N \neq 0$ are obtained using OUT- and IN-systems with at least one N_{IN} or N_{OUT} greater than zero. For the convergence of Cases 1, 2, and 3 see method of [3, 11]. For convergence of Taylor expansions see [10].

The set of the above behaviors obtained in [3–6, 8], coincides with the set of expansions obtained by power geometry, summarized in [2].

12.3 Parameterization in Terms of Monodromy Data

Every solution $y(x)$ is associated to a point in the space of monodromy data of (12.1). The power of the method of isomonodromic deformations is that it allows to parameterize the free parameters (integration constants) of $y(x)$ in terms of monodromy data. For a given point in the space of monodromy data, namely, for a given $y(x)$, the parameterization can be done at each of the critical points $x = 0, 1, \infty$. In this way the connection problem is solved. In order to achieve the goal, we need to compute the monodromy data of (12.1). This is done as follows. *Once the matching* $\Psi_{\mathrm{OUT}} \leftrightarrow \Psi_{\mathrm{IN}}$ *in* (12.4) *has been completed*, we have to match Ψ_{OUT} with a fundamental solution Ψ of (12.1) at $\lambda = \infty$, and we have to match Ψ_{IN} with *the same* Ψ in another region of the λ-plane, typically around $\lambda = 0$ or x. If this is done, then M_1 of Ψ coincides with M_1^{OUT} of Ψ_{OUT}, while M_0 and M_x coincide with M_0^{IN} and M_x^{IN} of Ψ_{IN}. The crucial point is that we are able to compute the monodromy matrices M_1^{OUT}, M_0^{IN} and M_x^{IN}

exactly, i.e., that we are able to solve the systems in terms of linear special functions. We cannot write the rather long parameterization formulas here. The reader may want to see [1, 4–6, 8]

Bibliography

[1] P. Boalch, From Klein to Painlevé, via Fourier, Laplace and Jimbo, *Proc. London Math. Soc.* **90**(1): 167–208, 2005.

[2] A. D. Bruno, I. V. Goryuchkina, Asymptotic expansions of solutions to the sixth Painleve equation, *Transactions of Moscow Math. Soc.* **71**: 1–105 (= *Tr. MMO*, **71**: 6–118, 2010, 2010).

[3] D. Guzzetti, The Elliptic Representation of the General Painlevé VI Equation, *Comm. Pure Appl. Math.,* **LV**: 1280–1363, 2002.

[4] D. Guzzetti, Matching procedure for the sixt Painlevé equation, *J. Phys. A: Math. Gen.*, **39**: 11973–12031, 2006.

[5] D. Guzzetti, The Logaritmic asymptotics of the sixt Painlevé equation, *J. Phys. A: Math. Theor.,* **41**: 205201, 2008.

[6] D. Guzzetti, *Solving the sixth Painlevé equation: towards the classification of all the critical behaviors and the connection formulae*, International Mathematics Research Notices 2011 rnr071, doi:10.1093/imrn/rnr071.

[7] D. Guzzetti, Poles Distribution of PVI transcendets close to a critical point, arXiv 1104.5066[math.CA] (2011)

[8] M. Jimbo,Monodromy Problem and the Boundary Condition for Some Painlevé Trascendents, *Publ. RIMS, Kyoto Univ.*, **18**: 1137–1161, 1982.

[9] M. Jimbo, T. Miwa, K. Ueno, Monodromy Preserving Deformations of Linear Ordinary Differential Equations with Rational Coefficients (I), *Physica,* **D2**: 306–352, 1981.

[10] K. Kaneko, Painlevé VI transcendents which are meromorphic at a fixed singularity, *Proceedings of the Japan Academy*, **82**(Ser. A No. 5): 71–76 , 2006.

[11] S. Shimomura, A Family of Solutions of a Nonlinear Ordinary Differntial Equation and its Application to Painlevé Equations (III), (V), (VI), *J. Math. Soc. Japan*, **39**: 649–662, 1987.

Author Information

Davide Guzzetti, KIAS, Seoul – SISSA, Trieste, South Korea – Italy.
E-mail: davide_guzzetti@yahoo.com

Chapter 13

Isomonodromy Deformation of the Heun Class Equation

Alexander Ya. Kazakov

Abstract. The confluent Heun equation with an additional two apparent singularities (*Heunc2*) can be considered as a reduction of the linear system of differential equations. The gauge transform of such system leads to the isomonodromic deformation of the *Heunc2*. Explicit relations describing isomonodromic deformation of *Heunc2* are presented.

Keywords. Heun Class Equation, Gauge Transform, Isomonodromy Deformation.

2010 Mathematics Subject Classification. Primary: 34M35; Secondary: 34M03, 33C60.

13.1 Introduction

The Heun equation

$$\frac{d^2w}{dz^2} + \left[\frac{\alpha}{z} + \frac{\beta}{z-1} + \frac{\gamma}{z-\tau}\right]\frac{dw}{dz} + \frac{\sigma z + \chi}{z(z-1)(z-\tau)}w = 0 \qquad (13.1)$$

is a Fuchsian differential equation with four regular singularities on complex plane z. This equation itself and other reduced and confluent forms of it play an important role in applied mathematics [1]. Most of the "linear" special functions, such as Mathieu functions, spheroidal functions, Coulomb spheroidal functions, etc., are solutions of equations of the Heun class. In some applications, Heun class equations with added apparent singularities arise. Thus, the Heun equation with added single apparent singularity is connected to the theory of Painleve equations [2]. In this chapter the confluent Heun equation with two added apparent singularities will be discussed, We will call this equation *Heunc*2 (see below for its explicit description). Necessary definitions and the results of the analytical theory of linear differential equations can be found in [2,3].

The most important of its applications is to give information about solutions, connection matrices, the monodromy group of concrete equation in dependence of parameters of the equation. Such analytic information concerning Heun class equations is very poor. Solutions of these equations cannot be expressed as a contour integral, in contrast to hypergeometric-type equations. Thus, it is necessary to derive another analytical tools for studying Heun class equations. In the present chapter we use gauge transforms. Closely related topics were discussed in [2, 4]. The initial point of these works was the study of the isomonodromic deformations of linear differential equations. It was shown that if "true" singularities are fixed, such deformation can be realized by the corresponding gauge transform [2, Theorem 1.11.1] realized with the help of a matrix rationally depending on z. In [4] these transforms were discussed as a potential tool for the "removing" of the apparent singularities. Our main aim here is the explicit description (in terms of parameters of equation) of the gauge transform of the Heun class equation. Due to extreme awkwardness of the corresponding calculations we will consider the situation where the gauge matrix is constant, which is a more special case then in [2,4].

This paper is organized as follows. In the next section the gauge transforms of the linear differential equations are discussed. Then as a first application the gauge transform of the hypergeometric equation with one added apparent singularity is considered. In section 13.4, the isomonodromic deformation of *Heunc*2 is described in explicit terms. Section 13.5 ontains some concluding remarks.

The transforms of equations considered here are very cumbersome, and the corresponding analytic calculations were performed with the help of the computer algebra system Maple.

13.2 Gauge Transforms of Linear Differential Equations

Let us consider a system of linear differential equations

$$D(z)\frac{dW(z)}{dz} = R(z)W(z), \tag{13.2}$$

where $D(z)$ is a scalar polynomial, $R(z)$ – 2×2 – matrix valued polynomial, $W(z) = (w_1(z), w_2(z))^T$ is a 2-vector-valued function, z takes its values on a complex plane. It is known that system (13.2) can be reduced to the second-order linear differential equation by excluding either $w_1(z)$ or $w_2(z)$. The singular points of this equation will be zeroes of the $D(z)$ and, may be, $z = \infty$, these singularities can be regular or irregular, depending on the parameters of the equation. It is not too difficult to show that roots of $R_{21}(z)$ (or, accordingly, $R_{12}(z)$ will be apparent singularities of of equation (see [3, Lemma 6.1.2]). Singular points which are not apparent will be called below "true" singularities.

Let $W(z) = UV(z)$, where U is a constant nonsingular 2×2-matrix. Then the vector-valued function $V(z)$ is a solution of system

$$D(z)\frac{dV(z)}{dz} = \widetilde{R}(z)V(z), \tag{13.3}$$

$\widetilde{R}(z) = U^{-1}R(z)U$. Note, that $\widetilde{R}(z)$ is a matrix-valued polynomial of the same degree as $R(z)$. So, systems (13.2) and (13.3) are connected by *a gauge transform.* System (13.3) can also be reduced to a scalar linear differential equation, and *gauge transform of scalar linear differential equations* arises. Here we consider such transforms for the hypergeometric class of equations and for the Heun class equations. But first we will discuss some properties of these transforms.

Let reductions of the systems to the equations be realized in the same manner, by excluding of the first component of vector-values functions. Then

$$w_1(z) = \frac{D(z)w_2'(z) - R_{22}(z)w_2(z)}{R_{21}(z)}, \tag{13.4}$$

$$v_1(z) = \frac{D(z)v_2'(z) - \widetilde{R}_{22}(z)v_2(z)}{\widetilde{R}_{21}(z)}, \tag{13.5}$$

and $(w(z) = w_2(z), v(z) = v_2(z))$

$$w(z) = U_{21}v_1(z) + U_{22}v_2(z) = \frac{U_{21}D(z)v'(z) + \left[U_{22}\widetilde{R}_{21}(z) - U_{21}\widetilde{R}_{22}(z)\right]v(z)}{\widetilde{R}_{21}(z)}. \tag{13.6}$$

This relation describes the solution of the initial equation in terms of the final equation in explicit terms.

Proposition 13.1. *Let apparent singularities not coincide with "true" singularities for the initial and final equations. Then initial and final equations have the same "true" singularities and corresponding characteristic exponents coincide.*

Proof. Let $z = z_0$ be a regular singularity of the final equation, $v(z)$ is a solution of the corresponding equation with fixed behavior in the vicinity of this point, $v(z) = 1 \cdot (z - z_0)^\rho [1 + O(z - z_0)]$, where ρ is characteristic exponent related to this point. Substituting this solution in (13.6), one can obtain the solution of the initial equation $w(z) = \text{const} \cdot (z - z_0)^\rho [1 + O(z - z_0)]$, which has the same characteristic exponent in the same singular point (it follows from our assumptions that $\widetilde{R}_{21}(z)$ does not root in this point). Irregular singularity can similarly be considered . □

The key role in the various applications of the linear differential equations is played by connection matrices. Let z_1, z_2 be regular singularities of the equation. In what follows we mark by index $k = 1, 2$ the objects which correspond to singularity z_k, by index $m = 1, 2$ the objects related to the different solutions with fixed behavior in these points, $D_k(z) = D(z)(z - z_k)^{-1}$. Let $\rho_m^{(k)}$ be a characteristic exponents for these points, $w_m^{(k)}(z)$ be the corresponding normalized solutions, $w_m^{(k)} = 1 \cdot (z - z_k)^{\rho_m^{(k)}} [1 + O((z - z_k))]$. Let L be some way of analytic continuation from the vicinity of point z_1 into vicinity of point z_2. Then in accordance with analytic theory of linear differential equations, the result of analytic continuation of $w_{1,2}^{(1)}(z)$ along L can be expressed with help of linear combinations of $w_{1,2}^{(2)}(z)$:

$$w_m^{(1)}(z) = C_{m1}w_1^{(2)}(z) + C_{m2}w_2^{(2)}(z). \tag{13.7}$$

Matrix C is called a *aconnection matrix*; it can be defined for each pair of regular singularities and for any L. *Monodromy* matrices as a special case of connection matrices, when the initial and final singularities coincide. The next result describes the relationship between connection matrices of the initial and final equations.

Proposition 13.2. *Let apparent singularities and "true" singularities of the initial and final equations not coincide. Then the next relation holds:*

$$C = \operatorname{diag}\left(\frac{1}{\zeta_{11}}, \frac{1}{\zeta_{12}}\right) \widetilde{C} \operatorname{diag}(\zeta_{21}, \zeta_{22}), \tag{13.8}$$

$$\zeta_{km} = U_{22} + U_{21} \frac{\rho_m^{(k)} D_k(z_k) - \widetilde{R}_{22}(z_k)}{\widetilde{R}_{21}(z_k)}. \tag{13.9}$$

Proof. Consider with help of (13.6) the relationship between normalized solutions of the initial and transformed equations. If we substitute $v_m^{(k)}(z)$,

$$v_m^{(k)}(z) = 1 \cdot (z - z_k)^{\rho_m^{(k)}}, \quad k, m = 1, 2,$$

in the right-hand side of (13.6), then the corresponding solution of the initial equation

$$w(z) = \zeta_{km} \cdot (z - z_k)^{\rho_m^{(k)}} (1 + O(z - z_k)).$$

So, the gauge transform converts $v_m^{(k)}(z)$ into $\zeta_{km} w_m^{(k)}(z)$. Respectively, relation (13.6) can be rewritten as

$$\zeta_{km} w_m^{(k)}(z) = \frac{U_{21} D(z)(v_m^{(k)}(z))' + \left[U_{22}\widetilde{R}_{21}(z) - U_{21}\widetilde{R}_{22}(z)\right] v_m^{(k)}(z)}{\widetilde{R}_{21}(z)}. \tag{13.10}$$

Let us rewrite the last relation for $k = 1$ and continue it analytically along L with help of (13.7) for the transformed equation, taking into account the analyticity of the functions in the right-hand side of (13.10) except but solutions $v_m^{(k)}(z)$:

$$v_m^{(1)}(z) = \widetilde{C}_{m1} v_1^{(2)}(z) + \widetilde{C}_{m2} v_2^{(2)}(z). \tag{13.11}$$

Applying once more relation (13.6), we conclude:

$$\zeta_{1m} w_m^{(1)}(z) = \widetilde{C}_{m1} \zeta_{21} w_1^{(2)}(z) + \widetilde{C}_{m2} \zeta_{22} w_2^{(2)}(z). \tag{13.12}$$

Finally,

$$C_{mk} = \frac{\zeta_{2k}}{\zeta_{1m}} \widetilde{C}_{mk}, \tag{13.13}$$

which is the same as (13.8). □

Remark 13.3. One can obtain analogous results when the apparent singularity of either the initial or the transformed equation coincide with "true" singularity.

Remark 13.4. Let the apparent singularities of the initial equation not coincide with "true" singularity. Generally speaking, choosing parameters of the gauge matrix U, one can coalescence apparent singularity with "true" singularity of the transformed equation. Thus, one can "remove" apparent singularity. We consider this fact in what follows for the concrete equations.

Remark 13.5. Results analogous to those described in Propositions 13.1 and 13.2 have already been considered in [4]. The initial purpose of this paper was to study the isomonodromic deformations of the scalar linear differential equations, such deformations which do not change the set of "true" singularities and the monodromy group. As was shown in [4], such transforms coincide with gauge transforms whose matrices $U = U(z)$ rationally depend on z. In the present chapter the constant gauge matrix is only considered if all calculations can be preformed in explicit terms.

Let the pair of the normalized solutions be fixed with a prescribed behavior in the vicinity of the regular singularity. Monodromy matrices describe the results of analytic continuation along different closed ways. How does gauge transform influence monodromy matrices? As follows from the preceding considerations, monodromy matrices of the transformed equation are similar to the corresponding monodromy matrices of the initial equation, $\widetilde{M} = K^{-1}MK$, where matrix K is the same for all monodromy matrices. So, gauge transfoms do not change the monodromy group of equations; in other words, they realize isomonodromic deformation of equation.

13.3 Gauge Transforms of the Hypergeometric Class Equations

Consider as a first example the hypergeometric class equation. This equation was discussed (from a slightly different point of view) in [4]. Let us set $D(z) = z(z-1)$, $R(z) = Fz + E$ in the system (13.2), where F, E are 2×2-constant matrices. Here the items of the matrix $R(z)$ are (in general) polynomials of the first order on z. Respectively, reduction of the system (13.2) is a hypergeometric equation with one added apparent singularity (perhaps up to the elementary factor). This equation can be considered to be a Heun equation when one regular singularity is the apparent one. But, as it follows from the results of [4,5], it is much closer to the hypergeometric equation from an analytical point of view. We choose the next parametrization of this equation:

$$w''(z) + \left[\frac{\alpha}{z} + \frac{\beta}{z-1} - \frac{1}{z-\lambda}\right] w'(z) + \frac{1}{z(z-1)}\left[\frac{\chi}{z-\lambda} + \frac{L}{\lambda(\lambda-1)}\right] w(z) = 0. \tag{13.14}$$

Here $\alpha, \beta, \chi, \lambda$ are free parameters of the equation, α, β determine the behavior of solutions in the vicinity of singularities $z = 0, 1$, parameter λ determines the ocation of the apparent singularity. L is not a free parameter, its value is determined by the condition that $z = \lambda$ is the apparent singularity, $L = -(\beta\lambda - 2\lambda + \alpha\lambda + 1 + \chi - \alpha)\chi$.

Our parametrization differs from the conventional one, in that usually parameter $\rho = L\lambda^{-1}(\lambda-1)^{-1}$ takes as a free parameter (it determines the behavior of solutions at infinity). Our choice leads to more simple relations for the transformed values.

As it follows from our definitions, the result of gauge transform will be the same equation with different parameters. Parameters α, β, ρ keep their values, only parameters λ and χ, respectively, L, change (but $\widetilde{L\lambda^{-1}(\lambda-1)^{-1}} = \widetilde{\rho} = \rho$). Let the gauge matrix be

$$U = \begin{pmatrix} u_1 & u_2 \\ u_3 & u_4 \end{pmatrix},$$

and $detU = 1, u_3 = \zeta u_1$. Then direct calculations lead to the next expressions:

$$\widetilde{\lambda} = \frac{P(\zeta)}{Q(\zeta)}, \quad \widetilde{\chi} = \frac{N(\zeta)}{Q(\zeta)}, \tag{13.15}$$

where

$$\begin{aligned} P(\zeta) &= \lambda(\lambda-1)(\lambda\alpha+\lambda\beta-2\lambda+\chi)(\lambda\alpha+\lambda\beta-2\lambda+\chi-\zeta) \\ &\quad \times (\lambda\alpha+\lambda\beta-2\lambda+\chi-\zeta+1-\alpha), \\ N(\zeta) &= \chi(\lambda\alpha+\lambda\beta-2\lambda+\chi-\alpha+1)(-2\lambda^2-\zeta\lambda+2\lambda-\lambda\beta+\beta\lambda^2 \\ &\quad -\chi+\lambda\chi-\alpha\lambda+\alpha\lambda^2)(\lambda\alpha+\lambda\beta-2\lambda+\chi-\zeta), \\ Q(\zeta) &= -\lambda\chi\zeta^2-\lambda(\lambda-1)(\beta+\alpha-2)(\lambda\alpha+\lambda\beta-2\lambda+\chi)\zeta \\ &\quad +(\lambda-1)\times(\lambda\alpha+\lambda\beta-2\lambda+\chi)^2(\lambda\alpha+\lambda\beta-2\lambda+\chi-\alpha+1). \end{aligned}$$

Thus, parameters of the transformed equation depends only on $\zeta = u_3/u_1$, other parameters of the matrix U have no influence on the transformed equation. It follows from these relations that, choosing ζ, one can get either $\widetilde{\lambda} = 0$ or $\widetilde{\lambda} = 1$ and, consequently, reduce the transformed equation to the hypergeometric one (up to the elementary factor). At the same time, all objects related to the initial equation (solutions, connections, and monodromy matrices) can be expressed in terms of corresponding objects of the transformed equation. Such "removing" of the apparent singularity was already discussed in [4] and realized with the help of concrete information about the monodromy matrices of related equations. Explicit relations for the connection matrices of equation (13.14) was obtained in [5] by means of a quite different approach. Note that $\widetilde{\lambda}(\zeta), \widetilde{\chi}(\zeta)$ are rational functions, at $\zeta = 0$ one gets $\widetilde{\lambda} = \lambda, \widetilde{\chi} = \chi$.

13.4 Gauge Transform of Heun Class Equations

13.4.1 Formulation of the Problem

Consider the system (13.2), supposing $D(z) = z(z-1)$, $R(z) = Hz^2 + Fz + E$, where H, F, E are constant 2×2-matrices. Roots of the function $D(z)$, $z = 0, z = 1$ will be regular singularities, $z = \infty$ will be an irregular singular point (in the general situation). Roots of the $R_{21}(z)$ will be apparent singularities, so that we obtain 2 apparent singularities in general. So, generally the situation system (13.2) produces after reduction of the confluent Heun equation with 2 added apparent singularities, equation *Heunc*2. We will use the next parametrization of this equation:

$$w''(z) + \left[a + \frac{\beta}{z} + \frac{\gamma}{z-1} - \frac{2z+S}{z^2+Sz+P}\right] w'(z) + \frac{1}{z(z-1)}\left[\sigma z + N + \frac{Lz+\chi}{z^2+Sz+P}\right] w(z) = 0. \quad (13.16)$$

Here $a, \beta, \gamma, \sigma, \chi, S, P$ are free parameters of (13.16). Parameters a, σ determine the behavior of solutions at infinity, parameters β, γ determine the behavior solution near regular singularities $z = 0, z = 1$, respectively, parameters S, P determine the location of the apparent singularities. Parameters L, N are not free; their values are defined by the condition that *Heunc*2 has two apparent singularities. Parameter χ is an *accessory* parameter of *Heunc*2. This parameter cannot be defined from the local behavior of the solutions near singularities (hypergeometric class equations do not have accessory parameters). Our parametrization, being nonstandard, leads to more simple final relations.

Gauge transform of the system (13.2) produces the transformed system (13.3), where $\widetilde{R}(z) = U^{-1}(Hz^2 + Fz + E)U$. Note that the matrix-valued function $\widetilde{R}(z)$ is polynomial of the second order and, consequently, reduction of the system (13.3) also yields equation *Heunc*2. Thus, we are interested in the explicit description of the orbit of the gauge transforms on the equations *Heunc*2 (with different parameters).

In order to obtain an explicit description of the action of gauge transforms on parameters of the equation *Heunc*2, one has to relate parameters of equation (13.16) and system (13.2) which generates it. Then one has to get in explicit form the transformed equation *Heunc*2 which is the reduction of the system (13.3),

$$v''(z) + \left[\widetilde{a} + \frac{\widetilde{\beta}}{z} + \frac{\widetilde{\gamma}}{z-1} - \frac{2z+\widetilde{S}}{z^2+\widetilde{S}z+\widetilde{P}}\right] v'(z) + \frac{1}{z(z-1)}\left[\widetilde{\sigma} z + \widetilde{N} + \frac{\widetilde{L}z+\widetilde{\chi}}{z^2+\widetilde{S}z+\widetilde{P}}\right] v(z) = 0. \quad (13.17)$$

13.4.2 Initial System of Equations and Equation *Heunc*2

First, let us get the relations between parameters of the initial system

$$z(z-1)W'(z) = (Hz^2 + Fz + E)W(z) \quad (13.18)$$

and the parameters of equation (13.16). Note, that here we exclude the first component of the vector value function $W(z)$, and reduce the system to the equation for the component $w(z) = w_2(z)$. It is enough to set

$$H = \begin{pmatrix} h & h_2 \\ 1 & h_2/h_1 \end{pmatrix}, \quad F = \begin{pmatrix} f_1 & f_2 \\ f_3 & f_4 \end{pmatrix}, \quad E = \begin{pmatrix} 0 & 0 \\ e_3 & e_4 \end{pmatrix}.$$

Further, we take $f_3 = S, e_3 = P$, so that $R_{21}(z) = (z^2 + Sz + P)$. This choice leads to the necessary location of the apparent singularities. Condition

$$f_4 = \frac{hh_2P - f_1e_4h - h^2e_4 - f_1h_2 + hh_2S + hf_2(1 + S + P)}{h(h + f_1)}$$

guarantees the absence of the poles of the second order in the coefficients of the equation. Comparing residues in the coefficient at $w'(z)$ in the point $z = 0$, one obtains: $e_4 = \beta - 1$. Calculating the constant term in this coefficient, one concludes $h_2 = -h(h + a)$. Residue of this coefficient in the point $z = 1$ yields:

$$f_2 = \frac{h + h^2(S + P) + ha(1 + S + P) - 2f_1h - f_1^2 + f_1 - \gamma(h + f_1)}{1 + S + P}.$$

Calculating χ in terms of (13.18), one gets:

$$f_1 = \frac{S - S\beta + SPa + SPh + 4P + \chi + 2Pa - 2P(\beta + \gamma)}{2P}.$$

Calculating σ, one finds a quadratic equation for h:

$$\begin{aligned}
&h^2P^2(S^2 - P) + 2hP\big[S^2(\beta - 1) + P^2(2\beta - 4a - Sa - 4 + 2\gamma) \\
&\quad + P(2S\beta + 2 - Sa - 3S + S\gamma - 2\beta) - \chi(2P + S)\big] + \chi^2 \\
&\quad + \chi\big[6P + 2S - 4P\beta + 2Pa - 2S\beta - 2P\gamma - 2P^2a\big] + 2P^2Sa\beta - 2S^2\beta \\
&\quad - 12P^2\beta - 2aSP\beta + 2P^2Sa\gamma + 2PS\beta\gamma + 4P^2\beta^2 \\
&\quad + S^2\beta^2 + 4P^2a + 6PS + 8P^2 + S^2 - 4P^2Sa - 4P^2\gamma + 2PSa \\
&\quad + 10PS\beta - 2PS\gamma - 4P^2a\beta + 4PS\beta^2 - 2P^2Sa^2 - P^2S^2a^2 \\
&\quad + 4P^2\beta\gamma - 8P^3a - 4P^3a^2 + 4P^3a\beta - 2P^3Sa^2 - 4P^2\sigma \\
&\quad - 4P^2S\sigma + 4P^3a\gamma - 4P^3\sigma = 0.
\end{aligned} \tag{13.19}$$

These formulas describe the relationship between the parameters of system (13.18) and equation (13.16). They are recurrent: parameter f_1 is determined by h, parameter f_2 is determined by f_1, and so on. They can be rewritten in a more direct manner, but we omit here the cumbersome corresponding details.

13.4.3 Parameters of the Transformed Equation

As it follows from Proposition 13.1, parameters a, β, γ, σ, which determine the behavior of the solutions in the "true" singularities, keep their values at gauge transforms, $\widetilde{a} = a, \widetilde{\beta} = \beta, \widetilde{\gamma} = \gamma, \widetilde{\sigma} = \sigma$. Gauge transforms have an affect on the location of the apparent singularities and value of accessory parameter, and so one must calculate $\widetilde{\chi}$, $\widetilde{S}$, $\widetilde{P}$. Set, as before, $detU = 1, u_3 = \zeta u_1$. Then direct calculations produce

$$\begin{aligned}\widetilde{S} = &\big[2P(\zeta h-1)(h\zeta+a\zeta-1)\big]^{-1}\big[2PS + \zeta(-2PSh+2S\beta-2PSa-4P-2S\\&+2P\gamma-2\chi-2Pa+2P\beta)+\zeta^2\big[h(4P-2P\beta+2Pa+2\chi+2S\\&+PSa-2S\beta-2P\gamma)-2Pa\beta+2Pa^2+a\chi+2P\sigma+PSa^2\\&+Sa-Sa\beta+4Pa-2Pa\gamma\big]\big],\end{aligned}\tag{13.20}$$

$$\widetilde{P} = \frac{P+\zeta(\beta-1)}{(\zeta h-1)(\zeta h+a\zeta-1)},\tag{13.21}$$

$$\widetilde{\chi} = \frac{b_2\chi^2+b_1\chi+b_0}{4P^2(1+S+P)(\zeta h+\zeta a-1)^2(\zeta h-1)},\tag{13.22}$$

where

$$\begin{aligned}b_2 = &-\zeta(\zeta Ph-\zeta\beta+\zeta Pa-2P),\\ b_1 = &\,\zeta^2(2P\gamma-4P^2h^2-2S\beta^2+4S\beta+10P\beta-4P\beta^2-2P^2Sh^2-2P\beta\gamma\\&-2P62a-8P^2ha-6P-2S-4^2a^2+2p^2\gamma h+2P^2a\gamma-2P^2h\\&-4P^2Sah-2P^2Sa^2)+4P\zeta(3P-2P\beta+PSh+PSa+2Ph\\&-P\gamma+2Pa+S-S\beta)-4P^2(1+S+P),\\ b_0 = &-\zeta(-1+Ph+\beta+Pa)[4P^2\beta-4P^2a+4P^2\gamma+4P^2S-8P^2-4P^2Sa\\&-2PS\beta\gamma\zeta-2P^2Sh\gamma\zeta-2P^2Sa\gamma\zeta-4P^2a\gamma\zeta+2PS\gamma\zeta-4P^2\beta\gamma\zeta\\&-4P^2h\gamma\zeta+2P^2Sh\zeta+2P^2Sa\zeta-2PS^2\beta+2P^2S^2ah\zeta\\&+4P^2ah\zeta-4P^3ah\zeta+4P^2Sah\zeta-8P^2\zeta+8P^3h+12P^2\beta\zeta\\&-4P^3\beta-4P^2\beta^2\zeta+2S^2\beta\zeta+8P^3-6PS\zeta+10PS\beta\zeta\\&+4P^3a-4P^3\gamma-2P^2S^2a-2P^2S^2h+4P^2h\zeta-4P^3h^2\zeta+4P^2a^2\zeta\\&-4PS\beta^2\zeta-4P^2S\beta+4P^2\gamma\zeta+P^2S^2h^2\zeta\\&-S^2\zeta+4P^2a\zeta-S^2\beta^2\zeta+2PS^2+4P^2Sa^2\zeta+P^2S^2a^2\zeta].\end{aligned}$$

Relations (13.20)–(13.22) are the main result of this chapter; they describe in explicit terms gauge transform of the equation *Heunc*2. As it follows from the preceding discussion and results [2,4], these transforms do not change the monodromy group of the equation. Therefore, relations (13.20)–(13.22) describe isomonodromic deformations of the equation *Heunc*2 in explicit terms. Note that the value ζ is a free parameter in these relations and functions $\widetilde{S}(\zeta)$, $\widetilde{P}(\zeta)$, $\widetilde{\chi}(\zeta)$ are rational, $\widetilde{S}(0) = S$, $\widetilde{P}(0) = P$, $\widetilde{\chi}(0) = \chi$. These relations make it possible to "remove" one apparent singularity of the equation *Heunc*2. Namely, one can choose the value ζ such that one apparent singularity of the transformed equation coincides with one from the "true" singularities, and he main objects (solutions, connection and monodromy matrices) of initial and transformed equations relate to each other by simple manner.

13.5 Conclusion

Gauge transforms of the systems of linear differential equations induce corresponding transforms of the scalar linear differential equations which are generated by reduction of the systems. Such transforms do not change the monodromy group of the scalar equations. In the present chapter we have discussed gauge transform of the linear differential equations of the second order at a constant gauge matrix. This restriction makes it possible to describe transforms of equations in an explicit manner, in terms of equation parameters. Such a description was obtained in two situations. Firstly, we considered a hypergeometric equation with one added apparent singularity. In this case gauge transform changes the location of the apparent singularity only. Secondly, we have examined the confluent Heun equation with two added apparent singularities. In this case gauge transform changes the location of the apparent singularities and the value of the accessory parameter. In both situations gauge transform does not change the monodromy group; in other words, this transform realizes isomonodromic deformation of the corresponding equation. Obtained results make it possible to "remove" one apparent singularity for the discussed equations (such a possibility for the hypergeometric equation with n added apparent singularities was previously discussed in [4]). Analogous results can also be obtained for the Heun equation with two added apparent singularities.

Bibliography

[1] S. Yu. Slavyanov and W. Lay, *Special functions: A unified theory based on singularities,* Oxford university press, Oxford, New York, 2000.

[2] Y. Sibuya, *Linear differential equations in the complex domain: problem of analytic continuation,* AMS, Providence, 1985.

[3] K. Iwasaki, H. Kimura, S. Shimomura, and M. Yosida, *From Gauss to Painleve: a modern theory of special functions,* Braunshweig, Vieweg, 1991.

[4] A. V. Shanin and R. V. Craster, Removing false singular points as a method of solving ordinary differential equations, *Euro. J. Appl. Math.,* **13**: 617–639, 2002.

[5] A. Ya. Kazakov, Euler integral symmetry and deformed hypergeometric equation, *Chapter 29, this volume.*

Author Information

Alexander Ya. Kazakov, St.-Petersburg State University of Aerospace Instrumentation,
B. Morskaya 67, 190000, S.-Petersburg, Russia.
E-mail: a_kazak@mail.ru

Painlevé Equations and Related Topics, 117–122

Chapter 14

Isomonodromy Deformations and Hypergeometric-Type Systems

Vladimir P. Leksin

Abstract. We consider some reduction of nonlinear upper-triangular Schlesinger systems of isomonodromic deformations of Fuchsian systems on Riemann sphere to a recurrent linear systems of nonhomogeneous Pfaff equations. In this case for two-dimensional Fuchsian systems we describe an integral representations of solutions of Schlesinger systems and obtain that Malgrange theta-divisors are empty and solutions have moderate growth.

Keywords. isomodromic Deformations of Fuchsian Systems, Schlesinger Systems, Jordan–Pochhammer Systems, Integral Representations of Solutions.

2010 Mathematics Subject Classification. 34M56, 34M03, 35C15.

14.1 Schlesinger Families of Fuchsian Systems

Let us consider a Fuchsian system of p equations

$$\frac{\mathrm{d}\,y(z)}{\mathrm{d}\,z} = \left(\sum_{i=1}^{n} \frac{B_i}{z - a_i^0}\right) y(z). \tag{14.1}$$

on Riemann sphere $\bar{\mathbb{C}}$ with $n+1$ singular points $a_1^0, \dots, a_n^0$, $a_{n+1}^0 = \infty$ and constant matrices B_i of the size $p \times p$. The Schlesinger family of Fuchsian systems for the initial system (14.1) is following family of Fuchsian systems

$$\frac{\mathrm{d}\,y(z,a)}{\mathrm{d}\,z} = \left(\sum_{i=1}^{n} \frac{B_i(a)}{z - a_i}\right) y(z,a). \tag{14.2}$$

with initial conditions $B_i(a^0) = B_i$, $i = 1, 2, \dots, n$, where $a^0 = (a_1^0, a_2^0, \dots, a_n^0) \in \mathbb{C}_*^n = \{(a_1, a_2, \dots, a_n) |\ a_i \neq a_j,\ i \neq j,\ i,j = 1, 2, \dots, n\}$.

The Schlesinger family (14.2) is an isomonodromic family for system (1) if for points a from enough small polydisk neighborhood $U(a^0) \subset \mathbb{C}_*^n$ of the point $a^0 = (a_1^0, a_2^0, \dots, a_n^0)$ there exist family of fundamental matrices of solutions $Y(z,a)$ such that their monodromy matrices G_i, $i = 1, \dots, n$ corresponding "small loops" going around singular points $a_1, \dots, a_n$ does not depend on a.

14.2 Schlesinger Systems

In small neighborhood $U(a^0) \subset \mathbb{C}^n_*$ of the point $a^0 = (a^0_1, a^0_2, \dots, a^0_n)$ the sufficient isomonodromic condition on the family (14.2) is the system of the Schlesinger equations on matrices $B_i(a)$, $i = 1, 2, \dots n$

$$\mathrm{d}\, B_i(a) = -\sum_{\substack{j=1 \\ j\neq i}}^{n} [B_i(a), B_j(a)] \frac{\mathrm{d}(a_i - a_j)}{a_i - a_j}. \tag{14.3}$$

Here $[B_i, B_j] = B_i B_j - B_j B_i$ denote the commutator of matrices.

We list some properties of the Schlesinger systems.

- Nonlinear Pfaff system (14.3) is integrable in Frobenius sense [1, 3] and consequently in a sufficiently small neighborhood of the a^0 there exist the local solution $B(a) = (B_1(a), B_2(a), \dots, B_n(a))$ of the Pfaff system (3) with every initial value $B(a^0) = (B_1(a^0) = B_1, B_2(a^0) = B_2, \dots, B_n(a^0) = B_n)$.
- It is well known [1] that all eigenvalues of all matrices $B_i(a)$ are constants, that is, do not depend on a.
- The sum
$$\sum_{i=1}^{n} B_i(a) = -B_\infty$$
is a constant matrix.
- Malgrange proved [6] that the local solution $B(a)$ has meromorphic continuation on whole universal covering $\widetilde{C}^n_*$.
- In general, polar divisor Θ (theta-divisor Malgrange) of the solution $B(a)$ on $\widetilde{C}^n_*$ is nonempty and depends on the initial data $B(a^0)$.
- The Malgrange theta-divisor is defined by zeros of the Miwa tau-function $\tau(a)$, which is a solution of the equation
$$\mathrm{d} \log \tau(a) = \kappa \sum_{\substack{i\neq j \\ i,j=1}}^{n} \mathrm{tr}(B_i(a) B_j(a)) \frac{\mathrm{d}(a_i - a_j)}{a_i - a_j}. \tag{14.4}$$

14.3 Upper-Triangular Schlesinger Systems

We suppose that all matrices $B_i(a), i = 1, \dots, n$ from the Schlesinger family (2) are upper-triangular matrices $B_i(a) = \Lambda_i + U^1_i(a) + \cdots + U^{p-1}_i(a)$. Here

$$\Lambda_i = \begin{pmatrix} \lambda^1_i & 0 & \dots & 0 \\ 0 & \lambda^2_i & \dots & 0 \\ \vdots & \vdots & \ddots & \vdots \\ 0 & \dots & \dots & \lambda^p_i \end{pmatrix}$$

are constant diagonal matrices. Further

$$U_i^1(a) = \begin{pmatrix} 0 & u_i^{1,1}(a) & & \dots & 0 \\ 0 & 0 & u_i^{1,2}(a) & \dots & 0 \\ \vdots & \vdots & \ddots & \ddots & \vdots \\ 0 & 0 & \dots & 0 & u_i^{1,p-1}(a) \\ 0 & 0 & \dots & 0 & 0 \end{pmatrix}$$

have nonzero entries only in the first off-diagonal, respectively,

$$U_i^2(a) = \begin{pmatrix} 0 & 0 & u_i^{2,1}(a) & \dots & \dots & 0 \\ 0 & 0 & 0 & u_i^{2,2}(a) & \dots & 0 \\ \vdots & \vdots & \ddots & & \ddots & \vdots \\ 0 & 0 & \dots & \ddots & 0 & u_i^{2,p-2}(a) \\ 0 & 0 & \dots & \dots & 0 & 0 \\ 0 & 0 & \dots & \dots & 0 & 0 \end{pmatrix}$$

have nonzero entries only in the second off-diagonal, and so on. Then the corresponding Schlesinger system (14.3) has the following special form

$$\mathrm{d}\,\Lambda_i = 0,$$

$$\mathrm{d}\,U_i^1 = -\sum_{\substack{j=1\\ j\neq i}}^{n} \left([\Lambda_i, U_j^1] + [U_i^1, \Lambda_j]\right)\frac{\mathrm{d}(a_i - a_j)}{a_i - a_j},$$

$$\mathrm{d}\,U_i^{k+1} = -\sum_{\substack{j=1\\ j\neq i}}^{n} \left([\Lambda_i, U_j^k] + [U_i^k, \Lambda_j] + \sum_{r+s=k+1} [U_i^r, U_j^s]\right)\frac{\mathrm{d}(a_i - a_j)}{a_i - a_j},$$

$$k = 2, \dots, p-1.$$

For entries of matrices $\Lambda^i, U_k^i, i = 1, \dots, n, k = 1, \dots, p-1$, we obtain the recurrent linear nonhomogeneous system of meromorphic differential equations

$$\mathrm{d}\,\lambda_i^m = 0, \quad i = 1, \dots, n; \quad m = 1, \dots, p,$$

$$\mathrm{d}\,u_i^{1,m} = -\sum_{\substack{j=1\\ j\neq i}}^{n} (\lambda_i^{m,m+1} u_j^{1,m} - \lambda_j^{m,m+1} u_i^{1,m})\frac{\mathrm{d}(a_i - a_j)}{a_i - a_j},$$

$$i, j = 1, \dots, n; \quad m = 1, \dots, p-1,$$

where $\lambda_i^{m,m+1} = \lambda_i^m - \lambda_i^{m+1}, i = 1,\dots,n; m = 1,\dots,p-1.$

$$\mathrm{d}\,u_i^{2,m} = -\sum_{\substack{j=1\\j\neq i}}^{n}\Big[(\lambda_i^{m,m+2}u_j^{2,m} - \lambda_j^{m,m+2}u_i^{2,m}) + (u_i^{1,m}u_j^{1,m+1} - u_j^{1,m}u_i^{1,m+1})\Big]\frac{\mathrm{d}(a_i-a_j)}{a_i-a_j},$$

where $\lambda_i^{m,m+2} = \lambda_i^m - \lambda_i^{m+2}, i, j = 1,\dots,n; m = 1,\dots,p-2,$

$$\mathrm{d}\,u_i^{k,m} = -\sum_{\substack{j=1\\j\neq i}}^{n}[(\lambda_i^{m,m+k}u_j^{k,m} - \lambda_j^{m,m+k}u_i^{k,m}) + \sum_{r+s=k}(u_i^{r,m}u_j^{s,m+r} - u_j^{s,m}u_i^{r,m+s})]\frac{\mathrm{d}(a_i-a_j)}{a_i-a_j},$$

$$i,j=1,\dots,n; \quad m = 1,\dots,p-k, \quad k = 1,\dots,p-1,$$

where $\lambda_i^{m,m+k} = \lambda_i^m - \lambda_i^{m+k}$.

In particular, for $p = 2$ and upper-triangular matrices $B_i(a)$ Schlesinger system (3) are the following linear meromorphic Pfaff system

$$\begin{aligned} &\mathrm{d}\,\lambda_i^m = 0, && i = 1,\dots,n; && m = 1,2;\\ &\mathrm{d}\,u_i = -\textstyle\sum_{\substack{j=1\\j\neq i}}^{n}(\lambda_i u_j - \lambda_j u_i)\frac{\mathrm{d}(a_i-a_j)}{a_i-a_j}, && && i,j = 1,\dots,n, \end{aligned} \tag{14.5}$$

where $\lambda_i = \lambda_i^1 - \lambda_i^2, i = 1,\dots n.$

14.4 Jordan–Pochhammer Systems

We will consider the class of Fuchsian systems on linear complex spaces $\mathbb{C}^n$

$$\mathrm{d}\,y(z) = \Bigg(\sum_{1\leqslant i<j\leqslant n} J_{ij}(\lambda)\frac{\mathrm{d}(z_i-z_j)}{z_i-z_j}\Bigg)y(z). \tag{14.6}$$

Here $J_{ij}(\lambda)$ are the following matrices of the size $n\times n$:

$$J_{ij}(\lambda) = \begin{pmatrix} 0 & \dots & 0 & \dots & 0 & \dots & 0\\ \vdots & \ddots & \vdots & & \vdots & & \vdots\\ 0 & \dots & \lambda_j & \dots & -\lambda_i & \dots & 0\\ \vdots & & \vdots & \ddots & \vdots & & \vdots\\ 0 & \dots & -\lambda_j & \dots & \lambda_i & \dots & 0\\ \vdots & & \vdots & & \vdots & \ddots & \vdots\\ 0 & \dots & 0 & \dots & 0 & \dots & 0 \end{pmatrix} \begin{matrix} \\ \\ i\\ \vdots\\ j\\ \\ \\ \end{matrix}$$

The symbol λ denotes an ordered collection of complex numbers $\lambda = (\lambda_1, \lambda_2, \dots, \lambda_n)$. The matrices $J_{ij}(\lambda)$ satisfy the relations

$$[J_{ij}(\lambda), J_{ik}(\lambda) + J_{jk}(\lambda)] = 0, \quad 1 \le i < j < k \le n; \tag{14.7}$$

$$[J_{ij}(\lambda), J_{kl}(\lambda)] = 0, \quad \{i, j\} \cap \{k, l\} = \emptyset. \tag{14.8}$$

The relations (14.7) and (14.8) are equivalent to the Frobenius condition of the integrability of system (14.6) (i.e., the system (14.6) is integrable) (see [4,5]).

Then any fundamental matrix of solutions $Y(z)$ of the (14.6) has a holomorphic continuation on the universal covering and deck transformations corresponding to elements $g \in \pi_1(\mathbb{C}^n_*, a^0)$ act on $Y(z)$ by the rule $Y(g^*z) = Y(z)M(g)$ (see [1,4]). The constant matrix $M(g)$ is called the monodromy matrix for $g \in \pi_1(\mathbb{C}^n_*, a^0)$.

The entries y_{ij} of the fundamental matrix of solutions of the system (14.6) have the following integral representations (see [5,7]):

$$y_{ij}(a_1, \dots, a_n) = \lambda_i \int_{\gamma_j} (t - a_1)^{\lambda_1} \cdots (t - a_n)^{\lambda_n} \frac{dt}{t - a_i}, \tag{14.9}$$

where γ_j, $j = 1, \dots, n$ is a basis in homology group $H_1(\mathbb{C}P^1 \backslash \{a_1, \dots, a_n, \infty\}, \mathcal{L}_\chi)$ with local coefficients $\mathcal{L}_\chi$. The local system $\mathcal{L}_\chi$ is defined by a representation χ : $\pi_1(\mathbb{C}P^1 \setminus \{a_1, \dots, a_n, \infty\}, t_0) = F_n \to \mathbb{C}^*$ of the fundamental group $\pi_1(\mathbb{C}P^1 \setminus \{a_1, \dots, a_n, \infty\}, t_0) = F_n$, that maps generators $x_1, \dots, x_n$ of this free group F_n to nonzero complex numbers $q_1 = e^{-2\pi i \lambda_1}, \dots, q_n = e^{-2\pi i \lambda_n}$.

14.5 The Basic Result

Now we make the following observation:

Theorem 14.1. *For upper-triangular matrices $B_i(a)$ of the size 2×2, the Schlesinger system* (14.5) *coincides with the Jordan–Pochhammer system in the form* (14.6) *under suitable choice parameters λ_i, $i = 1, \dots, n$. Hence, upper-triangular solutions of the Schlesinger system* (14.3) *have only hypergeometric-type solutions with entries of the form* (14.9) *and among such solution entries of solutions does not have a new transcendent.*

At the present moment it is not known whether or not there exists integral representations of the type (14.9) for entries of any solutions of arbitrary upper-triangular Schlesinger systems.

Corollary 14.2. *Under the conditions of the Theorem 14.1, the Malgrange divisor of the Schlesinger system* (14.3) *is the empty set.*

This corollary is the consequence of the linearity and integrability of Jordan–Pochhammer system and the properties of their solutions pointed out above.

Corollary 14.3. *In fixed singular points $a \in \mathcal{H} = \cup_{1 \le i < j \le n} H_{ij}$, all solutions of the upper-triangular 2×2-Schlesinger system* (14.3) *have moderate growth, i.e., at the most, power growth.*

In [2] R. Gontsov found sufficient conditions on an initial Fuchsian system of order two on the Riemann sphere that have only upper-triangular Schlesinger deformations up to the conjugation on a constant matrix. For the formulation this result we suppose that coefficients of the system (14.1) have traces equal to zero, i.e., $B_i \in sl_2(\mathbb{C}), i = 1, \dots, n$. Denote $\pm\lambda_i, i = 1, \dots, n$ eigenvalues of matrices B_i and $\pm\lambda_\infty$ eigenvalue of matrix $B_\infty = -\sum_{i=1}^n B_i$. We suppose also that B_∞ is a diagonal matrix.

Theorem 14.4 (R. Gontsov). *Let the monodromy representation of the Fuchsian system* (14.1) *be reducible. If for some choice of signs $\varepsilon_i = \pm 1$, $i \in \{1, 2, \dots, n, \infty\}$ the sum of eigenvalues $\sum_{i=1}^m \varepsilon_i \lambda_i + \varepsilon_\infty \lambda_\infty$ is equal to zero, then after conjugation $CB_i(a)C^{-1}$ by a constant matrix C of any solution $B_i(a)$, $i = 1, 2, \dots, n$ the Schlesinger system* (14.3) *is reduced to the upper-triangular form.*

Bibliography

[1] A. A. Bolibruch, *Inverse monodromy problems in analytic theory of differential equations*, MCCME, Moscow, 2009 (in Russian).

[2] R. R. Gontsov, On movable singularities of Garnier systems, *Math. Notes.*, **88**(5–6), 806–818, 2010.

[3] K. Aomoto, Une remarque la solution des équations de L.Schlesinger et Lappo–Danilevski, *Jour. Fac. Sci. Univ. Tokyo*, **17**(1–2): 341–354, 1970.

[4] T. Kohno, Linear representations of braid groups and classical Yang–Baxter equations, *Contemporary Math.,* **78**: 339–363, 1988.

[5] M. Kapovich and J. Millson, Quantization of bending deformations of polygons in $\mathbb{E}^3$, hypergeometric integrals and the Gassner representation, *Canad. Math. Bull.*, **44**: 36–60, 2001.

[6] B. Malgrange, *Sur les déformations isomonodromiques.I: Singularités régulières.* Progress in Math. 37, pp. 401–426, Birkhäuser, 1983.

[7] A. Varchenko, Special functions, KZ type equations, and representation theory, *CBMS regional conference series in Mathematics*, **98**, AMS-NSF, 2003.

Author Information

Vladimir P. Leksin, Moscow State Regional Social Humanity Institute, Zelenaya, 30, Kolomna, Moscow Region, 140411, Russia.
E-mail: `lexine@mccme.ru`

Chapter 15

A Monodromy Problem Connected with P_6

Dmitrii P. Novikov

Abstract. We consider an integral equation connected with the problem of construction of a linear equation with a given monodromy group.

Keywords. Isomonodromic Deformations, Belavin–Polyakov–Zamolodchikov Equations.

2010 Mathematics Subject Classification. Primary: 34M56; Secondary: 34M50, 34M55, 45F05.

15.1 Preliminaries I

Let L be a nonselfintersecting closed curve passing through all ramification points (t_1, $t_2, \ldots, t_m$, clockwise) of functions $y_1, \ldots, y_n$, that form the basis of solutions to some Fuchsian equation. We assume L to be smooth. Let D be the domain bounded by L and let $y^+ = (y_1^+, \ldots, y_n^+)$ be branches of these functions in D. The condition that $y^- = y^+$ on the arc (t_m, t_1) of L uniquely defines the analytical continuation y^- of y^+ onto $D^- = \overline{\mathbb{C}} \backslash \overline{D}$. It follows that $y^+ = y^- M_1$ on (t_1, t_2), $y^+ = y^- M_1 M_2$ on (t_2, t_3), etc., where M_i are the monodromy matrices. Thus, we get $y^+ = y^- C$ on L, where $C(\zeta)$ is a piecewise constant matrix. This is a particular case of the Riemann–Hilbert (or linear conjugation) problem: given C, find analytic vector functions y^+ in D and y^- in D^- such that $y^+ = y^- C$ on L. The following integral equations (the Sokhotskij–Plemelji formulas)

$$\frac{1}{2} y^+(x) = \frac{1}{2\pi i} \int_{\partial D} \frac{y^+(\zeta)}{\zeta - x}\, d\zeta, \qquad x \in \partial D,$$

$$\frac{1}{2} y^-(x) = \frac{-1}{2\pi i} \int_{\partial D} \frac{y^-(\zeta)}{\zeta - x}\, d\zeta + y^-(\infty)$$

imply the singular equation for the function $\varphi = (y^- + y^+)/2$:

$$\varphi(x) = \frac{1}{\pi i} \int_{\partial D} \varphi(\zeta) \frac{C(\zeta) - E}{C(\zeta) + E} \frac{d\zeta}{\zeta - x} + y^-(\infty). \tag{15.1}$$

In [1], Its and Harnad considered equation (15.1) with

$$(C(\zeta) - E)^2 = 0, \quad C(\zeta) = E + 2\pi i\, f(\zeta) g(\zeta),$$

$f(\zeta)$ and $g^T(\zeta)$ are $n \times l$ matrices ($l \leq n$) such that $g(\zeta)f(\zeta) = 0$. Suppose that C satisfies the following condition: for any nonzero $y^-(\infty)$ the equation (15.1) is soluble. Then the problem $Y^+ = Y^- C$, $Y^-(\infty) = E$, reduces to the equations on the matrix $\Phi = (Y^- + Y^+)/2$:

$$\Phi(x) = E + \int \Phi(\zeta) \frac{f(\zeta)g(\zeta)}{\zeta - x}\, d\zeta \left[= E + \int \frac{F(\zeta)g(\zeta)\, d\zeta}{\zeta - x} \right],$$

$$\Phi^{-1}(x) = E - \int \frac{f(\zeta)g(\zeta)}{\zeta - x} \Phi^{-1}(\zeta)\, d\zeta \left[= E - \int \frac{f(\zeta)G(\zeta)\, d\zeta}{\zeta - x} \right].$$

Here $F = \Phi f$ and $G = g\Phi^{-1}$ are solutions of integral equations whose kernel has no singularity on the diagonal:

$$\begin{cases} F(x) - \int F(\zeta)K(\zeta, x)\, d\zeta = f(x), \\ G(x) - \int K(x, \zeta)G(\zeta)\, d\zeta = g(x), \end{cases} \quad K(\zeta, x) = \frac{g(\zeta)f(x)}{\zeta - x}. \tag{15.2}$$

The resolvent is defined by the equation

$$R(x, z) - \int K(x, \zeta)R(\zeta, z)\, d\zeta = K(x, z)$$

and can be written explicitly (the proof is stated in [1]):

$$R(x, z) = \frac{G(x)F(z)}{x - z}. \tag{15.3}$$

15.2 Preliminaries II

The following case of the monodromy problem was considered in [1,2]. L is the system of nonintersected intervals, the number of ramification points is even ($m = 2k$), matrix $C = E + 2\pi i\, f_j g_j$ on the arcs (t_{2j-1}, t_{2j}), f_j and g_j^T are $n \times l$ constant matrices ($l \leq n$), and $g_j f_j = 0$. On the remainder of L it is assumed that $C = E$. Then the problem reduces (as it was explained in Preliminaries I) to the integral equations (15.2)

$$F(x) - \int_{t_1}^{t_2} \frac{F(\zeta)g_1 f_j}{\zeta - x}\, d\zeta - \cdots - \int_{t_{2k-1}}^{t_{2k}} \frac{F(\zeta)g_k f_j}{\zeta - x}\, d\zeta = f_j, \tag{15.4}$$

where $x \in (t_{2j-1}, t_{2j})$, $j = \overline{1, k}$; the equations for $G(x)$ are analogous. Here, the kernel is continuous. If the matrices f_i, g_i are independent of t_j and equations (15.2), (15.4) have a unique solution, then, differentiating on variables t_j, we get [3] the system

$$\frac{\partial F(x)}{\partial t_j} = (-1)^j F(t_j)R(t_j, x), \qquad \frac{\partial G(x)}{\partial t_j} = (-1)^j R(x, t_j)G(t_j).$$

Using formulas (15.3), one can transform the first equation to

$$\frac{\partial F(x)}{\partial t_j} = -\frac{A_j}{x - t_j} F(x), \qquad A_j = (-1)^j F(t_j) G(t_j). \tag{15.5}$$

It follows from the integral equation that $F(x)$ is homogeneous on x, t_j and holomorphic on x near infinity:

$$\frac{\partial F}{\partial x} + \frac{\partial F}{\partial t_1} + \cdots + \frac{\partial F}{\partial t_m} = 0, \qquad x\frac{\partial F}{\partial x} + t_1\frac{\partial F}{\partial t_1} + \cdots + t_m\frac{\partial F}{\partial t_m} = 0.$$

Thus, if the matrices M_i satisfy conditions $(M_i - E)^2 = 0, i = \overline{1, 2k}$, and $M_{2i-1} = M_{2i}^{-1}, i = \overline{1, k}$, and are independent of t_j, then they are generators of the monodromy group of the Fuchsian system

$$\frac{\partial F}{\partial x} = AF, \qquad A = \sum_{i=1}^{m} \frac{A_i}{x - t_i}. \tag{15.6}$$

The Schlesinger system, which follows from (15.6), (15.5),

$$\frac{\partial A_j}{\partial t_i} = \frac{[A_i, A_j]}{t_i - t_j}, \quad i \neq j, \qquad \frac{\partial A_i}{\partial t_i} = -\sum_{j \neq i} \frac{[A_i, A_j]}{t_i - t_j} \tag{15.7}$$

gives the isomonodromic deformations equations for A_j.

15.3 Main Result

The τ-function was introduced in [4] in connection with the system (15.7). Its construction involves the closed form

$$d \ln \tau = \sum_{i=1}^{m} H_i \, dt_i, \qquad H_i = \frac{1}{2} \operatorname*{res}_{x=t_i} (\mathrm{tr} A^2) \equiv \sum_{j \neq i}^{m} \frac{\mathrm{tr} A_j A_i}{t_i - t_j}.$$

In what follows, we describe the results of [5]. We assume that A_i are 2×2 matrices and $\mathrm{tr} A_i = 0$, $\det A_i = -\Delta_i$, then

$$A^2 = \sum_{i=1}^{m} \left[\frac{H_i}{x - t_i} + \frac{\Delta_i}{(x - t_i)^2} \right] \begin{pmatrix} 1 & 0 \\ 0 & 1 \end{pmatrix}.$$

According to (15.6), (15.5), we have

$$\sum_{i=1}^{m} \frac{1}{x - t_i} \frac{\partial F}{\partial t_i} = \frac{\partial^2 F}{\partial x^2} - \sum_{i=1}^{m} \left[\frac{H_i}{x - t_i} + \frac{\Delta_i}{(x - t_i)^2} \right] F.$$

For $\Phi = \tau F$ this read as

$$\sum_{i=1}^{m} \frac{1}{x - t_i} \frac{\partial \Phi}{\partial t_i} = \frac{\partial^2 \Phi}{\partial x^2} - \sum_{i=1}^{m} \frac{\Delta_i}{(x - t_i)^2} \Phi; \tag{15.8}$$

(15.8) is a special case of the equation in the conformal field theory

$$\sum_{i=1}^{m} \frac{1}{x - t_i} \frac{\partial \Phi}{\partial t_i} = \kappa \frac{\partial^2 \Phi}{\partial x^2} - \sum_{i=1}^{m} \frac{\Delta_i}{(x - t_i)^2} \Phi,$$

(the Virasoro algebra have central charge $c = -6\kappa + 13 - 6\kappa^{-1}$, formula (5.17) in [6]; we get (15.8) setting $c = 1$).

The Neumann series for (15.4) gives an example of a solution to (15.8) (with $\Delta_i = 0$) expressed through the polylogarithms. We apply linear superposition of solutions of (15.8) to construct solutions to systems (15.6), (15.7). This is only one aspect of the problem of the complete integration of system (15.7) in special functions.

15.4 Example

We demonstrate the reduction to P_6 in the particular case $m = 2k = 4$ in (15.4). The monodromy matrices satisfy the relations

$$M_1 = \begin{pmatrix} 1 & 2\pi i \gamma \\ 0 & 1 \end{pmatrix} = M_2^{-1}, \qquad M_3 = \begin{pmatrix} 1 & 0 \\ 2\pi i \gamma & 1 \end{pmatrix} = M_4^{-1}.$$

We choose

$$f_1 = \begin{pmatrix} 1 \\ 0 \end{pmatrix}, \ f_2 = \begin{pmatrix} 0 \\ 1 \end{pmatrix}, \qquad g_1 = (0\ \gamma),\ g_2 = (\gamma\ 0).$$

The equation for $F(x)$ follows from (15.4):

$$F(x) - \int_{t_3}^{t_4} \frac{\gamma F(z)}{z - x}\, dz = \begin{pmatrix} 1 \\ 0 \end{pmatrix}, \qquad F(z) - \int_{t_1}^{t_2} \frac{\gamma F(\zeta)}{\zeta - z}\, d\zeta = \begin{pmatrix} 0 \\ 1 \end{pmatrix},$$

here $x \in (t_1, t_2)$, $z \in (t_3, t_4)$. Let $F^*(x) = \begin{pmatrix} 1 & 0 \\ \gamma^2 \ln(-t_4) & -\gamma \end{pmatrix} F(x)$. Suppose that $t_1 = 1$, $t_2 = t$, $t_3 = 0$, and $t_4 \to \infty$. Then

$$F^*(x) - \lambda \int_1^t \frac{\ln x - \ln \zeta}{x - \zeta} F^*(\zeta)\, d\zeta = \begin{bmatrix} 1 \\ \lambda \ln x \end{bmatrix}, \tag{15.9}$$

where $x \in (1,t)$, $\lambda = \gamma^2$. Due to (15.3), (15.5), we get

$$F^*(x) = \begin{bmatrix} \psi_1(x;t,\lambda) \\ \psi_2(x;t,\lambda) \end{bmatrix}, \qquad R(x,z) = \frac{\psi_1(x)\psi_2(z) - \psi_2(x)\psi_1(z)}{x-z},$$

$$\frac{\partial F^*(x)}{\partial t} = \frac{-A_t}{x-t} F^*(x), \qquad A_t = \begin{pmatrix} -ab & a^2 \\ -b^2 & ab \end{pmatrix}, \qquad \begin{matrix} a = \psi_1(t;t,\lambda), \\ b = \psi_2(t;t,\lambda). \end{matrix}$$

Set $\widetilde{a} = \psi_1(1;t,\lambda)$, $\widetilde{b} = \psi_2(1;t,\lambda)$. Since $\psi_1(tx;t,\lambda) = \psi_1(x;t^{-1},-\lambda)$, and

$$\psi_2(tx;t,\lambda) = \psi_2(x;t^{-1},-\lambda) + \psi_1(x;t^{-1},-\lambda)\ln t$$

we have a Fuchsian system

$$\frac{\partial F^*(x)}{\partial x} = \frac{1}{x}\left[\begin{pmatrix} 0 & 0 \\ 1 & 0 \end{pmatrix} + \frac{tA_t}{x-t} + \frac{A_1}{x-1}\right] F^*(x), \qquad A_1 = \begin{pmatrix} -\widetilde{a}\widetilde{b} & \widetilde{a}^2 \\ -\widetilde{b}^2 & \widetilde{a}\widetilde{b} \end{pmatrix}.$$

Eliminating $F^*(x)$, we get

$$a' = \frac{f\widetilde{a}}{t(t-1)}, \qquad b' = \frac{a}{t} + \frac{f\widetilde{b}}{t(t-1)}, \qquad \widetilde{a}' = \frac{fa}{1-t}, \qquad \widetilde{b}' = \frac{fb}{1-t},$$

$f = b\widetilde{a} - a\widetilde{b}$, with integral $f^2 + a^2 + \widetilde{a}^2 = 0$. It follows that the function $u = -\widetilde{a}^2 a^{-2}$ is a solution to the Painlevé VI equation

$$u'' = \frac{1}{2}\left(\frac{1}{u} + \frac{1}{u-1} + \frac{1}{u-t}\right)u'^2 - \left(\frac{1}{t} + \frac{1}{t-1} + \frac{1}{u-t}\right)u'.$$

The σ-form (see [2] and the equation $\mathrm{E_{VI}}$ in [7]) of P_6 follows for the function $\sigma = ta^2 + \widetilde{a}^2 = t(t-1)R(t,t)$

$$\left[t(t-1)\sigma''\right]^2 = 4\sigma'(t\sigma' - \sigma)\big((1-t)\sigma' + \sigma\big). \tag{15.10}$$

Let δ be the Fredholm determinant (an entire function on λ) for (15.9):

$$K(x,\zeta) = \frac{\ln x - \ln\zeta}{x - \zeta},$$

$$\delta = 1 - \lambda\int_1^t K(x,x)\,dx + \frac{\lambda^2}{2!}\int_1^t\int_1^t \begin{vmatrix} K(x_1,x_1) & K(x_1,x_2) \\ K(x_2,x_1) & K(x_2,x_2) \end{vmatrix} dx_1\,dx_2 + \cdots.$$

Using the identity $R(t,t) = (\ln\delta)'$, we obtain $\sigma = t(t-1)(\ln\delta)'$. One can check that we get solutions to P-VI by substitution to (15.10)

$$\sigma = -\lambda(t-1) - \lambda^2 t(t-1)\int_1^t \frac{(\ln x - \ln t)^2}{(x-t)^2}\,dx + \cdots.$$

Moreover, for the function $\phi = \delta\psi_1$ we have the equation ([8])

$$\frac{t(t-1)}{x(x-1)(x-t)}\frac{\partial\phi}{\partial t} = \frac{\partial^2\phi}{\partial x^2} + \left(\frac{1}{x} + \frac{1}{x-1}\right)\frac{\partial\phi}{\partial x}. \tag{15.11}$$

The Fredholm theory, for the integral equation (15.9), gives the expansion $\phi = 1 + \lambda\phi_1 + \lambda^2\phi_2 + \cdots$ of the solution to equation (15.11). Coefficients at the powers of λ satisfies to (15.11), for example,

$$\phi_1 = \int_1^t [K(x,z) - K(z,z)]\,dz = \int_1^t \left(\frac{\ln x - \ln z}{x-z} - \frac{1}{z}\right) dz.$$

The complexity of ϕ_i grows quickly as i increases. Noting that all branches of ϕ_i satisfy (15.11), one can get simpler solutions; in particular, ϕ_1 generates solutions $\ln(x-1) - \ln(x-t)$ and $\ln x - \ln(x-t)$.

Bibliography

[1] J. Harnad and A. R. Its, Integrable Fredholm operators and dual isomonodromic deformations, *Commun. Math. Phys.*, **226**: 497–530, 2002.

[2] A. M. Borodin, *Isomonodromic deformations and the Painlev'e equations in random-matrix type problems,* preprint Nr. 04, 2005, POMI.

[3] M. G. Krein, Integral equations, that generate 2-nd order differential equations, *Dokl. AN SSSR.*, **97** (1954), 21–24.

[4] M. Sato, M. Jimbo, and T. Miwa, Holonomic quantum fields. II. The Riemann-Hilbert problem, *Publ. RIMS Kyoto Univ.*, **15**: 201–278, 1979.

[5] D. P. Novikov, The 2×2 matrix Schlesinger system and the Belavin–Polyakov–Zamolodchikov equation, *Teor. Mat. Fiz.*, **161**: 191–203, 2009.

[6] A. A. Belavin, A. M. Polyakov, and A. B. Zamolodchikov, Infinite conformal symmetry in two-dimensional quantum field theory, *Nucl. Phys. B*, **241**: 333–380, 1984.

[7] K. Okamoto, Polynomial hamiltonians associated with Painlevé equations. I, II, *Proc. Japan Acad.*, **56A**: 264–268; 367–371, 1980.

[8] B. I. Suleimanov, The Hamiltonian structure of Painlev'e equations and the method of isomonodromic deformations, in: *Asymptotic Properties of Solutions of Differential Equations* (A.M. Il'in, ed.), Akad. Nauk SSSR Ural. Otdel., Bashkir. Nauchn. Tsentr, Ufa (1988), pp. 93–102; (in Russian) Differential Equations **30**: 726–732, 1994.

Author Information

Dmitrii P. Novikov, Omsk State Technical University, Mira pr. 11, 644050, Omsk, Russia.
E-mail: `nvdmpr@mail.ru`

Painlevé Equations and Related Topics, 129–136

Chapter 16

Monodromy Evolving Deformations and Confluent Halphen's Systems

Yousuke Ohyama

Abstract. We study Halphen's confluent systems corresponding to the Whittaker, Bessel, Weber, and Airy functions. We show that Halphen's confluent systems are represented by monodromy evolving deformation found by Chakravarty and Ablowitz.

Keywords. Monodromy Evolving Deformations, Confluent Hypergeometric Equations.

2010 Mathematics Subject Classification. Primary: 34M55; Secondary: 33C15.

16.1 Introduction

In this chapter we represent quadratic differential systems of the Halphen type as monodromy evolving deformations (MED). We first explain quadratic systems of the Halphen type. In 1881, G. Halphen [10, 14] found a quadratic system of differential equations

$$\begin{aligned} X' + Y' &= 2XY, \\ Y' + Z' &= 2YZ, \\ Z' + X' &= 2ZX. \end{aligned}$$

These equations can be solved by theta constants:

$$X = 2\frac{d}{d\tau}\log\theta_2(0,\tau), \quad Y = 2\frac{d}{d\tau}\log\theta_3(0,\tau), \quad Z = 2\frac{d}{d\tau}\log\theta_4(0,\tau).$$

In the last thirty years, (16.1) has become popular in many mathematical fields. This equation is a reduction from the Bianchi IX cosmological models or the self-dual Yang–Mills equation [5, 6] and gives a special self-dual Einstein metric [3, 9]. If we set $y = 2(X + Y + Z)$, y satisfies the Chazy's equation

$$y''' = 2yy'' - 3(y')^2. \tag{16.1}$$

This work was supported by the Mitsubishi foundation and the JSPS Grant-in-Aid for Scientific Research (C) 21540217. The author also extends his thanks to the Newton Institute for their hospitality in September, 2011.

Chazy's equation appeared in his classification of the third-order Painlevé-type equation [7], but (16.1) does not have the Painlevé property, because generic solutions has a natural boundary.

In the same year Halphen found another system, called Halphen's second equation [11]

$$\begin{aligned} x_1' &= x_1^2 + a(x_1 - x_2)^2 + b(x_2 - x_3)^2 + c(x_3 - x_1)^2, \\ x_2' &= x_2^2 + a(x_1 - x_2)^2 + b(x_2 - x_3)^2 + c(x_3 - x_1)^2, \\ x_3' &= x_3^2 + a(x_1 - x_2)^2 + b(x_2 - x_3)^2 + c(x_3 - x_1)^2, \end{aligned} \tag{16.2}$$

which is less familiar. Halphen's second equation (16.2) appears in the study on hypercomplex structure [12]. Halphen's second equation is also studied in [2] as the DH-IX system (the Darboux–Halphen system of ninth order)

$$\frac{dM}{dt} = {}^t(\operatorname{adj} M) + {}^t M M - (\operatorname{Tr} M) M,$$

for a 3×3-matrix $M = M(t)$.

In the case $a = b = c = -1/8$, (16.2) is equivalent to (16.1) by the transform $2X = x_2 + x_3, 2Y = x_3 + x_1, 2Z = x_1 + x_2$. Since the solutions of (16.2) may have a natural boundary or moving branch points, (16.2) also does not have the Painlevé property.

We may consider confluent case of the Halphen equation [15]. The system

$$\begin{aligned} x_1' &= x_1^2 + \left(m^2 - \frac{1}{4}\right)(x_2 - x_1)^2 - k(x_2 - x_1)(x_3 - x_1) + \frac{1}{4}(x_3 - x_1)^2, \\ x_2' &= x_2^2 + \left(m^2 - \frac{1}{4}\right)(x_2 - x_1)^2 - k(x_2 - x_1)(x_3 - x_1) + \frac{1}{4}(x_3 - x_1)^2, \\ x_3' &= x_3^2 + \left(m^2 - \frac{1}{4}\right)(x_2 - x_1)^2 - k(x_2 - x_1)(x_3 - x_1) - \frac{3}{4}(x_3 - x_1)^2, \end{aligned}$$

is solved by the confluent hypergeometric functions $W_{k,m}(z)$. We can construct a system of Halphen's type from any second-order linear equation [15]. The Bessel type was studied also in [1], and the Airy type was studied in [8] independently. These confluent systems also do not have the Painlevé property in general.

In order to study such non-Painlevé-type equations, we may use monodromy evolving deformations, which do not preserve monodromy data. In 1996, Chakravarty and Ablowitz [4] showed that a fifth-order equation

$$\begin{aligned} \omega_1' &= \omega_2\omega_3 - \omega_1(\omega_2 + \omega_3) + \phi^2, \\ \omega_2' &= \omega_3\omega_1 - \omega_2(\omega_3 + \omega_1) + \theta^2, \\ \omega_3' &= \omega_1\omega_2 - \omega_3(\omega_1 + \omega_2) - \phi\theta, \\ \phi' &= \omega_1(\theta - \phi) - \omega_3(\theta + \phi), \\ \theta' &= -\omega_2(\theta - \phi) - \omega_3(\theta + \phi), \end{aligned} \tag{16.3}$$

can be represented by MED. The above system, called the Darboux–Halphen fifth equation (DH-V), arises in complex Bianchi IX cosmological models. The DH-V has two first integrals and is reduced to the third-order equation, which is a special case of Halphen's second equation.

The author has generalized the result of [4] and has shown that all of Halphen's second equation can be represented by MED. In this paper, we show that Halphen's confluent systems are also represented by MED.

16.2 Quadratic Systems and Nonassociative Algebras

We take a homogeneous quadratic system

$$\frac{dX_i}{dt} = \sum_{j,k=1}^{n} a_{jk}^i X_j X_k, \tag{16.4}$$

where $i = 1, 2, \dots, n$ and $a_{jk} = a_{kj}$. For (16.4), we define a commutative and non-associative algebra generated by $x_1, x_2, \dots, x_n$ with a multiplication table

$$x_j \cdot x_k = \sum_{i=1}^{n} a_{jk}^i x_i.$$

Theorem 16.1 ([16]). *We take a quadratic system with rank three. If and only if the associated algebra has a unit and its automorphism group is finite, the quadratic system is corresponding to one of hypergeometric type $HG(a, b, c)$, Whittaker type $W(k, m)$, Bessel type $B(\nu)$, Hermite–Weber type $HW(n)$ or Airy type Ai.*

If the associated algebra has a unit and its automorphism group is infinite, the algebra is isomorphic to J (Jordan algebra), $X(t)$, Y_1, Y_2, Y_3, or Y_4 defined below. For these cases the corresponding quadratic systems can be solved by elementary functions.

$HG(0, 0, 0)$, $X(0)$, Y_1, and Y_2 are associative. J and Y_3 are not associative but power-associative. $X(t)$, Y_1, Y_2, Y_3, and Y_4 has a nontrivial ideal.

We list three-dimensional commutative algebras with a unit e. We assume that the algebra is $\mathbb{C}e + \mathbb{C}f + \mathbb{C}g$ as a $\mathbb{C}$-vector space. We may take any base field whose characteristic is not two.

$HG(a, b, c)$: $\quad e = x + y + z$ and

$$x \cdot x = x + (b + c)(x + y + z), \quad x \cdot y = -c(x + y + z),$$

$$y \cdot y = y + (c + a)(x + y + z), \quad y \cdot z = -a(x + y + z),$$

$$z \cdot z = z + (a + b)(x + y + z), \quad z \cdot x = -b(x + y + z).$$

$$W(k,m): \qquad g\cdot g = m^2 e, \quad f\cdot f = e, \quad f\cdot g = -\frac{f}{2} - ke$$

$$B(\nu): \qquad g\cdot g = \frac{\nu^2}{4} e, \quad f\cdot f = 0, \quad f\cdot g = -\frac{f}{2} + e.$$

$$HW(n): \qquad g\cdot g = \frac{n^2}{4} e, \quad f\cdot f = 0, \quad f\cdot g = -\frac{f}{2} + e.$$

$$\mathrm{Ai}: \qquad w\cdot w = v, \quad v\cdot v = 0, \quad w\cdot v = e.$$

$$J: \qquad f\cdot f = e, \quad g\cdot g = e, \quad f\cdot g = 0.$$

$$X(t): \qquad f\cdot f = 0, \quad g\cdot g = te, \quad f\cdot g = f.$$

$$Y_j\ (j = 1,2,3,4): \quad f\cdot f = 0, \quad f\cdot g = 0,\ g\cdot g = 0,\ f, e \quad \text{or} \quad f + e.$$

Here, $a, b, c; k, m; \nu; n; t$ are complex parameters.

We call quadratic systems with the rank three associated with the algebras $W(k,m)$, $B(\nu)$, $HW(n)$, and Ai *Halphen's confluent system*. A quadratic system associated with $HG(a,b,c)$ is Halphen's second equation (16.2):

$$\begin{aligned}
x_1' &= x_1^2 + a(x_1 - x_2)^2 + b(x_2 - x_3)^2 + c(x_3 - x_1)^2,\\
x_2' &= x_2^2 + a(x_1 - x_2)^2 + b(x_2 - x_3)^2 + c(x_3 - x_1)^2,\\
x_3' &= x_3^2 + a(x_1 - x_2)^2 + b(x_2 - x_3)^2 + c(x_3 - x_1)^2.
\end{aligned}$$

Halphen's confluent system of type $W(k,m)$ is (16.3):

$$\begin{aligned}
x_1' &= x_1^2 + \left(m^2 - \frac{1}{4}\right)(x_2 - x_1)^2 - k(x_2 - x_1)(x_3 - x_1) + \frac{1}{4}(x_3 - x_1)^2,\\
x_2' &= x_2^2 + \left(m^2 - \frac{1}{4}\right)(x_2 - x_1)^2 - k(x_2 - x_1)(x_3 - x_1) + \frac{1}{4}(x_3 - x_1)^2,\\
x_3' &= x_3^2 + \left(m^2 - \frac{1}{4}\right)(x_2 - x_1)^2 - k(x_2 - x_1)(x_3 - x_1) - \frac{3}{4}(x_3 - x_1)^2.
\end{aligned}$$

Halphen's confluent system of type $B(\nu$:

$$\begin{aligned}
x_1' &= x_1^2 + (\nu^2 - 1)(x_2 - x_1)^2 - 2(x_2 - x_1)(x_3 - x_1),\\
x_2' &= x_2^2 + (\nu^2 - 1)(x_2 - x_1)^2 - 2(x_2 - x_1)(x_3 - x_1),\\
x_3' &= x_3^2 + (\nu^2 - 1)(x_2 - x_1)^2 - 2(x_2 - x_1)(x_3 - x_1) - (x_3 - x_1)^2,
\end{aligned}$$

which is solved by the Bessel function $J_\nu(2\sqrt{2x})$.

Halphen's confluent system of type $W(n)$:

$$\begin{aligned}
x_1' &= x_1^2 + (n^2 - 1)(x_2 - x_1)^2 - \frac{1}{4}(x_2 - x_1)(x_3 - x_1),\\
x_2' &= x_2^2 + (n^2 - 1)(x_2 - x_1)^2 - \frac{1}{4}(x_2 - x_1)(x_3 - x_1) - (x_2 - x_1)^2,
\end{aligned}$$

$$x_3' = x_3^2 + (n^2+1)(x_2-x_1)^2 - \frac{1}{4}(x_2-x_1)(x_3-x_1) - (x_3-x_1)^2,$$

which is solved by the parabolic cylinder function $D_n(x)$.

Halphen's confluent system of type Ai:

$$\begin{aligned} x_1' &= x_1^2 + (x_2-x_1)(x_3-x_1), \\ x_2' &= x_2^2 + (x_2-x_1)(x_3-x_2), \\ x_3' &= x_3^2 + (x_2-x_3)(x_3-x_1) + (x_2-x_1)^2, \end{aligned} \tag{16.5}$$

which is solved by the Airy function $\mathrm{Ai}(x)$. The system (16.5) is written as a single equation

$$x_1''' - 12x_1x_1'' + 48{x_1}^2x_1' - 6(x_1')^2 - 24{x_1}^4 = 0,$$

which is known from Clarskon and Olver [8]. The author does not know of any condition where a Halphen-type system can be written as a single equation.

16.3 Monodromy Evolving Deformations

We give a basic theory of monodromy evolving deformation (MED). In [4], Chakravarty and Ablowitz have shown that DH-V is given by MED. The author has shown that Halphen's second equation is given by MED [17]. As the same as monodromy preserving deformation, we consider

$$\frac{\partial Y}{\partial x} = A(x,t)Y, \quad \frac{\partial Y}{\partial t} = B(x,t)Y \tag{16.6}$$

where $A(x,t)$, $B(x,t)$ are $m\times m$-matrices and $A(x,t)$ is a rational function on x. When $B(x,t)$ is also rational on x, the monodromy data of the first equation in (16.6) is invariant for any t. When $B(x,t)$ is not rational, the monodromy data may be changed.

Assume that $A(x,t)$ has a singularity at $x=a$. Then Y can be developed as

$$Y \sim Y_0(x)\exp\left[\sum_{k=1}^{r} T_k(x-a)^{-k}\right](x-a)^L.$$

Here $Y_0(x)$ is a (formal) power series around $x=a$, r is the Poincaré rank at $x=a$ (when $x=a$ is regular singular, $r=0$), T_k $(k=1,2,\ldots,r)$, and L are commutative. We consider the following monodromy evolving deformation:

$$\frac{\partial L}{\partial t} = fI,$$

where I is a unit matrix. This deformation changes only the trace part of local exponent matrix and preserves projective monodromy [13]. In this case the deformation equation is given by

$$\frac{\partial Y}{\partial t} = \left[B_0(x,t) + f\log(x-a)\right]Y.$$

We review the result in [17]. We set $Q(x) = x^2 + a(x_1 - x_2)^2 + b(x_2 - x_3)^2 + c(x_3 - x_1)^2$, $P(x) = (x - x_1)(x - x_2)(x - x_3)$ and $R(x) = -(x + x_1 + x_2 + x_3)$. Halphen's second equation is nothing but $x_j' = Q(x_j)\,(j = 1, 2, 3)$. We take the following 2×2 linear system

$$\frac{\partial Y}{\partial x} = \left(\frac{\mu}{P} + \sum_{j=1}^{3} \frac{c_j S}{x - x_j}\right) Y, \tag{16.7}$$

$$\frac{\partial Y}{\partial t} = \left(\nu + \sum_{j=1}^{3} c_j x_j S\right) Y - Q(x)\frac{\partial Y}{\partial x}. \tag{16.8}$$

Here μ and c_j's are constants with $c_1 + c_2 + c_3 = 0$, and S is any traceless constant matrix. We assume

$$\frac{\partial \nu}{\partial x} = \frac{R}{P}\mu.$$

Theorem 16.2 ([17]). *The compatibility condition of* (16.7) *and* (16.8) *gives the Halphen's second equation.*

This theorem contains the result in [4].

We assume the local monodromy of $Y_j(x)$ around $x = x_j$ is $e^{2\pi i L_j}$. This deformation does not preserve the monodromy data. The local exponent L_j at $x = x_j$ evolves as

$$\frac{dL_j}{dt} = -\frac{R(x_j)}{\prod_{m \neq j}(x_j - x_m)}\mu.$$

We can eliminate the variables μ and ν in (16.7) and (16.8) by rescaling $Y = fZ$ for a scalar function $f = f(x,t)$. f satisfies the linear equations

$$\frac{\partial f}{\partial x} = \frac{\mu}{P(x)} f, \qquad \frac{\partial f}{\partial t} = \nu f - Q(x)\frac{\partial f}{\partial x}, \qquad \frac{\partial \nu}{\partial x} = \frac{R(x)}{P(x)}\mu. \tag{16.9}$$

Here $R(x) = -(x + x_1 + x_2 + x_3)$.

The integrability condition for f is

$$\frac{\partial P}{\partial t} + Q\frac{\partial P}{\partial x} - P\frac{\partial Q}{\partial x} + RP = 0, \tag{16.10}$$

which gives the Halphen equation. Therefore MED is essentially deformations of a scalar equation (16.10). Our problem is *to find* P, Q, *and* R *in* (16.9) *which gives Halphen's confluent equation.*

16.4 Halphen's Confluent Systems and Monodromy Evolving Deformations

We consider a nonlinear system (16.9). We list up P, Q, and R in (16.9) which gives Halphen's systems (16.10).

1) Hypergeometric type:

$$P(x) = (x - x_1)(x - x_2)(x - x_3), \qquad R(x) = -(x + x_1 + x_2 + x_3),$$
$$Q(x) = x^2 + a(x_1 - x_2)^2 + b(x_2 - x_3)^2 + c(x_3 - x_1)^2.$$

2) Whittaker type:

$$P(x) = (x - x_1)^2(x - x_2), \qquad R(x) = -x - 2x_1 - x_2,$$
$$Q(x) = x^2 + \left(m^2 - \frac{1}{4}\right)(x_1 - x_2)^2 - \frac{1}{4}(x_3 - x_1)^2 - k(x_2 - x_1)(x_3 - x_1).$$

3) Bessel type:

$$P(x) = (x - x_1)^2(x - x_2)^2, \qquad R(x) = -2(x + x_1 + x_2),$$
$$Q(x) = x^2 + \left(\frac{\nu^2 - 1}{4}\right)(x - x_2)^2 - 2(x_3 - x_1)(x_2 - x_1).$$

4) Weber type:

$$P(x) = (x - x_1)^3, \qquad R(x) = -x - 3x_1,$$
$$Q(x) = x^2 - \left(n + \frac{1}{2}\right)(x_2 - x_1)^2 + \frac{1}{4}(x_3 - x_1)^2.$$

5) Airy type:

$$P(x) = (x - x_1)^3, \qquad R(x) = -x - 3x_1,$$
$$Q(x) = x^2 + (x_3 - x_1)(x_2 - x_1).$$

Bibliography

[1] M. J. Ablowitz, S. Chakravarty, and R. Halburd, The generalized Chazy equation from the self-duality equations. *Stud. Appl. Math.*, **103**: 75–88, 1999.

[2] M. J. Ablowitz, S. Chakravarty, and R. Halburd, The Darboux–Halphen system and the singularity structure of its solutions, in "Mathematical and numerical aspects of wave propagation", 408–412, SIAM, 1998.

[3] M. Atiya and N. J. Hitchin, Low-energy scattering on non-Abelian magnetic monopoles, *Phil. Trans. R. Soc. Lond. A*, **315**: 459–469, 1985.

[4] S. Chakravarty and M. J. Ablowitz, Integrability, Monodromy Evolving Deformations and Self-Dual Bianchi IX Systems, *Phys. Rev. Lett.*, **76**: 857–860, 1996.

[5] S. Chakravarty, M. J. Ablowitz, and P. A. Clarkson, Reductions of Self-Dual Yang-Mills Fields and Classical Systems, *Phys. Rev. Let.*, **65**: 1085–1087, 1990.

[6] S. Chakravarty, M. J. Ablowitz, L. A. Takhtajan, Self-dual Yang-Mills equation and new special functions in integrable systems, *Nonlinear evolution equations and dynamical systems*, 3–11, World Sci. Publ., 1992.

[7] J. Chazy, Sur les équations différentielles du trousième ordre et d'ordre supérieur dont l'intégrale générale a ses points critiques fixes, *Acta Math.*, **34**: 317–385, 1911.

[8] P. A. Clarkson and P. J. Olver, Symmetry and the Chazy equation, *J. Diff. Eq.*, **124**: 225–246, 1996.

[9] G. W. Gibbons and C. N. Pope, The Positive Action Conjecture and Asymptotically Euclidean Metrics in Quantum Gravity, *Commun. Math. Phys.*, **66**: 267–290, 1979.

[10] G. Halphen, Sur une systém d'équations différentielles, *C. R. Acad. Sci., Paris*, **92**: 1101–1103, 1881.

[11] G. Halphen, Sur certains système d'équations différentielles, *C. R. Acad. Sci., Paris*, **92**: 1404–1406, 1881.

[12] N. Hitchin, Hypercomplex manifolds and the space of framings, in: *The Geometric Universe*, pp. 9–30, Oxford, 1996; Oxford, 1998.

[13] C. Mitschi and M. F. Singer, Projective Isomonodromy and Galois Groups, `arXiv:1002.2005`.

[14] Y. Ohyama, Differential Relations of Theta Functions, *Osaka J. Math.*, **32**: 431–450, 1995.

[15] Y. Ohyama, Systems of nonlinear differential equations related to second order linear equations, *Osaka J. Math.*, **33**: 927–949, 1997.

[16] Y. Ohyama, Hypergeometric functions and non-associative algebras, *CRM Proceedings and Lecture Notes*, **30**: 173–184, 2001.

[17] Y. Ohyama, Monodromy evolving deformations and Halphen's equation *CRM Proceedings and Lecture Notes*, **47**: 343–348, 2009.

Author Information

Yousuke Ohyama, Department of Mathematics, Osaka University, 1-1 Machikaneyama-machi, Toyonaka, 560-0043, Japan.
E-mail: `ohyama@ist.osaka-u.ac.jp`

Chapter 17

On the Gauge Transformation of the Sixth Painlevé Equation

Yoshikatsu Sasaki

Abstract. This article concerns the isomonodromic deformations of three linear ordinary differential equations which governed the sixth Painlevé equation. We show that there exists a 2-dimensional system of LODEs s.t. (i) by eliminating an entry, the system reduces to one of the 3 LODEs; (ii) the IMD of the system is equivalent to the IMD of one of the 3 LODEs with a certain condition; (iii) the system is obtained from a linearization of the sixth Painlevé equation by use of a gauge transformation.

Keywords. Sixth Painlevé Equation, Gauge Transformation.

2010 Mathematics Subject Classification. 34M55.

17.1 Linearizations of the Sixth Painlevé Equation

Definition 17.1. We say that a linear ordinary differential equation

$$\frac{\mathrm{d}^2 y}{\mathrm{d}x^2} + p_1 \frac{\mathrm{d}y}{\mathrm{d}x} + p_2 y = 0, \tag{17.1}$$

with rational functions p_1 and p_2, admits *the isomonodromic deformation* if and only if there exists a couple of rational functions $A = A(x,t)$ and $B = B(x,t)$ such that the following equations are completely integrable:

$$\frac{\partial^2 y}{\partial x^2} + p_1 \frac{\partial y}{\partial x} + p_2 y = 0, \tag{17.2a}$$

$$\frac{\partial y}{\partial t} = A \frac{\partial y}{\partial x} + By. \tag{17.2b}$$

Then (17.2) is called the deformation equation of (17.1).

Remark 17.2. Let (17.2) be completely integrable. $\vec{y}^{\gamma} = M(\gamma)\vec{y}$, $\vec{y}$: fundamental solution of (17.1), γ: a closed path on $\mathbb{CP}^1\backslash\{$singularities of the LODE$\}$, $M(\gamma) \in$

This study is supported by *Hiroshima University Grant-in-Aid for Exploratory Research.*

$GL(2,\mathbb{C})$. Differentiating both sides of the above,

$$\begin{aligned}\frac{\partial}{\partial t}\vec{y}^{\gamma} &= \frac{\partial M(\gamma)}{\partial t}\vec{y} + M(\gamma)\frac{\partial \vec{y}}{\partial t}\\ &= AM(\gamma)\frac{\partial \vec{y}}{\partial x} + \left(\frac{\partial M(\gamma)}{\partial t} + BM(\gamma)\right)\vec{y}.\end{aligned}$$

On the other hand, by analytic continuation of $\frac{\partial y}{\partial t} = A\frac{\partial y}{\partial x} + By$ along γ, we have

$$\frac{\partial}{\partial t}\vec{y}^{\gamma} = AM(\gamma)\frac{\partial}{\partial x}\vec{y} + BM(\gamma)\vec{y}.$$

Compare these two equations, and using the fact that y_1 and y_2 are linearly independent,

$$\frac{\partial M(\gamma)}{\partial t}\vec{y} = 0, \quad \text{i.e.,} \ \frac{\partial M(\gamma)}{\partial t} = 0L,$$

this means that the monodromy of $\vec{y}$ is independent of t.

17.1.1 LODE L_{VI}

$$\begin{aligned}L_{\mathrm{VI}}: \quad & \frac{\mathrm{d}^2 y}{\mathrm{d}x^2} + p_1\frac{\mathrm{d}y}{\mathrm{d}x} + p_2 y = 0,\ \kappa = \rho(\rho+\kappa_\infty)\ (= \rho(\bar{\rho}+1)),\\ & p_1 = \frac{1-\kappa_0}{x} + \frac{1-\kappa_1}{x-1} + \frac{1-\theta}{x-t} - \frac{1}{x-\lambda},\\ & p_2 = \frac{\kappa}{x(x-1)} - \frac{t(t-1)H}{x(x-1)(x-t)} + \frac{\lambda(\lambda-1)\mu}{x(x-1)(x-\lambda)},\end{aligned}$$

where $\rho := \{1-(\kappa_0+\kappa_1+\theta+\kappa_\infty)\}/2, \kappa := \rho(\rho+\kappa_\infty)$.

We assume that the regular singurarities of L_{VI} lies in generic locations; the parameters $\kappa_0, \kappa_1, \theta, \kappa_\infty \in \mathbb{C}\backslash\mathbb{Z}$ (then ρ is defined by Fuchsian rel.); L_{VI} has no logarithmic solutions (nonlogarithmic condition).

Riemann scheme of L_{VI} reads

$$\left\{\begin{matrix} x=0 & x=1 & x=t & x=\lambda & x=\infty \\ 0 & 0 & 0 & 0 & \rho \\ \kappa_0 & \kappa_1 & \theta & 2 & \rho+\kappa_\infty \end{matrix}\right\}.$$

The deformation equation of L_{VI},

$$\begin{aligned}\frac{\partial^2 y}{\partial x^2} + p_1\frac{\partial y}{\partial x} + p_2 y = 0,\\ \frac{\partial y}{\partial t} = A\frac{\partial y}{\partial x} + By,\end{aligned}$$

is equivalent to the Hamiltonian system as follows:

$$H_{\rm VI}: \quad \frac{d\lambda}{dt} = \frac{\partial H}{\partial \mu}, \quad \frac{d\mu}{dt} = -\frac{\partial H}{\partial \lambda},$$

$$H = \frac{1}{t(t-1)}\big[\lambda(\lambda-1)(\lambda-t)\mu^2 - \{\kappa_0(\lambda-1)(\lambda-t) + \kappa_1\lambda(\lambda-t) + (\theta-1)\lambda(\lambda-1)\}\mu + \kappa(\lambda-t)\big],$$

which is also equivalent to the sixth Painlevé equation

$$P_{\rm VI}: \quad \frac{d^2\lambda}{dt^2} = \frac{1}{2}\left\{\frac{1}{\lambda} + \frac{1}{\lambda-1} + \frac{1}{\lambda-t}\right\}\left(\frac{d\lambda}{dt}\right)^2 - \left\{\frac{1}{t} + \frac{1}{t-1} + \frac{1}{\lambda-t}\right\}\frac{d\lambda}{dt} + \frac{\lambda(\lambda-1)(\lambda-t)}{t^2(t-1)^2}\left[\alpha + \frac{\beta t}{\lambda^2} + \frac{\gamma(t-1)}{(\lambda-1)^2} + \frac{\delta t(t-1)}{(\lambda-t)^2}\right].$$

So the deformation equation is also called the linearization of $P_{\rm VI}$. (See [1].)

17.1.2 LODE $\bar{L}_{\rm VI}$

$$\bar{L}_{\rm VI}: \quad \frac{d^2y}{dx^2} + p_1\frac{dy}{dx} + p_2 y = 0, \ \bar{\kappa} = \rho(\rho+\kappa_\infty-1) \ (= \rho(\bar{\rho}+1)),$$

$$p_1 = \frac{1-\kappa_0}{x} + \frac{1-\kappa_1}{x-1} + \frac{1-\theta}{x-t} - \frac{2}{x-\bar{\lambda}},$$

$$p_2 = \frac{\bar{\kappa}}{x(x-1)} - \frac{t(t-1)\bar{H}}{x(x-1)(x-t)} + \frac{\bar{\lambda}(\bar{\lambda}-1)\bar{\mu}}{x(x-1)(x-\bar{\lambda})},$$

We assume that the regular singurarities of $\bar{L}_{\rm VI}$ lies in generic locations; the parameters $\kappa_0, \kappa_1, \theta, \kappa_\infty \in \mathbb{C}\backslash\mathbb{Z}$ (then ρ is defined by Fuchsian rel.); $\bar{L}_{\rm VI}$ has no logarithmic solutions (nonlogarithmic condition).

Riemann scheme of $\bar{L}_{\rm VI}$ reads

$$\left\{\begin{matrix} x=0 & x=1 & x=t & x=\bar{\lambda} & x=\infty \\ 0 & 0 & 0 & 0 & \rho \\ \kappa_0 & \kappa_1 & \theta & 3 & \rho+\kappa_\infty-1 \end{matrix}\right\}$$

T. Kimura studied this equation in [2]. He wrote that the isomonodromic deformation of $\bar{L}_{\rm VI}$ replaced κ_∞ by $\kappa_\infty + 1$ is equivalent to that of $L_{\rm VI}$ by construction of the Schlesinger transformation from $L_{\rm VI}$ to $\bar{L}_{\rm VI}$, and obtained an Hamiltonian system

$$\bar{H}_{\rm VI}: \quad \frac{d\bar{\lambda}}{dt} = \frac{\partial \bar{H}}{\partial \bar{\mu}}, \quad \frac{d\bar{\mu}}{dt} = -\frac{\partial \bar{H}}{\partial \bar{\lambda}},$$

which governs the isomonodromic deformation of $\bar{L}_{\rm VI}$. He did not write down the form of the Hamiltonian, but, however, it is easy to do. (In his papers, only the necessary condition of equiavalency is checked, but the sufficient condition is not.)

Theorem 17.3. *The Hamiltonian of the system $\bar{H}_{\mathrm{VI}}$ is as follows:*

$$\bar{H} = \frac{1}{t(t-1)}\left[\frac{\bar{\lambda}(\bar{\lambda}-1)(\bar{\lambda}-t)}{4}\bar{\mu}^2 + \bar{\kappa}(\bar{\lambda}-t) - \frac{\Phi}{f}\right],$$
$$- \{\kappa_0(\bar{\lambda}-1)(\bar{\lambda}-t) + \kappa_1\bar{\lambda}(\bar{\lambda}-t) + \theta\bar{\lambda}(\bar{\lambda}-1) - t(\bar{\lambda}-1)\}\frac{\bar{\mu}}{2},$$

where

$$f := \bar{\mu} - \frac{\kappa_0}{\bar{\lambda}} - \frac{\kappa_1}{\bar{\lambda}-1} - \frac{\theta}{\bar{\lambda}-t},$$
$$\Phi := \frac{(\bar{\lambda}^2 - 2\bar{\lambda}t)}{4}\bar{\mu}^2 - \left\{(\kappa_0+\theta)(\bar{\lambda}-1) + \frac{\kappa_1(\bar{\lambda}^2 - 2\bar{\lambda}+t)}{\bar{\lambda}-1}\right\}\frac{\bar{\mu}}{2} + \bar{\kappa}.$$

17.1.3 LODE $\tilde{L}_{\mathrm{VI}}$

$$\tilde{L}_{\mathrm{VI}}\colon\ \frac{\mathrm{d}^2y}{\mathrm{d}x^2} + p_1\frac{\mathrm{d}y}{\mathrm{d}x} + p_2 y = 0,\ \tilde{\kappa} = \rho(\rho+\kappa_\infty-1)\ (= \rho(\bar{\rho}+1)),$$
$$p_1 = \frac{1-\kappa_0}{x} + \frac{1-\kappa_1}{x-1} + \frac{1-\theta}{x-t} - \sum_{k=1,2}\frac{1}{x-\lambda_k},$$
$$p_2 = \frac{\tilde{\kappa}}{x(x-1)} - \frac{t(t-1)\tilde{H}}{x(x-1)(x-t)} + \sum_{k=1,2}\frac{\lambda_j(\lambda_k-1)\mu_k}{x(x-1)(x-\lambda_k)},$$

We assume that the regular singurarities of $\tilde{L}_{\mathrm{VI}}$ lies in generic locations; the parameters $\kappa_0, \kappa_1, \theta, \kappa_\infty \in \mathbb{C}\backslash\mathbb{Z}$ (then ρ is defined by Fuchsian rel.); $\tilde{L}_{\mathrm{VI}}$ has no logarithmic solutions (nonlogarithmic condition).

The Riemann scheme of $\tilde{L}_{\mathrm{VI}}$ reads

$$\left\{\begin{matrix} x=0 & x=1 & x=t & x=\lambda_1 & x=\lambda_2 & x=\infty \\ 0 & 0 & 0 & 0 & 0 & \rho \\ \kappa_0 & \kappa_1 & \theta & 2 & 2 & \rho+\kappa_\infty-1 \end{matrix}\right\}.$$

H. Ishikawa studied this equation in his master's thesis [3]. He obtained by construction of the Schlesinger transformation from $\tilde{L}_{\mathrm{VI}}$ to L_{VI} that

Theorem 17.4 ([3]). *The isomonodromic deformation of $\tilde{L}_{\mathrm{VI}}$ is equivalent to that of L_{VI} and it is governed by a Hamiltonian system with a certain first integral as follows:*

$$\tilde{H}_{\mathrm{VI}}:\ \frac{\mathrm{d}\lambda_k}{\mathrm{d}t} = \frac{\partial\tilde{H}}{\partial\mu_k},\quad \frac{\mathrm{d}\mu_k}{\mathrm{d}t} = -\frac{\partial\tilde{H}}{\partial\lambda_k},\quad (k=1,2)$$

$$\tilde{H} = -\frac{(t-\lambda_1)(t-\lambda_2)}{t(t-1)} \sum_{k=1,2} \frac{\tilde{\lambda}_k(\tilde{\lambda}_k - 1)}{(\tilde{\lambda}_k - \tilde{\lambda}_{k+1})} \left[\tilde{\mu}_k^2 - \left\{ \frac{\kappa_0}{\tilde{\lambda}_k} + \frac{\kappa_1}{\tilde{\lambda}_k - 1} + \frac{\theta - 1}{\tilde{\lambda}_k - t} \right\} \tilde{\mu}_k \right]$$

with an integral

$$\begin{aligned} &\tilde{\lambda}_1(\tilde{\lambda}_1 - 1)(\tilde{\lambda}_1 - t) \left[\tilde{\mu}_1^2 - \left\{ \frac{\kappa_0}{\tilde{\lambda}_1} + \frac{\kappa_1}{\tilde{\lambda}_1 - 1} + \frac{\theta}{\tilde{\lambda}_1 - t} \right\} \tilde{\mu}_1 \right] \\ &-\tilde{\lambda}_2(\tilde{\lambda}_2 - 1)(\tilde{\lambda}_2 - t) \left[\tilde{\mu}_2^2 - \left\{ \frac{\kappa_0}{\tilde{\lambda}_2} + \frac{\kappa_1}{\tilde{\lambda}_2 - 1} + \frac{\theta}{\tilde{\lambda}_2 - t} \right\} \tilde{\mu}_2 \right] \\ &\qquad + \bar{\kappa}(\tilde{\lambda}_1 - \tilde{\lambda}_2) = 0, \end{aligned}$$

where the index $k = 3$ *means* $k = 1$.

However, his work was not published.

IMD (isomonodromic deformation) of these equations have been studied separately [1–3]. In this article, we observe these three equations from a unific viewpoint.

Put

$$\begin{aligned} p_1 &= \frac{1-\kappa_0}{x} + \frac{1-\kappa_1}{x-1} + \frac{1-\theta}{x-t} - \frac{1}{x-\lambda}, \\ p_2 &= \frac{c_0}{x} + \frac{c_1}{x-1} + \frac{-H}{x-t} + \frac{\mu}{x-\lambda}, \\ q_1 &= \frac{1-\kappa_0}{x} + \frac{1-\kappa_1}{x-1} + \frac{1-\theta}{x-t} - \frac{2}{x-\bar{\lambda}}, \\ q_2 &= \frac{\bar{c}_0}{x} + \frac{\bar{c}_1}{x-1} + \frac{-\bar{H}}{x-t} + \frac{\bar{\mu}}{x-\bar{\lambda}}, \end{aligned}$$

where

$$\begin{aligned} c_0 &= -\kappa - (t-1)H + (\lambda - 1)\mu, \qquad & c_1 &= \kappa + tH - \lambda\mu, \\ \bar{c}_0 &= -\bar{\kappa} - (t-1)\bar{H} + (\bar{\lambda} - 1)\bar{\mu}, \qquad & \bar{c}_1 &= \bar{\kappa} + t\bar{H} - \bar{\lambda}\bar{\mu}. \end{aligned}$$

Then L_{VI} is $y'' + p_1 y' + p_2 y = 0$, and $\bar{L}_{\mathrm{VI}}$ is $y'' + q_1 y' + q_2 y = 0$.

Theorem 17.5. *If* $A^{21}(x)$ *has a double zero, in order that* $\bar{H}_{\mathrm{VI}}$ *be equivalent to* H_{VI}*, the necessary and sufficient condition (NSC) is that the parameters satisfy the five algebraic relations* $A_{\bar{\lambda}}, C_0, C_1, C_{\bar{\lambda}}, D_\lambda = 0$*, where*

$$\begin{aligned} A_{\bar{\lambda}} &= -2s_1'(\bar{\lambda}) + 2p_1(\bar{\lambda})s_1(\bar{\lambda}) + 2\bar{\mu}s_1(\bar{\lambda}) - 2s_2(\bar{\lambda}), \\ C_0 &= -2s_1'(0)c_0 + (1-\kappa_0)s_2'(0) + \frac{tc_0}{-\lambda} + s_2(0)(\bar{c}_0 - c_0), \end{aligned}$$

$$C_1 = -2s_1'(1)c_1 + (1-\kappa_1)s_2'(1) + \frac{(t-1)\tilde{c}_1}{\lambda} + s_2(1)(c_1 - c_1),$$
$$C_{\bar{\lambda}} = -2s_2'(\bar{\lambda}) + 2p_2(\bar{\lambda})s_1(\bar{\lambda}) + 2\bar{\mu}s_2(\bar{\lambda}),$$
$$D_\lambda = \lambda(\lambda-1)(\lambda-t)\{\mu^2 + (p_1 - \frac{-1}{x-\lambda})|_{x=\lambda}\cdot\mu + (p_2 - \frac{\mu}{x-\lambda})|_{x=\lambda}\}.$$

Remark 17.6. This is a criterion of whether $\bar{H}_{\mathrm{VI}}$ is equivalent to H_{VI} or not. Only one relation out of the five is redundant. So, we can choose either $\bar{\lambda}(t_0)$ or $\bar{\mu}(t_0)$ freely, but the other is determined by NSC.

Theorem 17.7. *There exists a 2-dimensional system of LODEs:*

$$S^{\sigma}_{\mathrm{VI}}: \quad \frac{d\vec{y}}{dx} = A(x)\vec{y},\ \vec{y} = \begin{pmatrix} y_1 \\ y_2 \end{pmatrix}, \quad A(x): a\ 2\times 2\ \textit{matrix with function entries}.$$

s.t.

(i) *by eliminating y_1 or y_2, S^{σ}_{VI} reduces to L_{VI}, $\bar{L}_{\mathrm{VI}}$ or $\tilde{L}_{\mathrm{VI}}$;*

(ii) *IMD of S^{σ}_{VI} is equivalent to IMD of L_{VI}, $\bar{L}_{\mathrm{VI}}$ or $\tilde{L}_{\mathrm{VI}}$ if $A^{21}(x)$ has two simple poles, or $A^{21}(x)$ has a double zero and parameters satisfy NSC;*

(iii) *S^{σ}_{VI} is obtained by a gauge transformation of the linearization of P_{VI} given in [4] (also see [5]).*

17.1.4 Schlesinger System with Symmetric Gauge

As is well known, the sixth Painlevé equation, P_{VI}, has another linearization which is the deformation equation of a system of differential equations of Schlesinger type:

$$S_{\mathrm{VI}}: \quad \frac{d\vec{y}}{dx} = A(x)\vec{y},$$

where

$$A(x) = \frac{A_0}{x} + \frac{A_1}{x-1} + \frac{A_t}{x-t} = \begin{pmatrix} A^{11}(x) & A^{12}(x) \\ A^{21}(x) & A^{22}(x) \end{pmatrix},$$

$$A_\xi = \begin{pmatrix} z_\xi + \theta_\xi & -u_\xi z_\xi \\ u_\xi^{-1}(z_\xi + \theta_\xi) & -z_\xi \end{pmatrix} \quad \text{for } \xi = 0, 1, t$$

$$-(A_0 + A_1 + A_t) = A_\infty = \begin{pmatrix} \alpha & \\ & \beta \end{pmatrix}.$$

$\vec{y} = {}^t(y_1, y_2)$. For given $\theta_0, \theta_1, \theta_t, \theta_\infty \notin \mathbb{Z}$, α, β is determined by $\alpha + \beta = -(\theta_0 + \theta_1 + \theta_t)$, $\alpha - \beta = \theta_\infty$. Correspondence of the parameters of S_{VI} $(\theta_0, \theta_1, \theta_t, \theta_\infty) = (\kappa_0, \kappa_1, \theta, 1 - \kappa_\infty)$.

R. Garnier showed in 1926 that the isomonodromic deformation of $S_{\rm VI}$ is reduced to that of $L_{\rm VI}$, and, in [5], that $P_{\rm VI}$ is obtained directly from the isomonodromic deformation of an equation similar to $S^{\sigma}_{\rm VI}$.

17.1.5 Schlesinger System with Asymmetric Gauge

Representation of the monodromy group has an ambiguity as

$$\vec{y} :\Rightarrow G\vec{\eta}, \quad A_\xi :\Rightarrow G^{-1}A_\xi G,$$

where ":⇒" means replacement. This replacement is called gauge transformation. Usually the gauge of $S_{\rm VI}$ is fixed such that A_∞ is diagonal, then each of $A^{12}(x)$, $A^{21}(x)$ has a simple pole.

By a simple gauge transformation with

$$G = \begin{pmatrix} 1 & \sigma/(\alpha-\beta) \\ & 1 \end{pmatrix},$$

$S_{\rm VI}$ is changed into

$$S^{\sigma}_{\rm VI} : \quad \frac{{\rm d}\vec{y}}{{\rm d}x} = A(x)\vec{y},$$

where

$$A(x) = \frac{A_0}{x} + \frac{A_1}{x-1} + \frac{A_t}{x-t} = \begin{pmatrix} A^{11}(x) & A^{12}(x) \\ A^{21}(x) & A^{22}(x) \end{pmatrix},$$

$$A_\xi = \begin{pmatrix} z_\xi + \theta_\xi & -u_\xi z_\xi \\ u_\xi^{-1}(z_\xi+\theta_\xi) & -z_\xi \end{pmatrix} \quad \text{for } \xi = 0, 1, t$$

$$-(A_0 + A_1 + A_t) = A_\infty = \begin{pmatrix} \alpha & \\ \sigma & \beta \end{pmatrix}.$$

In [5] Mahoux studied the case where A_ξ's are traceless, and he pointed out that $P_{\rm VI}$ is obtained by the isomonodromic deformation with asymmetric gauge at $x = \infty$.

The theorems are proved by straightforward computing. Therefore, in the following section we only explain how to construct the Schlesinger transformations and how to obtain NSC.

17.2 Schlesinger Transformation $\tilde{L}_{\rm VI} \to L_{\rm VI}$.

In this section, we review the unpublished results of Ishikawa [3]. By construction of the Schlesinger transformation $L_{\rm VI} \to \bar{L}_{\rm VI}$ in a similar way, we obtain NSC in Theorem B.

Rational functions s_1, s_2

$$L_{\mathrm{VI}}: \quad \frac{\mathrm{d}^2 z}{\mathrm{d}x^2} + q_1 \frac{\mathrm{d}z}{\mathrm{d}x} + q_2 z = 0,\ \kappa = \rho(\rho + \kappa_\infty)\ (= \rho(\bar{\rho} + 1)),$$

$$\begin{aligned} q_1 &= \frac{1-\kappa_0}{x} + \frac{1-\kappa_1}{x-1} + \frac{1-\theta}{x-t} - \frac{1}{x-\lambda}, \\ q_2 &= \frac{\kappa}{x(x-1)} - \frac{t(t-1)H}{x(x-1)(x-t)} + \frac{\lambda(\lambda-1)\mu}{x(x-1)(x-\lambda)} \\ &= \frac{c_0}{x} + \frac{c_1}{x-1} + \frac{-H}{x-t} + \frac{\mu}{x-\lambda}, \end{aligned}$$

where

$$\begin{aligned} c_0 &= -\kappa - (t-1)H + (\lambda - 1)\mu, \\ c_1 &= \kappa + tH - \lambda\mu. \end{aligned}$$

$$\tilde{L}_{\mathrm{VI}}: \quad \frac{\mathrm{d}^2 y}{\mathrm{d}x^2} + p_1 \frac{\mathrm{d}y}{\mathrm{d}x} + p_2 y = 0,\ \bar{\kappa} = \rho(\rho + \kappa_\infty - 1)\ (= \rho\bar{\rho}),$$

$$\begin{aligned} p_1 &= \frac{1-\kappa_0}{x} + \frac{1-\kappa_1}{x-1} + \frac{1-\theta}{x-t} - \sum_{k=1,2} \frac{1}{x-\lambda_k}, \\ p_2 &= \frac{\bar{\kappa}}{x(x-1)} - \frac{t(t-1)\tilde{H}}{x(x-1)(x-t)} + \sum_{k=1,2} \frac{\lambda_k(\lambda_k - 1)\mu_k}{x(x-1)(x-\lambda_k)} \\ &= \frac{\tilde{c}_0}{x} + \frac{\tilde{c}_1}{x-1} + \frac{-\tilde{H}}{x-t} + \sum_{k=1,2} \frac{\mu_k}{x-\lambda_k}, \end{aligned}$$

where

$$\begin{aligned} \tilde{c}_0 &= -\bar{\kappa} - (t-1)\tilde{H} + \sum_{k=1,2} (\lambda_k - 1)\mu_k, \\ \tilde{c}_1 &= \bar{\kappa} + t\tilde{H} - \sum_{k=1,2} \lambda_k \mu_k. \end{aligned}$$

L_{VI} and $\tilde{L}_{\mathrm{VI}}$ have the same monodromy if and only if a pair of rational functions satisfying $z = s_1 y' + s_2 y$ exists. Then, s_1, s_2 satisfies $F_1 = F_2 = 0$, where

$$\begin{aligned} F_1 &:= s_1'' + (q_1 - 2p_1)s_1' + 2s_2' + (p_1^2 - p_1 q_1 + q_2 - p_2 - p_1')s_1 + (q_1 - p_1)s_2, \\ F_2 &:= s_2'' - 2p_2 s_1' + q_1 s_2' + (p_1 p_2 - q_1 p_2 - p_2')s_1 + (q_2 - p_2)s_2. \end{aligned}$$

Rough forms of s_1, s_2. First, we determine the rough forms of the rational functions s_1, s_2.

At a regular singularity $x = \xi(\neq \infty)$, with the local parameter $u := x - \xi$:

- solutions of $\tilde{L}_{\mathrm{VI}}$ are $y_1 = \varphi_1$, $y_2 = u^v \varphi_2$;
- solutions of L_{VI} are $z_1 = a\psi_1 + bu^w \psi_2$, $z_2 = c\psi_1 + du^w \psi_2$, where $\varphi_1, \varphi_2, \psi_1, \psi_2$: regular at $u = 0$, $a, b, c, d \in \mathbb{C}$. $' := \mathrm{d}/\mathrm{d}x$.

$$\begin{aligned} s_1 &= \begin{vmatrix} z_1 & y_1 \\ z_2 & y_2 \end{vmatrix} \Big/ \begin{vmatrix} y_1' & y_1 \\ y_2' & y_2 \end{vmatrix} \\ &= [au\psi_1\varphi_2 + bu^{w+1}\psi_2\varphi_2 - cu^{-v+1}\psi_1\varphi_1 - du^{w-v+1}\psi_2\varphi_1]/f_1, \\ s_2 &= \begin{vmatrix} y_1' & z_1 \\ y_2' & z_2 \end{vmatrix} \Big/ \begin{vmatrix} y_1' & y_1 \\ y_2' & y_2 \end{vmatrix} \\ &= \big[a\psi_1(v\varphi_2 + u\varphi_2') + bu^w\psi_2(v\varphi_2 + u\varphi_2') \\ &\quad - cu^{-v+1}\psi_1\varphi_1' - du^{w-v+1}\psi_2\varphi_1'\big]/f_2, \end{aligned}$$

where f_1, f_2: regular and not vanish at $u = 0$.

At regular singularity $x = \infty$, with the local parameter $u := 1/x$:

- solutions of $\tilde{L}_{\mathrm{VI}}$ are $y_1 = u^\rho \varphi_1$, $y_2 = u^{\rho+\kappa_\infty - 1}\varphi_2$;
- solutions of L_{VI} are $z_1 = au^\rho\psi_1 + bu^{\rho+\kappa_\infty}\psi_2$, $z_2 = cu^\rho\psi_1 + du^{\rho+\kappa_\infty}\psi_2$, where $\varphi_1, \varphi_2, \psi_1, \psi_2$: regular at $u = 0$, $a, b, c, d \in \mathbb{C}$. $\dot{} := \mathrm{d}/\mathrm{d}u$.

$$\begin{aligned} s_1 &= \begin{vmatrix} z_1 & y_1 \\ z_2 & y_2 \end{vmatrix} \Big/ \begin{vmatrix} y_1' & y_1 \\ y_2' & y_2 \end{vmatrix} \\ &= [au^{-1}\psi_1\varphi_2 + bu^{\kappa_\infty - 1}\psi_2\varphi_2 - cu^{-\kappa_\infty}\psi_1\varphi_1 - d\psi_2\varphi_1]/f_1, \\ s_2 &= \begin{vmatrix} y_1' & z_1 \\ y_2' & z_2 \end{vmatrix} \Big/ \begin{vmatrix} y_1' & y_1 \\ y_2' & y_2 \end{vmatrix} \\ &= \big[a\psi_1\{(\rho + \kappa_\infty - 1)\varphi_2 + u\dot{\varphi}_2\} + bu^{\kappa_\infty}\psi_2\{(\rho + \kappa_\infty - 1) + u\dot{\varphi}_2\} \\ &\quad - cu^{-\kappa_\infty + 1}\psi_1(\rho\varphi_1 + u\dot{\varphi}_1) - du\psi_2(\rho\varphi_1 + u\dot{\varphi}_1)\big]/f_2, \end{aligned}$$

where f_1, f_2: regular and not vanish at $u = 0$.

By the above, the caracteristic exponents of solutions and properties of the rational functions s_1, s_2 reads

reg.sing.	char.exp.	$s_1(x)$	$s_2(x)$
$\xi = 0, 1, t$	$v - w \notin \mathbb{Z}$	simple zero	regular
$\xi = \lambda_k (k = 1, 2)$	$v = 2, w = 1$	simple pole	simple pole
$\xi = \lambda$	$v = 1, w = 2$	regular	regular
$\xi = \infty$	$\rho, \rho + \kappa_\infty$	siple pole	regular

So, the rough forms of s_1, s_2 are

$$s_1(x) = A\frac{x(x-1)(x-t)}{(x-\lambda_1)(x-\lambda_2)} = A\left\{x + d_0 + \frac{d_1}{x-\lambda_1} + \frac{d_2}{x-\lambda_2}\right\},$$

$$s_2(x) = B_0 + \frac{B_1}{x-\lambda_1} + \frac{B_2}{x-\lambda_2},$$

where $A, B_0, B_1, B_2 \in \mathbb{C}$,

$$\begin{aligned} d_0 &= \lambda_1 + \lambda_2 - t - 1, \\ d_1 &= \lambda_1(\lambda_1 - 1)(\lambda_1 - t)/(\lambda_1 - \lambda_2), \\ d_2 &= \lambda_2(\lambda_2 - 1)(\lambda_2 - t)/(\lambda_2 - \lambda_1). \end{aligned}$$

Determination of s_1, s_2. Substituting the rough forms of s_1, s_2 into two equations $F_1 = 0$, $F_2 = 0$, we determine $A, B_0, B_1, B_2 \in \mathbb{C}$. Now we list the data used in calculation for convenience.

$$\begin{aligned} F_1 &:= s_1'' + (q_1 - 2p_1)s_1' + 2s_2' + (p_1^2 - p_1q_1 + q_2 - p_2 - p_1')s_1 + (q_1 - p_1)s_2, \\ F_2 &:= s_2'' - 2p_2s_1' + q_1s_2' + (p_1p_2 - q_1p_2 - p_2')s_1 + (q_2 - p_2)s_2. \end{aligned}$$

$$\begin{aligned} s_1 &= A\{x + d_0 + d_1/(x-\lambda_1) + d_2/(x-\lambda_2)\} \\ &= Ax(x-1)(x-t)/(x-\lambda_1)(x-\lambda_2), \\ s_1' &= A\{1 - d_1/(x-\lambda_1)^2 - d_2/(x-\lambda_2)^2\}, \\ s_1'' &= A\{2d_1/(x-\lambda_1)^3 + 2d_2/(x-\lambda_2)^3\}, \\ s_2 &= B_0 + B_1/(x-\lambda_1) + B_2/(x-\lambda_2), \\ s_2' &= -B_1/(x-\lambda_1)^2 - B_2/(x-\lambda_2)^2, \\ s_2'' &= 2B_1/(x-\lambda_1)^3 + 2B_2/(x-\lambda_2)^3, \end{aligned}$$

where $A, B_0, B_1, B_2 \in \mathbb{C}$.

$$\begin{aligned} F_1 =& s_1'' + (q_1 - 2p_1)s_1' + 2s_2' + (p_1^2 - p_1q_1 + q_2 - p_2 - p_1')s_1 + (q_1 - p_1)s_2. \\ F_2 =& s_2'' - 2p_2s_1' + q_1s_2' + (p_1p_2 - q_1p_2 - p_2')s_1 + (q_2 - p_2)s_2. \end{aligned}$$

In each term of F_1, F_2, the order of zero at regular singularities are as follows:

	$x=0$	$x=1$	$x=t$	$x=\lambda$	$x=\lambda_k$	$x=\infty$
s_1''	0	0	0	0	−3	3
$(q_1-2p_1)s_1'$	−1	−1	−1	−1	−3	1
$2s_2'$	0	0	0	0	−2	2
$p_1(p_1-q_1)s_1$	0	0	0	−1	−3	1
$(q_2-p_2)s_1$	0	0	0	−1	−2	1
$-p_1's_1$	−1	−1	−1	0	−3	1
$(q_1-p_1)s_2$	0	0	0	−1	−2	1

	$x=0$	$x=1$	$x=t$	$x=\lambda$	$x=\lambda_k$	$x=\infty$
s_2''	0	0	0	0	−3	3
$-2p_2s_1'$	−1	−1	−1	0	−3	2
q_1s_2'	−1	−1	−1	−1	−2	3
$p_2(p_1-q_1)s_1$	0	0	0	−1	−3	2
$-p_2's_1$	−1	−1	−1	0	−3	2
$(q_2-p_2)s_2$	−1	−1	−1	−1	−2	2

Put $x=1/u$. Investigate $u\to 0$.

$$\begin{aligned}
s_1'' &= O(u^3).\\
(q_1-2p_1)s_1' &= A(\kappa_0+\kappa_1+\theta)u+O(u^2).\\
2s_2' &= O(u^2).\\
p_1(p_1-q_1)s_1 &= A(\kappa_0+\kappa_1+\theta-1)u+O(u^2).\\
(q_2-p_2)s_1 &= A\rho u+O(u^2).\\
-p_1's_1 &= -A(\kappa_0+\kappa_1+\theta-1)u+O(u^2).\\
(q_1-p_1)s_2 &= B_0u+O(u^2).
\end{aligned}$$

Therefore,

$$F_1=\{B_0+A(\kappa_0+\kappa_1+\theta+\rho)\}u+O(u^2)=(B_0-A\bar{\rho})u+O(u^2).$$

$$\begin{aligned}
s_2'' &= O(u^3).\\
-2p_2s_1' &= -2A\bar{\kappa}u^2+O(u^3).\\
q_1s_2' &= O(u^3).\\
p_2(p_1-q_1)s_1 &= -A\bar{\kappa}u^2+O(u^3).\\
-p_2's_1 &= 2A\bar{\kappa}u^2+O(u^3).\\
(q_2-p_2)s_2 &= B_0\rho u^2+O(u^3).
\end{aligned}$$

Therefore,

$$F_2 = \{B_0\rho - A\bar{\kappa}\}u^2 + O(u^3) = (B_0 - A\bar{\rho})\rho u^2 + O(u^3).$$

And then

$$-\operatorname*{Res}_{x=\infty} F_2 = 0, \quad -\operatorname*{Res}_{x=\infty} F_1 = -\operatorname*{Res}_{x=\infty} xF_2 = B_0 - A\bar{\rho}.$$

In the above, we used:

- the definitions of $\kappa, \bar{\kappa}$ gives us $\kappa - \bar{\kappa} = \rho(\bar{\rho}+1) - \rho\bar{\rho} = \rho$;
- Fuchsian relation: $\kappa_0 + \kappa_1 + \theta + \kappa_\infty + 2\rho = 1$.

Here, we determine the constants A, B_0, B_1, B_2, and the form of the rational functions s_1, s_2:

- $\mathrm{Res}_{x=\infty} F_1 = 0$ and $\mathrm{Res}_{x=\infty} xF_2 = 0$ imply $B_0 = A\bar{\rho}$. Without loss of generality, we can put $B_0 = \bar{\rho} = \rho + \kappa_\infty - 1$;
- residue of F_1 at $x = \xi \in \{0, 1, t\}$: $\mathrm{Res}_{x=\xi} F_1 = 0$;
- coefficient of $1/(x-\lambda_k)^3$ in F_1 at $x = \lambda_k$: $\mathrm{Res}_{x=\lambda_k}(x-\lambda_k)^2 F_1 = 0$;
- for coefficient of $(x-\lambda_k)^{-2}$ at $x = \lambda_k$ in F_1 and coefficient of $(x-\lambda_k)^{-3}$ at $x = \lambda_k$ in F_2 vanishes, we have $B_k = -d_k\mu_k$ $(k = 1, 2)$.

Now, constants A, B_0, B_1, B_2 are determined, and F_1, F_2 is of the form:

$$F_1 = \frac{A_\lambda}{x-\lambda} + \sum_{k=1,2} \frac{A_{\lambda_k}}{x-\lambda_k},$$

$$F_2 = \frac{C_0}{x} + \frac{C_1}{x-1} + \frac{C_t}{x-t} + \frac{C_\lambda}{x-\lambda} + \sum_{k=1,2}\left\{\frac{C_{\lambda_k}}{x-\lambda_k} + \frac{D_{\lambda_k}}{(x-\lambda_k)^2}\right\}.$$

If the pair of rational functions s_1, s_2 satisfying the relations $F_1 = 0$, $F_2 = 0$ exists, then the 11 relations $A_\xi = 0$ (3 rel.), $C_\xi = 0$ (6 rel.), $D_\xi = 0$ (2 rel.) must be satisfied. Inversely, if these are satisfied, with A, B_0, B_1, B_2 determined the above, the pair of rational functions s_1, s_2 satisfying the relations $F_1 = 0$, $F_2 = 0$ exists.

We want to express the six parameters $\lambda_k, \mu_k, \tilde{H}, H$ by λ, μ, t. However, we have 11 relations A_ξ, C_ξ, $D_\xi = 0$. In order that only six relations are independent, five others are reduced to the six. We must discuss which of the five are reduced and

- how the sum of residues of F_1, F_2, xF_2 vanishes,
- how $C_{\lambda_k} = 0$ is dependent to another. If we see the explicit forms of $A_{\lambda_k}, C_{\lambda_k}, D_{\lambda_k}$, we find that $C_{\lambda_k} = 0$ is reduced to $A_{\lambda_k} = 0$ and $D_{\lambda_k} = 0$.

So, five are reduced, and we have only six relations $A_{\lambda_1}, A_{\lambda_2}, C_t, C_\lambda, D_{\lambda_1}, D_{\lambda_2} = 0$. Therefore, if A^{21} has two simple poles, L_{VI} and $\tilde{L}_{\mathrm{VI}}$ have the same monodromy.

The explicit forms of A_ξ, C_ξ, D_ξ are as follows:

$$A_\lambda = -s_1'(\lambda) + p_1(\lambda)s_1(\lambda) + \mu s_1(\lambda) - s_2(\lambda).$$

$$\begin{aligned}
A_{\lambda_k} &= 2\Big(s_1 - \frac{d_k}{x-\lambda_k}\Big)'\Big|_{x=\lambda_k} - \Big(p_1 - \frac{-1}{x-\lambda_k}\Big)'\Big|_{x=\lambda_k} \cdot d_k \\
&\quad - \Big(s_1 - \frac{d_k}{x-\lambda_k}\Big)\Big|_{x=\lambda_k} \cdot \Big\{\Big(p_1 - q_1 - \frac{-1}{x-\lambda_k}\Big)\Big|_{x=\lambda_k} + \Big(p_1 - \frac{-1}{x-\lambda_k}\Big)\Big|_{x=\lambda_k}\Big\} \\
&\quad + \Big(p_1 - q_1 - \frac{-1}{x-\lambda_k}\Big)\Big|_{x=\lambda_k} \cdot \Big(p_1 - \frac{-1}{x-\lambda_k}\Big)\Big|_{x=\lambda_k} \cdot d_k \\
&\quad - \Big(s_1 - \frac{d_k}{x-\lambda_k}\Big)\Big|_{x=\lambda_k} \cdot \mu_k - \Big(q_1 - p_1 - \frac{1}{x-\lambda}\Big)\Big|_{x=\lambda_k} \cdot d_k\mu_k \\
&\quad + \Big(s_2 - \frac{-d_k\mu_k}{x-\lambda_k}\Big)\Big|_{x=\lambda_k} + \Big(q_2 - p_2 - \frac{-\mu_k}{x-\lambda_k}\Big)\Big|_{x=\lambda_k} \cdot d_k.
\end{aligned}$$

$$\begin{aligned}
C_0 &= -2s_1'(0)\tilde{c}_0 + (1-\kappa_0)s_2'(0) + \frac{t\tilde{c}_0}{\lambda_1\lambda_2} + s_2(0)(c_0 - \tilde{c}_0). \\
C_1 &= -2s_1'(1)\tilde{c}_1 + (1-\kappa_1)s_2'(1) + \frac{(1-t)\tilde{c}_1}{(\lambda_1-1)(\lambda_2-1)} + s_2(1)(c_1 - \tilde{c}_1). \\
C_t &= 2s_1'(t)\tilde{H} + (1-\theta)s_2'(t) - \frac{t(t-1)\tilde{H}}{(\lambda_1-t)(\lambda_2-t)} - s_2(t)(H - \tilde{H}). \\
C_\lambda &= -s_2'(\lambda) + s_1(\lambda)p_2(\lambda) + \mu s_2(\lambda).
\end{aligned}$$

$$\begin{aligned}
C_{\lambda_k} &= \Big(p_1 - \frac{-1}{x-\lambda_k}\Big)'\Big|_{x=\lambda_k} \cdot d_k\mu_k - 2\Big(s_1' - \frac{-d_k}{(x-\lambda_k)^2}\Big)\Big|_{x=\lambda_k} \cdot \mu_k \\
&\quad + \Big(s_1 - \frac{d_k}{x-\lambda_k}\Big)\Big|_{x=\lambda_k} \\
&\quad \cdot \Big\{\Big(p_1 - q_1 - \frac{-1}{x-\lambda_k}\Big)\Big|_{x=\lambda_k} \cdot \mu_k - \Big(p_2 - \frac{\mu_k}{x-\lambda_k}\Big)\Big|_{x=\lambda_k}\Big\} \\
&\quad + \Big(p_1 - q_1 - \frac{-1}{x-\lambda_k}\Big)\Big|_{x=\lambda_k} \cdot \Big(p_2 - \frac{\mu_k}{x-\lambda_k}\Big)\Big|_{x=\lambda_k} \cdot d_k \\
&\quad - \Big(s_2 - \frac{-d_k\mu_k}{x-\lambda_k}\Big)\Big|_{x=\lambda_k} \cdot \mu_k - \Big(q_2 - p_2 - \frac{-\mu_k}{x-\lambda_k}\Big)\Big|_{x=\lambda_k} \cdot d_k\mu_k.
\end{aligned}$$

$$D_{\lambda_k} = d_k\Big\{\mu_k^2 + \Big(p_1 - \frac{-1}{x-\lambda_k}\Big)\Big|_{x=\lambda_k} \cdot \mu_k + \Big(p_2 - \frac{\mu_k}{x-\lambda_k}\Big)\Big|_{x=\lambda_k}\Big\}.$$

So, for $k = 1, 2$, we have:

$$A_{\lambda_k}\mu_k + C_{\lambda_k} = \Big\{\Big(p_1 - q_1 - \frac{-1}{x-\lambda_k}\Big)\Big|_{x=\lambda_k} \cdot d_k - \Big(s_1 - \frac{d_k}{x-\lambda_k}\Big)\Big|_{x=\lambda_k}\Big\}\frac{D_{\lambda_k}}{d_k}.$$

Remark 17.8. If we try to construct the Schlesinger transformation $L_{\mathrm{VI}} \to \bar{L}_{\mathrm{VI}}$ in a similar way, F_1, F_2 is of the form $F_1 = A_{\bar{\lambda}}/(x-\bar{\lambda})$ and $F_2 = C_0/x + C_1/(x-1) + C_t/(x-t) + C_{\bar{\lambda}}/(x-\bar{\lambda}) + C_\lambda/(x-\lambda) + D_\lambda/(x-\lambda)^2$. Note that the sum of residues of F_2 vanishes, and $C_\lambda = 0$ is subject to $D_\lambda = 0$, i.e., $C_\lambda = D_\lambda\{(p_1 - q_1 - \frac{-1}{u})|_{u=0} + d_1/d_2\}$. So, two relations of seven are reduced to others.

Bibliography

[1] K. Okamoto, Isomonodromic deformation and Painlevé equations, and the Garnier system, *J. Fac. Sci. Univ. Tokyo Sect. IA Math.*, **33**: 575–618, 1986.

[2] T. Kimura, On the isomonodromic deformation for linear ordinary differential equations of the second order, I, *Proc. Japan Acad. Ser. A* **57**: 285–290, 1981; ibid., **58**: 294–297, 1982.

[3] H. Ishikawa, *On the monodromy preserving deformation of a Fuchsian equation with two apparent singularities*, Masters Thesis, The Univ. of Tokyo, 1999 (in Japanese).

[4] M. Jimbo and T. Miwa, Monodromy preserving deformation of linear ordinary differential equations with rational coefficients II, *Physica*, **2D**(3): 407–448, 1981.

[5] G. Mahoux, Introduction to the theory of isomonodromic deformations of linear ordinary differential equations with rational coefficients, *The Painlevé property*, ed. R. Conte, CRM Ser. Math. Phys., pp. 35–76, Springer, New York, 1999.

Author Information

Yoshikatsu Sasaki, Graduate School of Science, Hiroshima University, 1-3-1 Kagamiyama, Higashi-Hiroshima, 739-8526, Japan.
E-mail: `sasakiyo@hiroshima-u.ac.jp`

Chapter 18

Expansions for Solutions of the Schlesinger Equation at a Singular Point

Ilya V. Vyugin

Abstract. A local behavior of solutions of the Schlesinger equation is studied. We obtain expansions for these solutions which converge in the neighborhood of some singular point. As a corollary, a similar result for the sixth Painlevé equation was obtained. In our analysis, we use the isomonodromic approach to solve this problem.

Keywords. Isomonodromic Deformation, Fuchsian System, Schlesinger Equation, Bundle with Connection.

2010 Mathematics Subject Classification. 34M56, 34M55, 34M03.

18.1 Introduction

We study a local behavior of solutions of the Schlesinger equation. We present solutions of this equation in the form of power series or logarithmic-power series. These series converge in the neighborhood of some singular point. As a corollary we obtain a similar result for the description of the behavior of solutions of the sixth Painlevé equation in some sectorial neighborhood. We use the isomonodromic approach to solve this problem.

Let us consider the following system of analytical partial differential equations:

$$dB_i = -\sum_{j=1, j\neq i}^{n} \frac{[B_i, B_j]}{a_i - a_j} d(a_i - a_j), \qquad i = 1, \ldots, n, \tag{18.1}$$

where B_i $(i = 1, \ldots, n)$ – are analytical $p\times p$-matrix functions of the variable $a = (a_1, \ldots, a_n)$, $[B_i, B_j]$ denotes the commutator of matrices B_i and B_j. The matrix-functions $B_i(a)$ are defined and meromorphic (see B. Malgrange [1], R. Gontsov and I. Vyugin [4]) on the space

$$\Big\{a | a = (a_1, \ldots, a_n) \in \overline{\mathbb{C}}^n \setminus \bigcup_{i,j} A_{ij}\Big\}, \qquad A_{ij} = \{a \mid a_i = a_j\}.$$

The work is supported by Russian Federation presidential grants NSh-5998.2012.1 and MK-4270.2011.1 and by RFBR, Grant 11-01-00339.

This system is called the *Schlesinger equation* (for more on this, see A. A. Bolibruch [7]). Divisor of the Schlesinger equation is the set

$$\Omega = \bigcup_{i,j} A_{ij}.$$

We will describe a local form of solutions of Schlesinger equation (18.1) in a neighborhood of the point $a^0 = (a_1^0, \ldots, a_n^0)$, which belongs to the following singular set

$$a^0 \in \Omega' = \Omega \setminus \Big(\bigcup_{i,j,k} A_{ijk} \Big), \qquad A_{ijk} = \{a \mid a_i = a_j = a_k\}.$$

We obtain the local expansions of the solutions of the system (18.1) in the form of power and logarithmic-power series of $(a_s - a_r)$ (if $a^0 \in A_{sr}$), which converges in the neighborhood of the point $z = a^0$ (for the first version of these result, see [6]). These series have terms of complex degrees.

Theorem 18.1. *Any solution of a two-dimensional Schlesinger equation (18.1) can be represented in the neighborhood of a point $a^0 = (a_1^0, \ldots, a_n^0) \in \Omega'$, where $a_r^0 = a_s^0$, $r \neq s$, in one of two following forms:*

- $b_{kl}^i(a) = F_1^{kli}(a) + (a_s - a_r)^{\varphi} F_2^{kli}(a) + (a_s - a_r)^{-\varphi} F_3^{kli}(a)$, $\varphi \in \mathbb{C}$ *in the general case;*
- $b_{kl}^i(a) = F_1^{kli}(a) + F_2^{kli}(a) \ln(a_s - a_r) + F_3^{kli}(a) \ln^2(a_s - a_r)$ *in the degenerate case,*

where $F_1^{kli}(a)$, $F_2^{kli}(a)$, $F_3^{kli}(a)$ are meromorphic (holomorphic in the generic case) functions, $i = 1, \ldots, n$, and $k, l \in \{1, 2\}$.

The notions of "general case" and "nongeneral case" are explained below. Notice that the measure of the systems of nongeneral case is equal to zero.

Now consider the case $n = 4$, $p = 2$, which is equivalent to the case of the sixth Painlevé equation (18.2). Without loss of generality, let us fix three variables $a_1 = 0$, $a_2 = 1$, $a_3 = \infty$ and denote a_4 by t. We obtain the system of ordinary differential equations with variable t and unknown matrix-functions

$$B_i(t) = \begin{pmatrix} b_{11}^i(t) & b_{12}^i(t) \\ b_{21}^i(t) & b_{22}^i(t) \end{pmatrix}, \qquad i = 0, t, 1, \infty.$$

With the above restrictions the following corollary holds.

Corollary 18.2. *Any solution of the Schlesinger equation under the above constraints can be represented in the neighborhood of $t = 0$ in one of two forms:*

(i) $b_{kl}^i(t) = F_1^{kli}(t) + t^{\varphi} F_2^{kli}(t) + t^{-\varphi} F_3^{kli}(t)$, $\varphi \in \mathbb{C}$ *in the general case;*

(ii) $b_{kl}^i(t) = F_1^{kli}(t) + F_2^{kli}(t) \ln t + F_3^{kli}(t) \ln^2 t$ *in the degenerate case, where* $F_1^{kli}(t)$, $F_2^{kli}(t)$, $F_3^{kli}(t)$ *are meromorphic in* $t = 0$ *functions,* $i = 0, t, 1, \infty$, *and* $k, l \in \{1, 2\}$.

Note that in the well-known sixth Painlevé equation

$$\frac{d^2w}{dt^2} = \frac{1}{2}\left(\frac{1}{w} + \frac{1}{w-1} + \frac{1}{w-t}\right)\left(\frac{dw}{dt}\right)^2 - \left(\frac{1}{t} + \frac{1}{t-1} + \frac{1}{w-t}\right)\frac{dw}{dt}$$
$$+ \frac{w(w-1)(w-t)}{t^2(t-1)^2}\left(\alpha + \beta\frac{t}{w^2} + \gamma\frac{t-1}{(w-1)^2} + \delta\frac{t(t-1)}{(w-t)^2}\right), \qquad (18.2)$$

α, β, γ, $\delta \in \mathbb{C}$ is equivalent to the system (18.1), where

$$w(t) = \frac{tb_{12}^0}{(t+1)b_{12}^0 + tb_{12}^1 + b_{12}^t}. \qquad (18.3)$$

Corollary 18.2 and (18.3) give the power expansions for solutions of the sixth Painlevé equation. Different asymptotic forms and expansions for the sixth Painlevé equation were obtained in D. Guzzetti [2], A. Bruno and I. Goryuchkina [5], M. Mazzocco [3], and others.

For the sixth Painlevé equation, we have an analogue of Corollary 18.2.

Corollary 18.3. *Any solution* $w(t)$ *of the sixth Painlevé equation* (18.2) *in the intersection of the given sector for* t *sufficiently close to singular point* $t = 0, 1, \infty$ *can be represented as a converged power series or as a converged logarithmic-power series:*

- *if* $G_1 G_\infty$ *is digonalizable, then* $w(t) = S(t, t^\varphi, t^{-\varphi})$, *where*

$$\varphi = \varphi(\alpha, \beta, \gamma, \delta, t_0, w(t_0), w'(t_0))$$

  can be found approximately;
- *if* $G_1 G_\infty$ *is a Jordan block, then* $w(t) = S(t, \ln t, \ln^{-1} t)$.

Using the expressions for $b_{12}^i(t)$ we obtain the following expressions for $w(t)$:

$$w(t) = \frac{f_1(t) + t^\varphi f_2(t) + t^{-\varphi} f_3(t)}{g_1(t) + t^\varphi g_2(t) + t^{-\varphi} g_3(t)} \qquad (18.4)$$

and

$$w(t) = \frac{f_1(t) + f_2(t)\ln t + f_3(t)\ln^2 t}{g_1(t) + g_2(t)\ln t + g_3(t)\ln^2 t}, \qquad (18.5)$$

where f_j, g_j, $j = 1, 2, 3$ are meromorphic functions. The denominators of these ratios are not identically zero. We can express these ratios as power series with powers of t, $t^\varphi, t^{-\varphi}, \ln t, \ln^{-1} t$. These power series will be converged in the sectorial neighborhood with any angle ψ and with radius r, depending on ψ, $r = r(\psi)$. This sectorial neighborhood is described by the condition: the denominator of the ratio does not vanish.

18.2 Schlesinger Equation and Isomonodromic Deformations

In this section we give a description of the Schlesinger equation (18.1) as an isomonodromy condition for a family of Fuchsian systems. Let us consider a Fuchsian system

$$\frac{dy}{dz} = \left(\sum_{i=1}^{n} \frac{B_i^0}{z - a_i^0}\right) y, \qquad B_i^0 \in \mathrm{Mat}_{p\times p}(\mathbb{C}), \quad y(z) \in \mathbb{C}^p. \tag{18.6}$$

The family of such systems

$$\frac{dy}{dz} = \left(\sum_{i=1}^{n} \frac{B_i(a)}{z - a_i}\right) y \tag{18.7}$$

is called *isomonodromic* if the following conditions hold:

- $B_i(a)$ are continuous matrix-functions of $a = (a_1, \ldots, a_n)$;
- the Fuchsian system (18.7) with any fixed a has fixed monodromy representation

$$\chi : \pi_1(\overline{\mathbb{C}} \setminus \{a_1, \ldots, a_n\}, z_0) \longrightarrow GL(p, \mathbb{C}). \tag{18.8}$$

The *Schlesinger isomonodromic family* is a family defined by equation (18.1). An isomonodromic fundamental matrix $Y(z, a)$ of the Schlesinger isomonodromic family (18.7) satisfies the condition

$$Y(\infty, a) \equiv Y(\infty, a_0).$$

The initial data of such a family are the coefficients $B_i(a^0) = B_i^0, i = 1, \ldots, n$ of system (18.6). It is known that the solutions of theSchlesinger equation are meromorphic functions on the space $a \in \overline{\mathbb{C}}^n \setminus \Omega$.

Let us consider the Painlevé VI case ($n = 4, p = 2, a_1 = 0, a_2 = 1, a_3 = \infty$, $a_4 = t$). Usually the following family

$$\frac{dy}{dz} = \left(\frac{B_0(t)}{z} + \frac{B_t(t)}{z - t} + \frac{B_1(t)}{z - 1}\right) y \tag{18.9}$$

is considered, where

$$\operatorname{tr} B_0 = \operatorname{tr} B_t = \operatorname{tr} B_1 = \operatorname{tr} B_\infty = 0, \qquad B_\infty = -(B_0 + B_t + B_1),$$

and the matrices $B_0, B_t, B_1, B_\infty = \mathrm{diag}(\delta, -\delta)$ are diagonalizable.

Formula (18.3) gives a solution of the sixth Painlevé equation (18.2) with the constants

$$\alpha = \frac{(2\lambda_\infty - 1)^2}{2}, \quad \beta = -2\lambda_0^2, \quad \gamma = 2\lambda_1^2, \quad \delta = \frac{1}{2} - 2\lambda_t^2,$$

where $\lambda_0, \lambda_t, \lambda_1, \lambda_\infty$ are eigenvalues of matrices B_0, B_t, B_1, B_∞.

18.3 Sketch of the Proof

We begin by we studying the simplest isomonodromic family. Consider the family of Fuchsian systems

$$\frac{dy}{dz} = \left(\frac{B_0}{z} + \frac{B_t}{z-t}\right) y, \qquad B_0, B_t \equiv \text{const.} \tag{18.10}$$

It is easy to see that this is an isomonodromic family. The systems of this family are mutually equivalent by a linear mapping of z. This family is non-Schlesinger if $B_\infty = -B_0 - B_t \neq 0$, but it can be transformed to a Schlesinger family by the gauge transformation $\tilde{y} = t^{B_\infty} y$. The transformed family has the form

$$\frac{dy}{dz} = \left(\frac{B_0'(t)}{z} + \frac{B_t'(t)}{z-t}\right) y, \qquad B_i' = t^{B_\infty} B_i t^{-B_\infty}, \quad i = 0, t. \tag{18.11}$$

We call this family the *canonical normalized family*.

Let us write the coefficients B_0', B_t' explicitly. There are two cases: in the first case when the matrix B_∞ is diagonalizable, and in the second case when B_∞ is a Jordan block:

- first case: $B_\infty = \begin{pmatrix} \lambda_1 & 0 \\ 0 & \lambda_2 \end{pmatrix}$ and $B_i' = \begin{pmatrix} b_{11}^i & b_{12}^i t^{\lambda_1-\lambda_2} \\ b_{21}^i t^{\lambda_2-\lambda_1} & b_{22}^i \end{pmatrix}$;
- second case: $B_\infty = \begin{pmatrix} \lambda & 1 \\ 0 & \lambda \end{pmatrix}$ and

$$B_i' = \begin{pmatrix} b_{11}^i + b_{21}^i \ln t & b_{12}^i + (b_{22}^i - b_{11}^i)\ln t - b_{21}^i \ln^2 t \\ b_{21}^i & b_{22}^i - b_{21}^i \ln t. \end{pmatrix}.$$

Now we study the limit of the family (18.11) as $t \to 0$. We would like to find a limit Fuchsian system. For the existence of this limit, we impose the following condition on the real part of $\lambda_1 - \lambda_2$

$$|\mathrm{Re}(\lambda_1 - \lambda_2)| < 1. \tag{18.12}$$

The condition (18.12) implies the equation (see [7])

$$\begin{aligned} \frac{B_0'(t)}{z} + \frac{B_t'(t)}{z-t} &= t^{B_\infty}\left(\frac{B_0}{z} + \frac{B_t}{z-t}\right) t^{-B_\infty} \\ &= t^{B_\infty}\left(\frac{B_0 + B_t}{z} + O(t)\right) t^{-B_\infty} \\ &= \frac{B_\infty}{z} + o(1). \end{aligned}$$

The limit system as $t \to 0$ of the family (18.11) under the condition (18.12) is

$$\frac{dy}{dz} = \frac{B_\infty}{z} y. \tag{18.13}$$

The proof is similar to that given by A.A. Bolibruch [7].

The Riemann–Hilbert theory gives that for almost all monodromy representations (18.8) with generators G'_0, G'_1, G'_∞ there exists a Fuchsian system (18.10) with this monodromy data and given asymptotics (see [8]). In all other cases we can construct such a system with one regular singular point (see [9]). These cases were called in Theorem 18.1 the "general case" and the "nongeneral case".

Now let us consider the nongeneral case. It is the case, when there isn't a system (18.10) having the given monodromy (18.8), and exponents in the points $z = 0, t, \infty$. The results of I. Vyugin and R. Gontsov [9] states that there exist the regular system

$$\frac{dy}{dz} = \left(\frac{B_{-r}}{z^{r+1}} + \ldots + \frac{B_0}{z} + \frac{B_t}{z - t_0}\right) y,$$

having the given monodromy (18.8), and exponents β_t^1, β_t^2, β_∞^1, β_∞^2 in points $z = t, \infty$, and $r < 3\max(\beta_t^1 - \beta_t^2, \beta_\infty^1 - \beta_\infty^2)$. Note that the family

$$\frac{dy}{dz} = \left(t^r \frac{B'_{-r}}{z^{r+1}} + t^{r-1}\frac{B'_{-(r-1)}}{z^r} + \ldots + \frac{B'_0}{z} + \frac{B'_t}{z - t}\right) y, \tag{18.14}$$

where $B'_i = t^{B_\infty} B_i t^{-B_\infty}$, $B_\infty = -(B_0 + B_t)$, is an isomonodromic. The limit (18.14) at $t \to 0$ is (18.13).

We will use the family (18.11) for the proof of the Theorem 18.1.

Let us consider a family of holomorphic vector bundles with logarithmic connection having the following description

$$(F_t, \nabla_t) = (D_0, D_\infty, g^t_{0\infty}(z), \omega^t_0, \omega_\infty),$$

where D_0, D_∞ – circles with centers 0 and ∞, which has a nonempty intersection $K = D_0 \cap D_\infty$, $g^t_{0\infty}(z)$ is a holomorphic cocycle $g^t_{0\infty}(z) : K \to \mathrm{GL}(p, \mathbb{C})$ and ω^t_0, ω_∞ is a differential 1-forms of logarithmic connection ∇_t.

Define the pairs (F_t, ∇_t) by the following description:

(i) ω_∞ is a 1-form of coefficients of the initial system (18.9), when $t = t_0$, which has the monodromy representation (18.8) and generators G_0, G_t, G_1, G_∞;

(ii) ω^t_0 is a 1-form of coefficients of system (18.11), with monodromy (18.8) with generators G'_0, G'_t, G'_∞ ($G'_0 = G_0$, $G'_t = G_t$, $G'_\infty = G_1 G_\infty$);

(iii) cocycle $g^t_{0\infty}(z)$ is a ratio $g^t_{0\infty}(z) = Y^t_0(z) Y^{-1}_\infty(z)$, where $Y^t_0(z)$ and $Y_\infty(z)$ are fundamental matrices of the systems

$$dy = \omega^t_0 y, \qquad dy = \omega_\infty y,$$

normalized in $z = \infty$.

Proposition 18.4. *Assume that the family*

$$(F_t, \nabla_t) = \left(D_0, D_\infty, g^t_{0\infty}(z), \omega^t_0, \omega^t_\infty\right),$$

holomorphically depends on the valuable t. If the limit

$$(F_0, \nabla_0) = \left(D_0, D_\infty, g^0_{0\infty}(z), \omega^0_0, \omega^0_\infty\right), \quad t \to 0$$

exists and it is a trivial bundle with trivialization $V^0(z)$, $W^0(z)$, then bundles F_t are trivial bundles, for sufficiently small t, and their trivializations $V^t(z)$, $W^t(z)$ have the limits

$$\lim_{t\to 0} V^t(z) = V^0(z), \quad \lim_{t\to 0} W^t(z) = W^0(z),$$

which are uniform for $z \in D_0 \cap D_\infty$

The family (18.9) can be represented in the neighborhood D_0 as

$$\frac{dy}{dz} = \left(V(t,z)\omega^t_0 V^{-1}(t,z) + \frac{dV}{dz}V^{-1}\right)y, \quad V(t,z) = V_0(t) + V_1(t)z + \dots, \tag{18.15}$$

and in the neighborhood D_∞ as

$$\frac{dy}{dz} = \left(W(t,z)\omega_\infty W^{-1}(t,z) + \frac{dW}{dz}W^{-1}\right)y, \quad W(t,z) = W_0(t) + W_1(t)z + \dots. \tag{18.16}$$

We obtain that in the general case the matrix coefficients of the system (18.9) are

$$B_i = V_0(t)B'_i(t)V_0^{-1}(t), \qquad i = 0, t,$$

and

$$B_i = W_0(t)B^0_i(t)W_0^{-1}(t), \qquad i = 1, \infty.$$

From these formulas and from the form (18.11) we obtain Theorem 18.1 and Corollary 18.2 in the general case.

Bibliography

[1] B. Malgrange, Sur le déformations isomonodromiques. Singularités régulières. *Progr. Math.*, **37**, 401–426, 1983.

[2] D. Guzzetti, The logarithmic asymptotics of the sixth Painlevé equation. *J. Phys. A: Math. Theor.*, **41**: 205201 (46 p.), 2008.

[3] M. Mazzocco, Picard and Chazy Solutions of the Painlevé VI equation. *Math. Ann.*, **321**(1): 157–195, 2001.

[4] R. R. Gontsov, I. V. Vyugin, Apparent singularities of Fuchsian equations, and the Painleve VI equation and Garnier systems, arXiv:0905.1436.

[5] A. D. Bruno and I. V. Goryuchkina, Asymptotic expansions of solutions of the sixth Painlevé equation. *Trans. Moscow Math. Soc.*, **71**: 1–105, 2010 (= *Tr. MMO*, **71**: 6–118, 2010).

[6] I. V. Vyugin, Isomonodromic confluentions of sindular points and singulaities of Schlesinger equation. *Proc. of the conference ITIS-09.*, pp. 447–452 (in Russian).

[7] A. A. Bolibruch, On isomonodromic confluences of Fuchsian singularities, *Proc. of the Steklov Inst. of Math.*, **221**: 117–132, 1998.

[8] D. V. Anosov and A. A. Bolibruch, *The Riemann–Hilbert Problem,* Vieweg, Braunschweig/Wiesbaden, 1994.

[9] I. V. Vyugin and R. R. Gontsov, Additional parameters in inverse problems of monodromy. *Sbornik: Mathematics*, **197**(12): 1753–1773, 2006.

Author Information

Ilya V. Vyugin, Institute for Information Transmission Problems, Bol'shoy Karetny Per. 19, 127994, Moscow, Russia.
E-mail: vyugin@gmail.com

Part IV

Painlevé Property

Chapter 19

Painleve Analysis of Lotka–Volterra Equations

Pantelis A. Damianou

Abstract. We examine a class of Lotka–Volterra equations in three and four dimensions which satisfy the Kowalevski–Painlevé property. We restrict our attention to Lotka–Volterra systems defined by a skew symmetric matrix. We obtain a complete classification of such systems. The classification is obtained using Painlevé analysis and, more specifically, by the use of Kowalevski exponents. The imposition of certain integrality conditions on the Kowalevski exponents gives the necessary conditions. We also show that the conditions are sufficient. In the four-dimensional case we also examine the Liouville integrability of the systems.

Keywords. Lotka–Volterra Equations, Kowalevski Exponents, Painlevé Analysis.

2010 Mathematics Subject Classification. 34G20, 34M55, 37J35.

The Lotka–Volterra model is a basic model of predator-prey interactions. The model was developed independently by Alfred Lotka (1925), and Vito Volterra (1926). It forms the basis for many models used today in the analysis of population dynamics. In three dimensions it describes the dynamics of a biological system where three species interact.

The most general form of Lotka–Volterra equations is

$$\dot{x}_i = \varepsilon_i x_i + \sum_{j=1}^{n} a_{ij} x_i x_j , \quad i = 1, 2, \dots, n . \tag{19.1}$$

We consider Lotka–Volterra equations without linear terms ($\varepsilon_i = 0$), and where the matrix of interaction coefficients $A = (a_{ij})$ is skew-symmetric. The special case of the Kac–van Moerbeke system (KM-system) was used to describe population evolution in a hierarchical system of competing individuals. The KM-system is closely connected to the Toda lattice. Lotka–Volterra equations have been studied by many authors in its various aspects, e.g., complete integrability [7] Poisson and bi-Hamiltonian formulation [5,10,12,15], stability of solutions and Darboux polynomials [9,17]. In [11] we examined such Lotka–Volterra equations in three dimensions satisfying the Kowalevski–Painlevé property. The basic tools for the required classification are the use of Painlevé analysis, the examination of the eigenvalues of the Kowalevski matrix, and other standard Lax pair and Poisson techniques. Kowalevski exponents are useful in establishing integrability or nonintegrability of Hamiltonian systems (see [1,2,6,13,14,16,18,19]). The first step is to impose certain conditions on the exponents, i.e., we require that all the Kowalevski exponents be integers for every solution of the indicial equation. This gives a finite list of values of the parameters satisfying such conditions. This step requires some elementary number theoretic techniques as is usual with this type of classification.

The second step is to check that the leading behavior of the Laurent series solutions agrees with the weights of the corresponding homogeneous vector field defining the dynamical system. In our case the weights are all equal to one, and we must therefore exclude the possibility that some of the Laurent series have leading terms with poles of order greater than one. To accomplish this step we use old-fashioned Painlevé analysis, i.e., Laurent series. The application of Painlevé analysis and especially of the ARS algorithm (see [3, 4, 7, 8, 14]) is useful in calculating the Laurent solution of a system and for check if there are $(n-1)$ free parameters.

In this classification of Lotka–Volterra systems we discover, as expected, some well-known integrable systems like the open and periodic Kac–van Moerbeke systems and systems associated with simple Lie algebras.

We must also point out that our classification is given up to isomorphism. In other words, if one system is obtained from another by an invertible linear change of variables, we do not consider them to be different. Modulo this identification, we obtain only six classes of solutions.

The Lotka–Volterra system can be expressed in hamiltonian form as follows. Define a quadratic Poisson bracket by the formula

$$\{x_i, x_j\} = a_{ij} x_i x_j\,, \quad i, j = 1, 2, \ldots, n\,. \tag{19.2}$$

Then the system can be written in the form $\dot{x}_i = \{x_i, H\}$, where $H = \sum_{i=1}^{n} x_i$. The Louville integrability in the three-dimensional case can be easily established.

In dimension three the system is defined by a matrix of the form

$$A = \begin{pmatrix} 0 & a & b \\ -a & 0 & c \\ -b & -c & 0 \end{pmatrix},$$

where a, b, c are constants. We use the notation (a, b, c) to denote this system. It turns out that the Lotka–Volterra systems which possess the Kowalevski–Painlevé property fall either into two infinite families or four exceptional cases:

Theorem 1. *The Lotka–Volterra equations in three dimensions satisfy the Kowalevski–Painlevé property if and only if (a, b, c) is in the class of*

(l_2) $(1, 0, 1)$

(l_3) $(1, -1, 1)$

(l_4) $(1, -1, 2)$

(l_6) $(1, -2, 3)$

(l_λ) $(1, 1, \lambda)$ $\lambda \in \mathbf{Z} \setminus 0.$
(l_0) $(1, 1 + \mu, \mu)$ $\mu \in \mathbf{R}.$

The analogous classification in four dimensions is the subject of the Ph.D. thesis of K. Constandinides. In the four-dimensional case, the situation is much more complicated than in the three-dimensional case. The system is not automatically Liouville integrable. We distinguish two different cases. First, the case where the Poisson bracket is of rank 2. Then we have two Casimirs, and therefore in this case the system is integrable (since the Hamiltonian is a constant of motion). The interesting case is when the rank is full, i.e., four. In the classification of the systems which satisfy the Painleve–Kowalevski condition we obtained over 100 such cases. For example, there are 117 cases which satisfy this condition in the case of full rank. The analysis of Kowalevski exponents indicate that these systems should be integrable. We analyze some of these examples and show their integrability. Special cases include the KM-system (Volterra lattice) and some systems associated with simple and affine Lie algebras.

Bibliography

[1] M. Adler, P. van Moerbeke, Kowalewski's asymptotic method, Kac–Moody Lie algebras and regularization, *Comm. Math. Phys.*, **83**: 83–106, 1982.

[2] M. Adler, P. Van Moerbeke, P. Vanhaecke, *Algebraic integrability, Painlevé geometry and Lie algebras,* Ergebnisse der Mathematik und ihrer Grenzgebiete 47, Springer-Verlag, Berlin/Heidelberg, 2004.

[3] M. J. Ablowitz, A. Ramani, and H. Segur, A connection between nonlinear evolution equations and ordinary differential equations of P-type I, *J. Math. Phys.*, **21**: 715–721, 1980.

[4] M. J. Ablowitz, A. Ramani, and H. Segur, A connection between nonlinear evolution equations and ordinary differential equations of P-type II, *J. Math. Phys.*, **21**: 1006–1015, 1980.

[5] B. Hernández–Bermejo, V. Fairen, Hamiltonian structure and Darboux theorem for families of generalized Lotka-Volterra systems, *J. Math. Phys.*, **39**(11): 6162–6174, 1998.

[6] A. V. Borisov, A. V. Tsygvintsev, Kowalewski exponents and integrable systems of classic dynamics, I, II, *Regul. Chaotic. Dyn.*, **1**: 15–37, 1996.

[7] T. C. Bountis, M. Bier, and J. Hijmans, On the integrability of some generalized Lotka–Volterra systems, *Phys. Lett. A*, **97**: 11–14, 1983.

[8] T. C. Bountis, What can complex time tell us about real time dynamics? *Internat. J. Bifur. Chaos Appl. Sci. Engrg.*, **2**: 217–232, 1992.

[9] L. Cairo and J. Llibre, Darboux integrability for 3D Lotka–Volterra systems, *J. Phys. A: Math. Gen.*, **33**: 2395–2406, 2000.

[10] C. Cronström and M. Noga, Multi-Hamiltonian Structure of Lotka–Volterra and quantum Volterra models, *Nucl. Phys. B*, **445**: 501–515, 1995.

[11] K. Constandinides and P. A. Damianou, Lotka-Volterra equations satisfying the Kowalevski-Painleve property, *Regul. and Chaotic Dyn.*, **16**(3-4), 311–329, 2011.

[12] P. A. Damianou, The Volterra model and its relation to the Toda lattice, *Phys. Lett. A*, **155**: 126–132, 1991.

[13] A. Goriely, A brief history of Kovalevskaya exponents and modern developments, *Regul. Chaotic Dyn.*, **5**: 3–15, 2000.

[14] A. Goriely, *Integrability and Nonintegrability of Dynamical Systems,* Advanced Series in Nonlinear Dynamics 19, World Scientific Publishing Company, Singapore, 2001.

[15] E. H. Kerner, Comment on Hamiltonian structures for the n-dimensional Lotka–Volterra equations, *J. Math. Phys.*, **38**: 1218–1223, 1997.

[16] V. V. Kozlov, Tensor invariants of quasihomogeneous systems of differential equations and the asymptotic method of Kovalevskaya–Lyapunov, *Mat. Zametki,* **51**: 46–52, 1992.

[17] J. Moulin–Ollagnier, Polynomial first integrals of the Lotka–Volterra system, *Bull. Sci. Math.*, **121**: 463–476, 1997.

[18] S. T. Sadetov, Resonances on the Kovalevskaya exponent, *Math. Notes*, **54**(3–4): 1081–1082, 1994 (= *Mat. Zametki,* **54**: 152–154, 1993).

[19] H. Yoshida, Necessary condition for the existence of algebraic first integrals. I. Kowalevski's exponents, *Celestial Mech.*, **31**: 363–379, 1983.

Author Information

Pantelis A. Damianou, Department of Mathematics and Statistics, University of Cyprus, P.O. Box 20537, 1678, Nicosia, Cyprus.
E-mail: `damianou@ucy.ac.cy`

Chapter 20

Painlevé Test and Briot–Bouquet Systems

Evgenii Gricuk and Valerii Gromak

Abstract. In this chapter, for differential equations $w^{(n)} = R(w^{(n-1)}, \dots, w, z)$, where $R(w^{(n-1)}, \dots, w, z)$ is a rational function of the first n arguments whose coefficients are analytic with respect to z in the domain $G \subset \mathbb{C}$, we clarify the relationship between the resonance method, which is also referred to as the Painlevé formal test, and the representation of these equations in the form of Briot–Bouquet systems.

Keywords. Painlevé Formal Test, Briot–Bouquet System.

2010 Mathematics Subject Classification. Primary: 34M55.

Hierarchies of ordinary differential equations with Painlevé property and their solutions are a topic of great current interest. For higher-order differential equations, the Painlevé classifcation problem proved to be very difficult, and the most complete results have only been obtained for equations of polynomial type.

In this chapter, for the differential equations

$$w^{(n)} = R\big(w^{(n-1)}, \dots, w, z\big), \tag{20.1}$$

where $R(w^{(n-1)}, \dots, w, z)$ is a rational function of the first n arguments whose coefficients are analytic with respect to z in the domain $G \subset \mathbb{C}$, we clarify the relationship between the resonance method, which is also referred to as the Painlevé formal test, and the representation of these equations in the form of Briot–Bouquet systems.

The test itself is in brief the successive performance of the following steps.

We find the least possible power of the variable t in the formal expansion of a solution of equation (20.1) of the form

$$w = \sum_{j=0}^{\infty} c_j t^{j-k}, \quad t = z - z_0. \tag{20.2}$$

If $c_0 \neq 0$ and the numbers k are positive integers, then the first step is complete, and the analysis of the equation can be continued. Note that the case of negative integers k can be reduced to the previous case by the replacement $w \to w^{-1}$, and in the case $k = 0$ the point $z = 0$ is not singular. The second step of the test is related to finding the Fuchs indices for each pair $(c_0; k)$. The Fuchs indices, or resonances, are defined as the indices j_r of the coefficients c_j in the expansion of the solution (20.2) for which the coefficients c_j remain arbitrary. The Fuchs indices are found from the polynomial equation $Q(r) = 0$, which is referred to as the resonance equation. If, for each pair $(c_0; k)$, all roots of the resonance equation are simple and roots other than -1 and possibly zero are positive integers, then the second step of the formal test is considered to be complete. If all coefficients c_j in the expansion (20.2) can be determined, and the number of arbitrary coefficients among them is equal to $n - 1$, then we assume that the third step is complete for the differential equation (20.1), and, therefore, the whole Painlevé formal test is complete.

Therefore, the passage of the Painlevé formal test for the differential equation (20.1) eliminates the presence of movable algebraic singular points of solutions of that equation. In this connection, obviously, it is assumed that there can exist movable singular points of the logarithmic type or movable essentially singular points.

The expansion (20.2) used in the test described above is formal; the proof of its convergences required additional investigations. It turns out that, to prove the formal expansion (20.2) in the general case, one can use the theory of Briot–Bouquet systems

$$tu'_j = f_j(t, u_1, \dots, u_n), \quad j = 1, \dots, n \tag{20.3}$$

with functions f_j analytic in some neighborhood of the point $t = u_1 = \dots = u_n = 0$ and with the conditions $f_j(0, 0, \dots, 0) = 0$.

First we prove that for all pairs $(c_0; k)$; $k \in \mathbb{N}$; $c_0 \neq 0$; one root of the resonance equation $Q(r) = 0$ is equal to -1. The main result is

Theorem 1. *If equation* (20.1) *passes the Painlevé formal test, then, in a neighborhood of a singular point, it can be reduced to a Briot–Bouquet system* (20.3), *and expansions of the form* (20.2) *for equation* (20.1) *are convergent.*

We also consider the inverse result to Theorem 1 and the case of rational k. In the last case we looking for the algebraic solutions of equation (20.1).

Author Information

Evgenii Gricuk, Belarusian State University, pr. Nezavisimosti, 4, 220030, Minsk, Belarus.
E-mail: `gricuk_e@tut.by`

Valerii Gromak, Belarusian State University, pr. Nezavisimosti, 4, 220030, Minsk, Belarus.
E-mail: `vgromak@gmail.com`

Chapter 21

Solutions of the Chazy System

Valerii Gromak

Abstract. In the present paper, we give a derivation of the system $(\mathcal{A})$–$(\mathcal{F})$, the solution of system $(\mathcal{A})$ expressed in expanded form via the parameters a_k, and the solution of the system $(\mathcal{B})$–$(\mathcal{F})$ in the case of constant a_k as well as in the general case under some restrictions, which permit the solution to be written out in closed form. All computations have been carried out with the Mathematica computer algebra system.

Keywords. Painlevé Property, Third Order Equation, Chazy System.

2010 Mathematics Subject Classification. 34M55.

The six Painlevé equations, were discovered by Painlevé and Gambier through the classification problem for ordinary differential equations, whose solutions have no movable branch points. Today this property is referred to as the Painlevé property. For higher-order equations, the Painlevé classification problem proves to be very complicated, and the most complete results have so far been obtained only for higher-order polynomial equations. The Chazy paper [1] is one of the first papers on the classification of higher-order equations with respect to the Painlevé property; it deals with the analysis of the Painlevé property of the equation

$$y''' = \sum_{k=1}^{6} \frac{(y'-a_k')(y''-a_k'')}{y-a_k} + \sum_{k=1}^{6} \frac{A_k(y'-a_k')^3 + B_k(y'-a_k')^2 + C_k(y'-a_k')}{y-a_k} + Dy'' + Ey' + \prod_{k=1}^{6}(y-a_k)\sum_{k=1}^{6}\frac{F_k}{y-a_k}; \tag{1}$$

here the poles $a_k = a_k(z)$ are finite and distinct and, in general, are functions of the independent variable z. The paper [1] also presents the Chazy system of 31 algebraic and differential equations

$$\sum_{j=1}^{6} A_j = 0, \sum_{j=1}^{6} a_j A_j = -6, \sum_{j=1}^{6} a_j^2 A_j = -2\sum_{j=1}^{6} a_j, 2A_k^2 + \sum_{j=1}^{6} \frac{A_k - A_j}{a_k - a_j} = 0, \tag{$\mathcal{A}$}$$

where $k = 1, \dots, 6$ $(j \neq k)$,

$$\sum_{j=1}^{6}(B_k - B_j)\left(-\frac{A_k}{2} - \frac{1}{a_k - a_j}\right) + A_k' - \sum_{j=1}^{6}\frac{a_k' - a_j'}{a_k - a_j}(A_k - 3A_j)$$

$$-\frac{3}{2}A_k\sum_{i=1}^{6}a_i' A_i = 0, \qquad (\mathscr{B})$$

$$\left(-2A_kC_k - \sum_{j=1}^{6}\frac{C_k - C_j}{a_k - a_j}\right) + \sum_{j=1}^{6}\frac{3A_j(a_k' - a_j')^2 + +(2B_j - B_k)(a_k' - a_j') + a_k'' - a_j''}{a_k - a_j}$$

$$-B_k^2 + B_k' - B_k D + E = 0, \qquad (\mathscr{C})$$

$$2D + \sum_{j=1}^{6}(B_j - 3a_j' A_j) = 0, \qquad (\mathscr{D})$$

$$\sum_{j=1}^{6}F_j = 0, \sum_{j=1}^{6}a_j F_j = 0, \sum_{j=1}^{6}a_j^2 F_j = 0,$$

$$a_k''' - B_k C_k + C_k' + D(a_k'' - C_k) + E a_k' + F_k \prod_{j=1}^{6}(a_k - a_j)$$

$$+\sum_{j=1}^{6}\frac{A_j(a_k' - a_j')^3 + B_j(a_k' - a_j')^2 - (C_k - C_j)(a_k' - a_j') + (a_k' - a_j')(a_k'' - a_j'')}{a_k - a_j}$$

$$= 0, \qquad (\mathscr{F})$$

for 26 unknown functions $A_k = A_k(z)$, $B_k = B_k(z)$, $C_k = C_k(z)$, $D = D(z)$, $E = E(z)$, and $F_k = F_k(z)$, whose solution, as Chazy claims, determines necessary and sufficient conditions for Painleve property of (1).

The problem of solving the Chazy system remains open. Moreover, in some papers the equations of the system are written out in a form different from the original, and Chazy himself did not present the derivation of the system in [1].

In the present chapter, we give a derivation of system $(\mathscr{A})$–$(\mathscr{F})$, the solution of system $(\mathscr{A})$ expressed in expanded form via the parameters a_k [2], and the solution of system $(\mathscr{B})$–$(\mathscr{F})$ in the case of constant a_k as well as in the general case under some restrictions, which permit one to write out the solution in closed form. All computations have been carried out with the Mathematica computer algebra system.

Bibliography

[1] J. Chazy, Sur les équations différentielles du troisième ordre et d'ordre supérieur dont l'intégrale générale a ses points critiques fixes, *Acta Math.*, **34**: 317–385, 1911.

[2] V. I. Gromak, On Solutions of the Chazy System, *Differen. Equations*, **43**(5): 631–635, 2007.

Author Information

Valerii Gromak, Belarusian State University, pr. Nezavisimosti, 4, 220030, Minsk, Belarus.
E-mail: vgromak@gmail.com

Painlevé Equations and Related Topics, 171–184

Chapter 22

Third-Order Ordinary Differential Equations with the Painlevé Test

Yasin Adjabi and Arezki Kessi

Abstract. In this chapter we obtain the necessary conditions for the absence of movable critical singular points of solutions of some third equation with 16 parameters.

Keywords. Painlevé Property, Third-Order Equation.

2010 Mathematics Subject Classification. 34M55.

22.1 Introduction

Painlevé, Chazy, Garnier, and Bureau obtained partial results on the Painlevé classification of rational third-order differential equations of the form

$$y''' = R\left(y'', y', y, x\right). \tag{22.1}$$

In 1899, Painlevé established the necessary conditions for a third-order first degree ODE (22.1) satisfying the Painlevé property, when R is a rational function in y'', y', algebraic in y, and analytic coefficients in varaible x.

The *first necessary condition* is that the highest derivative y''', as a function of the next highest derivative y'', be a polynomial of degree at most two. We then write equation (22.1) as

$$y''' = A_2\left(y', y, x\right) y''^2 + A_1\left(y', y, x\right) y'' + A_0\left(y', y, x\right). \tag{22.2}$$

The *second necessary condition* states that, as a function of the second next highest derivative y', each coefficient A_j has for only singularities simple poles, the poles of A_1 and A_0 are among those of A_2, and the difference between the degrees of the numerator and denominator of A_j does not exceed $-1, 1, 3$ for, respectively, $j = 2, 1, 0$.

The *third necessary condition* is that the function A_2 be identical to one of the 11 types (see [5]), the case where $A_2 = 0$ is already studied. In this work we study the case where $A_2 \equiv (1 - 1/\eta)/(y' + \alpha y^2)$

$$y''' = \frac{1 - 1/\eta}{y' + \alpha y^2} y''^2 + A_1\left(y', y, x\right) y'' + A_0\left(y', y, x\right). \tag{22.3}$$

Third order equation (22.3), where $\eta \neq 0, -1$, was considered in [5, 8]; [8] starts with the following simplified equation, i.e., an equation which contains the leading terms with leading order $k = -1$ as $x \to x_0$ only

$$y''' = \beta \frac{(y'' - 2yy')^2}{y' - y^2} + a_1 yy'' + a_2 y'^2 + a_3 y^2 y' + a_4 y^4 + c_1 \frac{y'' y'}{y} + c_2 \frac{y'^3}{y^2}, \quad (22.4)$$

and investigates the values of a_i $(i = 1, \ldots, 4)$, c_1, c_2 and β such that the equation is of Painlevé type.

In this chapter, we obtain the necessary conditions for the absence of movable critical singular points of the solutions of equations

$$\begin{aligned} y''' \left(\delta y' + \alpha y^2\right) = {} & \beta y''^2 + c_1 \frac{y'' y'^2}{y} + c_2 \frac{y'^4}{y^2} + a_1 y'' y' y + a_2 y'' y^3 + a_3 y'^3 \\ & + a_4 y'^2 y^2 + a_5 y' y^4 + a_6 y^6 + a_7 y'' y^2 + a_8 y'^2 y \\ & + a_9 y^4 + a_{10} y' y^2 + a_{11} y^3. \end{aligned} \quad (22.5)$$

We are looking for such values of constants $\delta, \alpha, \beta, c_1, c_2$, and a_i $(i = 1, \ldots, 11)$, so that the equations pass the Painlevé test.

22.2 Simplified Equation

The simplified equation corresponding to (22.5) is easily determined by setting $(x \mapsto x_0 + \varepsilon x)$, where $\varepsilon \neq 0$ is a parameter. We then put $\varepsilon = 0$; if equation (22.5) is to be free of mobile critical points, it is necessary that y be uniform, so that it must be given by an equation of the form

$$y''' = \beta \frac{y''^2}{y'} + c_1 \frac{y'' y'}{y} + c_2 \frac{y'^3}{y^2}. \quad (22.6)$$

By making the change of variable

$$y' = vy, \quad (22.7)$$

in equation (22.6) , we get

$$v'' = \beta \frac{v'^2}{v} + a_{12} v v' + a_{22} v^3, \quad (22.8)$$

where $(a_{12}, a_{22}) = (c_1 - 3 + 2\beta, c_1 + c_2 - 1 + \beta)$, and β must be either 1 or $(\eta - 1)/\eta$, where η is either infinite or an integer distinct from $0, -1$.

It is known [4] that, for solutions of (22.6) to have only single-valued nonstationary polar singularities, it is necessary and sufficient that all nonstationary singularities of solutions of (22.8) be simple poles with residues that are integers. The substitution

$$vw = 1 \quad (22.9)$$

transforms equation (22.8) into an equation of the same form, showing that without loss of generality we may restrict η to being positive.

In order to determine whether (22.8) possesses the Painlevé property, we apply the algorithm from Ablowitz, Ramani, and Segur [1]. We seek a solution of (22.8) in the neighborhood of an arbitrary point x_0 in the form of a Laurent series

$$v = v_0\,(x - x_0)^r + v_1\,(x - x_0)^{(r+1)} + v_2\,(x - x_0)^{(r+2)} + \cdots, \qquad (22.10)$$

with v_0 being nonzero and r is the leading power that needs to be found. When we substitute this series into (22.8) , we see that $r = -1$, and the residue v_0 of any moving pole must be an integer root of the equation

$$a_{22}v_0^2 - a_{12}v_0 - 2 + \beta = 0. \qquad (22.11)$$

Moreover, the roots of the indicial equation corresponding to

$$-J^2 + (a_{12}v_0 + 3 - 2\beta)\,J + 3a_{22}v_0^2 - 2a_{12}v_0 + \beta - 2 = 0, \qquad (22.12)$$

must be distinct integers. where the constant β has the values

$$\{0;\ 1;\ 1 - 1/\eta,\ \eta > 1 \text{ an integer};\ 3/2;\ 4/3;\ 7/6\}.$$

By taking into account these requirements, we find that equation (22.6) is with fixed critical points only if the couple (c_1, c_2) takes one of the following values:

$$(c_1, c_2) \in \left\{\begin{array}{ll}
\eta = -6 & : \left(\frac{2(m-1)}{3m},\ -\frac{(m-1)(5m+1)}{6m^2}\right), \\
\eta = -3 & : \left(\frac{m-1}{3m},\ -\frac{(m-1)(2m+1)}{3m^2}\right);\ \left(2,\ -\frac{3(m-1)(m+1)}{2m^2}\right), \\
\eta = -2 & : \left(0,\ \frac{1-m^2}{2m^2}\right), \\
\eta = 2 & : \left(\frac{2m-1}{m},\ -\frac{(m-1)(3m+1)}{2m^2}\right);\ \left(2,\ -\frac{3(m-1)(m+1)}{2m^2}\right), \\
\eta \geqslant 2 & : \left(\frac{(m-1)(2+\eta)}{\eta m},\ -\frac{(m-1)(\eta m+m-1)}{\eta m^2}\right), \\
\eta = 3 & : \left(\frac{5m-2}{3m},\ -\frac{2(m-1)(2m+1)}{3m^2}\right), \\
\eta = 5 & : \left(\frac{7m-2}{5m},\ -\frac{2(m-1)(3m+2)}{5m^2}\right), \\
\eta = \infty & : \left(\frac{m-1}{m},\ \frac{1-m}{m}\right);\ \left(1,\ \frac{1}{m^2} - 1\right), \\
\eta \text{ integer} & : \left(\frac{2}{\eta} + 1,\ -\frac{1}{\eta} - 1\right),
\end{array}\right\},$$

where m ($m = v_0$) is an integer number different from 0.

22.3 Reduced Equations

Now consider the dominant behavior in the neighborhood of a critical point $x = x_0$,

$$y = y_0 (x - x_0)^k, \quad as \ x \to x_0, \ \ k \in \mathbb{Z}, \tag{22.13}$$

where k is the singularity order, y_0 is a coefficient and substitute equation (22.13) in equation (22.5). For certain values of the exponent k, two or more terms may balance, and the remainder of the terms can be ignored as $x \to x_0$. The terms which balance for each choice of the exponent k are called the dominant terms.

In the following sections, the reduced equations that retain only leading terms as $x \to x_0$ will be considered for $k = -1, -2, -3$, and -4, with distinct Fuchs indices (resonances).

22.3.1 Leading Order $k = -1$

If $a_6 \neq 0$ and/or one or more of $a_i \neq 0$, $i = 1, \ldots, 5$, then the ε-test,

$$x \to x_0 + \varepsilon x, \ \ y \to \varepsilon^{-1} y, \ \ \varepsilon \to 0, \tag{22.14}$$

reduces equation (22.5) to

$$\begin{aligned} y''' \left(\delta y' + \alpha y^2\right) = \beta y''^2 + c_1 \tfrac{y'' y'^2}{y} + c_2 \tfrac{y'^4}{y^2} + a_1 y'' y' y \\ + a_2 y'' y^3 + a_3 y'^3 + a_4 y'^2 y^2 + a_5 y' y^4 + a_6 y^6. \end{aligned} \tag{22.15}$$

The case where $\beta = 0$. This has already been treated in [3–5, 7, 8].

The case where $\beta \neq 0$. Consider the following cases:

Case a. The case where $\delta = 0$ was already studied by Bureau in 1972, Cosgrove and Scoufis in 1993 and Chazy [5].

Case b. Let $\delta \neq 0$, $\alpha \neq 0$ or $\alpha = 0$; we obtain an equation of third order nonpolynomial. Then (22.15) can be written as

$$\begin{aligned} y' y''' = \beta y''^2 + c_1 \frac{y'' y'^2}{y} + c_2 \frac{y'^4}{y^2} + a_1 y'' y' y + a_2 y'' y^3 + a_3 y'^3 \\ + a_4 y'^2 y^2 + a_5 y' y^4 + a_6 y^6. \end{aligned} \tag{22.16}$$

Equation (22.16) satisfies the Painlevé test if all its solutions (*general and singular*) can be expanded in the formal Laurent series:

$$y = (x - x_0)^{-1} \sum_{i=0}^{\infty} y_i (x - x_0)^i, \tag{22.17}$$

with y_0 being nonzero: the indicial equation of the equation (22.16) is

$$(\delta - \alpha y_0) r^3 + \left(a_2 y_0^2 + (6\alpha - a_1) y_0 + c_1 + 4\beta - 6\delta\right) r^2 + Q(y_0) r + P(y_0) = 0, \tag{22.18}$$

where

$$\begin{cases} Q(y_0) = a_5 y_0^3 - (3a_2 + 2a_4) y_0^2 + (3a_3 + 5a_1 - 11\alpha) y_0 \\ \qquad - 12\beta + 17\delta - 7c_1 - 4c_2, \\ P(y_0) = 6a_6 y_0^4 - 5a_5 y_0^3 + 4(a_4 + 2a_2) y_0^2 + 3(6\alpha - 2a_1 - a_3) y_0 \\ \qquad + 4c_1 + 2c_2 + 2(4\beta - 6\delta), \end{cases}$$

and y_0 is root of

$$a_6 y_0^4 - a_5 y_0^3 + (a_4 + 2a_2) y_0^2 + (6\alpha - 2a_1 - a_3) y_0 + 2c_1 + c_2 + (4\beta - 6\delta) = 0, \tag{22.19}$$

In order for equation (22.16) to be with fixed critical points, it is necessary that, for each values of y_0 solution of the equation (22.19), the solutions of equation (22.18) be distinct integer numbers.

There are two cases to consider, according to whether or not α is zero.

Case (b.i). Let $\alpha = 0$. According to Theorem of I. P. Martynov, the roots of (22.19) must be distinct. Let the roots be denoted y_1, y_2, y_3, and y_4, and let the corresponding products of the indice numbers be denoted $P(y_1)$, $P(y_2)$, $P(y_3)$ and $P(y_4)$ where

$$\begin{aligned} P(y_i) = {} & 6a_6 y_i^4 - 5a_5 y_i^3 + 4(a_4 + 2a_2) y_i^2 + 3(-2a_1 - a_3) y_i \\ & + 4c_1 + 2c_2 + 2(4\beta - 6\delta). \end{aligned}$$

From the expressions in terms of the y_i, we see that the $P(y_i)$ satisfy the Diophantine equation,

$$\sum_{i=1}^{i=N} \frac{1}{P(y_i)} = -\frac{1}{2c_1 + c_2 + (4\beta - 6)}, \tag{22.20}$$

where N is number of branches.

Equation with $a_6 = a_5 = 2a_2 + a_4 = 0$ and $(6\alpha - 2a_1 - a_3) \neq 0$. By applying the Painlevé test, we find that the list of equations with integer Fuchs's indices is as follows:

$$\begin{cases} y'''y' = \frac{1}{2}y''^2 - \frac{3}{2}y''y'y - \frac{1}{2}y''y^3 - y'^3 + y'^2y^2 + \dfrac{y''y'^2}{y}, \\ \quad y_{01} = 1, (r_{11}, r_{12}) = (1, 2), \end{cases}$$

$$\begin{cases} y'''y' = \frac{1}{2}y''^2 + 24y''y'y + 12y''y^3 - 52y'^3 - 24y'^2y^2 + 12\dfrac{y''y'^2}{y} - 24\dfrac{y'^4}{y^2}, \\ \quad y_{01} = 1, (r_{11}, r_{12}) = (1, 4), \end{cases}$$

$$\begin{cases} y'''y' = \frac{3}{4}y''^2 + 24y''y'y + 12y''y^3 - 51y'^3 - 24y'^2y^2 + 12\dfrac{y''y'^2}{y} - 24\dfrac{y'^4}{y^2}, \\ \quad y_{01} = 1, (r_{11}, r_{12}) = (1, 3). \end{cases}$$

Equation with $a_5 = a_6 = 0$. The reduced equations for these cases are as follows:

$$\begin{cases} y'''y' = \frac{4}{3}y''^2 + \frac{7}{3}y''y'y + \frac{10}{3}y''y^3 - 6y'^3 - \frac{28}{3}y'^2y^2 + \frac{2}{3}\frac{y''y'^2}{y} + \frac{2}{3}\frac{y'^4}{y^2}, \\ \quad y_{01} = 1, y_{02} = -\frac{1}{2}, (r_{11}, r_{12}) = (2, -2), (r_{21}, r_{22}) = (1, -2), \end{cases}$$

$$\begin{cases} y'''y' = \frac{7}{6}y''^2 - \frac{40}{3}y''y'y - \frac{64}{3}y''y^3 + 40y'^3 + \frac{160}{3}y'^2y^2 + \frac{4}{3}\frac{y''y'^2}{y} + \frac{4}{3}\frac{y'^4}{y^2}, \\ \quad y_{01} = 1, y_{02} = \frac{1}{4}, (r_{11}, r_{12}) = (1, -2), (r_{21}, r_{22}) = (1, 8), \end{cases}$$

$$\begin{cases} y'''y' = \frac{7}{6}y''^2 - \frac{4}{3}y''y'y + \frac{20}{3}y''y^3 - \frac{56}{3}y'^2y^2 + \frac{4}{3}\frac{y''y'^2}{y} + \frac{4}{3}\frac{y'^4}{y^2}, \\ \quad y_{01} = 1, y_{02} = -\frac{1}{2}, (r_{11}, r_{12}) = (1, -8), (r_{21}, r_{22}) = (-2, 2), \end{cases}$$

$$\begin{cases} y'''y' = \frac{3}{2}y''^2 - 4y'^3 - 8y'^2y^2 + 2\frac{y''y'^2}{y}, \\ \quad y_{01} = 1, y_{02} = -1, (r_{11}, r_{12}) = (3, -4), (r_{21}, r_{22}) = (-3, 2), \end{cases}$$

$$\begin{cases} y'''y' = \frac{1}{2}y''^2 + \frac{63}{8}y''y'y - \frac{745}{8}y''y^3 + \frac{21}{2}y'^3 + \frac{315}{8}y'^2y^2 + \frac{3}{2}\frac{y''y'^2}{y} - \frac{7}{8}\frac{y'^4}{y^2}, \\ \quad y_{01} = 1, y_{02} = -\frac{1}{15}, (r_{11}, r_{12}) = (2, 15), (r_{21}, r_{22}) = (1, 2), \end{cases}$$

$$\begin{cases} y'''y' = \frac{1}{2}y''^2 + \frac{15}{2}y''y'y + \frac{45}{4}y'^3 + \frac{225}{8}y'^2y^2 + \frac{3}{2}\frac{y''y'^2}{y} - \frac{7}{8}\frac{y'^4}{y^2}, \\ \quad y_{01} = 1, y_{02} = -\frac{1}{15}, (r_{11}, r_{12}) = (5, 6), (r_{21}, r_{22}) = (1, 2), \end{cases}$$

$$\begin{cases} y'''y' = \frac{2}{3}y''^2 + \frac{1}{2}y''y'y - \frac{3}{2}y''y^3 + 2y'^3 + \frac{15}{2}y'^2y^2 + \frac{4}{3}\frac{y''y'^2}{y} - \frac{5}{6}\frac{y'^4}{y^2}, \\ \quad y_{01} = 1, y_{02} = -\frac{1}{3}, (r_{11}, r_{12}) = (2, 3), (r_{21}, r_{22}) = (1, 2), \end{cases}$$

$$\begin{cases} y'''y' = y''^2 - \frac{1}{2}y''y'y - 3y''y^3 + 4y'^3 + \frac{21}{2}y'^2y^2 + \frac{1}{2}\frac{y''y'^2}{y} - \frac{1}{2}\frac{y'^4}{y^2}, \\ \quad y_{01} = 1, y_{02} = -\frac{1}{3}, (r_{11}, r_{12}) = (2, 3), (r_{21}, r_{22}) = (1, 2), \end{cases}$$

$$\begin{cases} y'''y' = y''^2 - \frac{9}{2}y''y^3 + 3y'^3 + \frac{27}{2}y'^2y^2 + \frac{1}{2}\frac{y''y'^2}{y} - \frac{1}{2}\frac{y'^4}{y^2}, \\ \quad y_{01} = 1, y_{02} = -\frac{1}{3}, (r_{11}, r_{12}) = (1, 6), (r_{21}, r_{22}) = (1, 2), \end{cases}$$

$$\begin{cases} y'''y' = \frac{3}{2}y''^2 - \frac{3}{2}y''y'y + \frac{3}{2}y''y^3 + 3y'^3 - 5y'^2y^2 + \frac{y''y'^2}{y}, \\ \quad y_{01} = 1, y_{02} = -1, (r_{11}, r_{12}) = (1, -4), (r_{21}, r_{22}) = (-2, 2), \end{cases}$$

$$\begin{cases} y'''y' = y''^2 + 4y''y^3 - 10y'^2y^2 + 2\frac{y''y'^2}{y}, \\ \quad y_{01} = 1, y_{02} = -1, (r_{11}, r_{12}) = (1, -4), (r_{21}, r_{22}) = (1, -4), \end{cases}$$

$$\begin{cases} y'''y' = y''^2 + \frac{10}{3}y''y'y + \frac{16}{3}y''y^3 - \frac{26}{3}y'^3 - \frac{44}{3}y'^2y^2 + 2\frac{y''y'^2}{y}, \\ \quad y_{01} = 1, y_{02} = -\frac{1}{2}, (r_{11}, r_{12}) = (2, -3), (r_{21}, r_{22}) = (1, -3), \end{cases}$$

$$\begin{cases} y'''y' = \frac{1}{2}y''^2 + \frac{1}{3}y''y'y - \frac{2}{3}y''y^3 + \frac{4}{3}y'^3 + \frac{16}{3}y'^2y^2 + \frac{y''y'^2}{y}, \\ \quad y_{01} = 1, y_{02} = -\frac{1}{2}, (r_{11}, r_{12}) = (2, 3), (r_{21}, r_{22}) = (1, 3), \end{cases}$$

$$\begin{cases} y'''y' = \frac{1}{2}y''^2 + \frac{35}{4}y''y'y - \frac{75}{4}y''y^3 + \frac{35}{4}y'^3 + \frac{525}{8}y'^2y^2 + \frac{3}{2}\frac{y''y'^2}{y} - \frac{7}{8}\frac{y'^4}{y^2}, \\ \quad y_{01} = 1, y_{02} = -\frac{1}{15}, (r_{11}, r_{12}) = (1, 30), (r_{21}, r_{22}) = (1, 2), \end{cases}$$

$$\begin{cases} y'''y' = \frac{3}{2}y''^2 + 3y''y^3 - 8y'^2y^2 + \frac{y''y'^2}{y}, \\ \quad y_{01} = 1, y_{02} = -1, (r_{11}, r_{12}) = (1, -4), (r_{21}, r_{22}) = (1, -4), \end{cases}$$

$$\begin{cases} y'''y' = \frac{3}{2}y''^2 - 2y''y'y - 4y''y^3 + 2\frac{y''y'^2}{y}, \\ \quad y_{01} = 1, y_{02} = -\frac{1}{2}, (r_{11}, r_{12}) = (4, -3), (r_{21}, r_{22}) = (3, -2), \end{cases}$$

$$\begin{cases} y'''y' = \frac{1}{2}y''^2 + y''y'y - 2y''y^3 + 8y'^2y^2 + \frac{y''y'^2}{y}, \\ \quad y_{01} = 1, y_{02} = -\frac{1}{2}, (r_{11}, r_{12}) = (1, 6), (r_{21}, r_{22}) = (1, 3). \end{cases}$$

Equation with $a_5 \neq 0$ and $a_6 = 0$. The reduced equations for these cases are as follows:

$$\begin{cases} y'''y' = \frac{1}{2}y''^2 + \frac{(1-5i\sqrt{2})}{9}y''y'y - \frac{(8+5i\sqrt{2})}{9}y''y^3 + \frac{(-2+4i\sqrt{2})}{3}y'^3 \\ \quad + \frac{(26+14i\sqrt{2})}{9}y'^2y^2 + \frac{(-4+2i\sqrt{2})}{9}y'y^4 + \frac{y''y'^2}{y}, \\ \quad y_{01} = 1, y_{02} = -1 - 2i\sqrt{2}, y_{03} = -1 + \frac{1}{2}i\sqrt{2}, \\ (r_{11}, r_{12}) = (1, 4), (r_{21}, r_{22}) = (-4, -5), (r_{31}, r_{32}) = (1, 5), \end{cases}$$

$$\begin{cases} y'''y' = \frac{3}{2}y''^2 - 2y''y'y - 2y''y^3 + 5y'^3 - y'y^4 + 2\frac{y''y'^2}{y}, \\ \quad y_{01} = 1, y_{02} = 4, y_{03} = -1, \ (r_{11}, r_{12}) = (2, -3), \\ \quad (r_{21}, r_{22}) = (3, 20), (r_{31}, r_{32}) = (-2, 5), \end{cases}$$

$$\begin{cases} y'''y' = \frac{3}{2}y''^2 - \frac{1}{3}y''y'y + \frac{2}{3}y''y^3 + \frac{4}{3}y'^3 - 4y'^2y^2 + \frac{2}{3}y'y^4 + 2\frac{y''y'^2}{y}, \\ \quad y_{01} = 1, y_{02} = -2, y_{03} = -3, \ (r_{11}, r_{12}) = (2, -4), \\ \quad (r_{21}, r_{22}) = (1, -4), (r_{31}, r_{32}) = (-2, -4), \end{cases}$$

$$\begin{cases} y'''y' = \frac{3}{2}y''^2 - 4y''y'y + 9y'^3 - 4y'^2y^2 - y'y^4 + 2\frac{y''y'^2}{y}, \\ \quad y_{01} = 1, y_{02} = 4, y_{03} = -1, \ (r_{11}, r_{12}) = (1, -6), \\ \quad (r_{21}, r_{22}) = (-5, -12), (r_{31}, r_{32}) = (-2, 5), \end{cases}$$

$$\begin{cases} y'''y' = \frac{3}{2}y''^2 + y''y'y + y''y^3 - 4y'^3 - 6y'^2y^2 + 2y'y^4 + 2\frac{y''y'^2}{y}, \\ \quad y_{01} = 1, y_{02} = -2, y_{03} = -1, \ (r_{11}, r_{12}) = (-4, 3), \\ \quad (r_{21}, r_{22}) = (-3, -4), (r_{31}, r_{32}) = (-4, 1), \end{cases}$$

$$\begin{cases} y'''y' = \frac{3}{2}y''^2 + 2y''y'y - 3y''y^3 - 6y'^3 + 2y'^2y^2 + 2y'y^4 + 2\frac{y''y'^2}{y}, \\ \quad y_{01} = 1, y_{02} = -2, y_{03} = -1, \ (r_{11}, r_{12}) = (-2, 6), \\ \quad (r_{21}, r_{22}) = (3, 4), (r_{31}, r_{32}) = (-2, 2), \end{cases}$$

$$\begin{cases} y'''y' = \frac{3}{2}y''^2 - \frac{11}{2}y''y'y + \frac{9}{2}y''y^3 + 9y'^3 - 13y'^2y^2 + 2y'y^4 + 2\frac{y''y'^2}{y}, \\ \quad y_{01} = 1, y_{02} = -2, y_{03} = -1, \ (r_{11}, r_{12}) = (1, -12), \\ \quad (r_{21}, r_{22}) = (-2, -6), (r_{31}, r_{32}) = (-2, 2), \end{cases}$$

$$\begin{cases} y'''y' = \frac{3}{2}y''^2 + \frac{31}{6}y''y'y + \frac{1}{3}y''y^3 - \frac{89}{12}y'^3 - \frac{3}{2}y'^2y^2 - \frac{1}{12}y'y^4 + 2\frac{y''y'^2}{y}, \\ \quad y_{01} = 1, y_{02} = 16, y_{03} = -3, \ (r_{11}, r_{12}) = (-1, 5), \\ \quad (r_{21}, r_{22}) = (19, 20), (r_{31}, r_{32}) = (1, -19). \end{cases}$$

Equation with $a_6 \neq 0$. The resonances and the reduced equations for these cases are as follows:

$$\begin{cases} y'''y' = \frac{4}{3}y''^2 + \frac{5}{3}y''y^3 - 5y'^2y^2 + \frac{1}{3}y^6 + \frac{y''y'^2}{y}, \\ \quad y_{01} = 1, y_{02} = -1, y_{03} = 2, y_{04} = -2, (r_{11}, r_{12}) = (1, -2), \\ (r_{21}, r_{22}) = (1, -2), (r_{31}, r_{32}) = (-2, -4), (r_{41}, r_{42}) = (-2, -4). \end{cases}$$

$$\begin{cases} y'''y' = \frac{1}{2}y''^2 + \frac{5}{2}y'^2y^2 - \frac{1}{2}y^6 + \frac{y''y'^2}{y}, \\ y_{01} = 1, y_{02} = -1, y_{03} = 2, y_{04} = -2, (r_{11}, r_{12}) = (1, 3), \\ (r_{21}, r_{22}) = (1, 3), (r_{31}, r_{32}) = (-2, 6), (r_{41}, r_{42}) = (-2, 6). \end{cases}$$

$$\begin{cases} y'''y' = \frac{2}{3}y''^2 + \frac{1}{3}y''y^3 + y'^2y^2 - \frac{1}{3}y^6 + \frac{y''y'^2}{y} \\ y_{01} = 1, y_{02} = -1, y_{03} = 2, y_{04} = -2, (r_{11}, r_{12}) = (1, 2), \\ (r_{21}, r_{22}) = (1, 2), (r_{31}, r_{32}) = (-2, 4), (r_{41}, r_{42}) = (-2, 4), \end{cases}$$

The general solution of the latter equation possesses four families, the differentiation of which with respect to x gives the fourth order equation

$$y'''' = 2\frac{y'''y'}{y} + \frac{3}{2}\frac{y''^2}{y} - 2\frac{y''y'^2}{y^2} + 5y''y^2 + \frac{5}{2}y'^2y - \frac{5}{2}y^5. \tag{22.21}$$

This equation passes the Painleve test and has four solution branches. These branches have two collections of the Fuchs indices that are

$$y_{01} = 1, \quad y_{02} = -1, \quad y_{03} = 2, \quad y_{04} = -2, \quad (r_{11}, r_{12}) = (1,3,5),$$
$$(r_{21}, r_{22}) = (1,3,5), \quad (r_{31}, r_{32}) = (-2,5,6), \quad (r_{41}, r_{42}) = (-2,5,6).$$

The canonical form of (22.21) was given in [7].

Case $(b.ii)$. Let $\alpha \neq 0$, without loss of generality. One can see that one of the roots of (22.19) may be taken equal to -1 (i.e., $y_0 = -1$) and the corresponding indicial equation of the equation (22.18) can be writen as

$$(r+1)\left((\alpha+1)r^2 + Mr + N\right) = 0, \text{ and } y_0 = -1,$$

where

$$\begin{cases} M = a_2 + a_1 - 7 - 7\alpha + c_1 + 4\beta, \\ N = 24 + 18\alpha - (a_5 + 2a_4 + 3a_3 + 4a_2 + 6a_1) - 16\beta - 8c_1 - 4c_2. \end{cases}$$

We will distinguish between two cases according to whether $(N, M) = (0,0)$, or $(N, M) \neq (0,0)$ (N and M are both nonzero).

The case where $\alpha = -1$ and $M = N = 0$. Equation (22.15) takes the form (22.4) which was already studied by Martynov [8].

Remark 22.1. In this case, equation (22.15) has the property that the corresponding equation (22.19) has a double root for y_0.

The resonances and the reduced equations for these cases are as follows:

$$\begin{cases} y''' = \frac{1}{2}\frac{(y''-2yy')^2}{y'-y^2} + 2yy'' + 8y^2y' - 8y^4 + \frac{y''y'}{y}, \\ \quad y_{01} = -1, \; y_{02} = \frac{1}{2}, \; y_{03} = \frac{-1}{2}, \; (r_{11}, r_{12}) = (-2,6), \\ \qquad (r_{21}, r_{22}) = (2,3), \; (r_{31}, r_{32}) = (1,2), \end{cases}$$

$$\begin{cases} y''' = \frac{1}{2}\frac{(y''-2yy')^2}{y'-y^2} - \frac{5}{2}yy'' - \frac{37}{24}y'^2 + \frac{125}{24}y^2y' + \frac{125}{24}y^4 + \frac{3}{2}\frac{y''y'}{y} - \frac{7}{8}\frac{y'^3}{y^2}, \\ \quad y_{01} = -1, y_{02} = \frac{9}{5}, y_{03} = \frac{1}{5}, (r_{11}, r_{12}) = (5,7), \\ \quad (r_{21}, r_{22}) = (-7,6), (r_{31}, r_{32}) = (1,2), \end{cases}$$

$$\begin{cases} y''' = \frac{1}{2}\frac{(y''-2yy')^2}{y'-y^2} + 7yy'' + \frac{265}{24}y'^2 - \frac{833}{24}y^2y' + \frac{343}{24}y^4 + \frac{y''y'}{y} - \frac{5}{8}\frac{y'^3}{y^2}, \\ \quad y_{01} = -1, \; y_{02} = \frac{-9}{7}, \; y_{03} = \frac{-1}{7}, \; (r_{11}, r_{12}) = (-7,1), \\ \qquad (r_{21}, r_{22}) = (-3, \; 2), \; (r_{31}, r_{32}) = (1,2), \end{cases}$$

$$\begin{cases} y''' = \frac{1}{2}\frac{(y''-2yy')^2}{y'-y^2} + \frac{20}{3}yy'' + \frac{118}{9}y'^2 - \frac{200}{9}y^2y' + \frac{4}{3}\frac{y''y'}{y} - \frac{8}{9}\frac{y'^3}{y^2}, \\ y_{01} = -1, \; y_{02} = -\frac{1}{10}, \; (r_{11}, r_{12}) = (-10,4), \; (r_{21}, r_{22}) = (1,2), \end{cases}$$

$$\begin{cases} y''' = \frac{1}{2}\frac{(y''-2yy')^2}{y'-y^2} - \frac{8}{3}yy'' - \frac{46}{9}y'^2 + \frac{128}{9}y^2y' + \frac{5}{3}\frac{y''y'}{y} - \frac{10}{9}\frac{y'^3}{y^2}, \\ y_{01} = -1,\ y_{02} = \frac{1}{8},\ (r_{11}, r_{12}) = (4, 8),\ (r_{21}, r_{22}) = (1, 2), \end{cases}$$

$$\begin{cases} y''' = \frac{4}{5}\frac{(y''-2yy')^2}{y'-y^2} + \frac{6}{5}yy'' - \frac{(125n^2-153)}{108}a_0y'^2 - 2a_0y^2y' + a_0y^4 \\ +\frac{6}{5}\frac{y''y'}{y} - \frac{4}{5}\frac{y'^3}{y^2}, \\ y_{01} = -1, y_{02} = \frac{5n-3}{6}, y_{03} = \frac{-5n-3}{6}, (r_{11}, r_{12}) = (1, 6), \\ (r_{21}, r_{22}) = (2, n), (r_{31}, r_{32}) = (-n, 2), a_0 = -\frac{216}{5(25n^2-9)}. \end{cases}$$

These reduced equations do not appear on Martynov's list [8].

The case when $\alpha \neq 0$ and $(N, M) \neq (0, 0)$. Equation (22.19) has a simple root for y_0, implying that, in general, there are four branches if $a_6 \neq 0$.

Remark 22.2. In this case, equation (22.15) has the property that one family is not maximal (see R. Conte et al. [6]), so we use it for the generation of the necessary stability conditions.

We obtained a list of equations with integer Fuchs's indices:

Equation with $a_5 = a_6 = 0$ and $(6\alpha - 2a_1 - a_3) \neq 0$. In this case y_0 takes a single value i.e., there is only one branch:

$$\begin{cases} y'''(y' - y^2) = y''^2 - 4y''y'y - y''y^3 + 4y'^3 + y'^2y^2 + \frac{y''y'^2}{y} - \frac{y'^4}{y^2}, \\ y_{01} = -1, (r_{11}, r_{12}) = (-1, 1), \end{cases}$$

$$\begin{cases} y'''(y' - y^2) = y''^2 - 3y''y'y - y''y^3 + 3y'^3 + \frac{1}{2}y'^2y^2 + 2\frac{y''y'^2}{y} - \frac{3}{2}\frac{y'^4}{y^2}, \\ y_{01} = -1, (r_{11}, r_{12}) = (-1, 3). \end{cases}$$

Equation with $a_5 \neq 0$ and $a_6 = 0$. The reduced equations for these cases are as follows:

$$\begin{cases} y'''(y' + y^2) = \frac{1}{2}y''^2 + \frac{3}{2}y''y'y - \frac{3}{2}y''y^3 + 3y'^2y^2 + y'y^4 + \frac{y''y'^2}{y}, \\ y_{01} = 1,\ y_{02} = -1,\ y_{03} = 2, (r_i, r_j) = (-1, -2),\ (1, 3),\ (1, 6), \end{cases}$$

$$\begin{cases} y'''(y' + y^2) = \frac{1}{2}y''^2 + \frac{5}{2}y''y'y - \frac{1}{2}y''y^3 - 2y'^3 + y'^2y^2 + y'y^4 + \frac{y''y'^2}{y}, \\ y_{01} = 1,\ y_{02} = -1,\ y_{03} = 2, (r_i, r_j) = (-1, -6),\ (1, 3),\ (2, 3), \end{cases}$$

$$\begin{cases} y'''(y'+y^2) = \frac{1}{2}y''^2 + 3y''y'y - y''y^3 - 2y'^3 + 2y'^2y^2 + \frac{2}{3}y'y^4 + 2\frac{y''y'^2}{y} - \frac{4}{3}\frac{y'^4}{y^2}, \\ y_{01} = 1,\ y_{02} = -1,\ y_{03} = 2, (r_i, r_j) = (-1, -2),\ (1, 2),\ (1, 4). \end{cases}$$

Equation with $a_6 \neq 0$. The resonances and the reduced equations for these cases are as follows:

$$\begin{cases} y'''\left(y'-y^2\right)=\frac{2}{3}y''^2-\frac{14}{3}y''y'y-\frac{1}{3}y''y^3+\frac{4}{3}y'^3+\frac{16}{3}y'^2y^2-\frac{10}{3}y'y^4 \\ \qquad +\frac{2}{3}y^6+\frac{7}{3}\frac{y''y'^2}{y}-\frac{4}{3}\frac{y'^4}{y^2}, \\ y_{01}=-1,\ y_{02}=-3,(r_{11},r_{12})=(1,4),\ (r_{21},r_{22})=(-3,4), \end{cases}$$

$$\begin{cases} y'''\left(y'-y^2\right)=\frac{3}{4}y''^2-\frac{13}{2}y''y'y+\frac{1}{2}y''y^3+6y'^3+\frac{3}{4}y'^2y^2-\frac{1}{4}y^6 \\ \qquad -\frac{1}{2}y'y^4+3\frac{y''y'^2}{y}-3\frac{y'^4}{y^2}, \\ y_{01}=-1,\ y_{02}=4,(r_{11},r_{12})=(1,5),(r_{21},r_{22})=(-5,4), \end{cases}$$

$$\begin{cases} y'''\left(y'-y^2\right)=\frac{1}{2}y''^2+5y''y'y-2y'^3-\frac{9}{2}y'^2y^2-\frac{1}{2}y^6 \\ \qquad +3y'y^4-3\frac{y''y'^2}{y}+2\frac{y'^4}{y^2}, \\ y_{01}=-1,\ y_{02}=-4,(r_{11},r_{12})=(1,3),(r_{21},r_{22})=(-4,6), \end{cases}$$

$$\begin{cases} y'''\left(y'-y^2\right)=\frac{1}{2}y''^2-3y''y'y-y''y^3-3y'^3+\frac{19}{2}y'^2y^2+\frac{3}{2}y^6 \\ \qquad -6y'y^4+2\frac{y''y'^2}{y}, \\ y_{01}=-1,\ y_{02}=-2,(r_{11},r_{12})=(1,3),(r_{21},r_{22})=(-2,3). \end{cases}$$

These equations are not considered in [8].

22.3.2 Leading Order $k = -2$

If $a_i = 0$, $(i = 1,\dots,5)$ and $a_6 = 0$, then it may be possible to construct analogous solutions of (22.5) which feature a leading term containing a pole of order $m = -2$; then the $\varepsilon-$ test

$$x = x_0 \to \varepsilon x,\quad y \to \varepsilon^{-2} y,\quad \varepsilon \to 0,$$

reduces equation (22.5) to

$$y'''y' = \beta y''^2 + c_1\frac{y''y'}{y} + c_2\frac{y'^3}{y^2} + a_7 y'' y^2 + a_8 y'^2 y + a_9 y^4. \tag{22.22}$$

If we substitute

$$y \cong y_0\left(x-x_0\right)^{(-2)} + \hat{\imath}\left(x-x_0\right)^{(r-2)} \tag{22.23}$$

into (22.22), then we obtain the following equations for the Fuchs indices r and y_0:

$$\begin{cases} 2r^2+(a_7y_0+12\beta-20+4c_1)\,r-(6a_7+4a_8)y_0+96-72\beta-48c_1-32c_2=0, \\ a_9y_0^2+(6a_7+4a_8)\,y_0-48+36\beta++24c_1+16c_2=0. \end{cases}$$

The reduced equations for these cases are as follows:

$$\begin{cases} y'''y' = \frac{1}{2}y''^2 + \frac{3}{2}\frac{y''y'}{y} - \frac{7}{8}\frac{y'^3}{y^2} + 10y'^2y - 32y^4. \\ y_{01} = 1, y_{02} = \frac{1}{4}, (r_{11}, r_{12}) = (-2, 6), (r_{21}, r_{22}) = (1, 3), \end{cases}$$

$$\begin{cases} y'''y' = \frac{1}{2}y''^2 + \frac{y''y'}{y} + 6y^4. \\ y_{01} = 1, y_{02} = -1, (r_{11}, r_{12}) = (2, 3), (r_{21}, r_{22}) = (2, 3), \end{cases}$$

$$\begin{cases} y'''y' = \frac{1}{2}y''^2 + 2\frac{y''y'}{y} - 12y''y^2 + 27y'^2y - 54y^4. \\ y_{01} = 1, y_{02} = -\frac{1}{3}, (r_{11}, r_{12}) = (-3, 12), (r_{21}, r_{22}) = (4, -3), \end{cases}$$

$$\begin{cases} y'''y' = \frac{1}{2}y''^2 + 2\frac{y''y'}{y} - 18y^4. \\ y_{01} = 1, y_{02} = -\frac{1}{3}, (r_{11}, r_{12}) = (-3, 12), (r_{21}, r_{22}) = (4, -3), \end{cases}$$

$$\begin{cases} y'''y' = \frac{1}{2}y''^2 + \frac{y''y'}{y} - \frac{5}{8}\frac{y'^3}{y^2} + 20y'^2y - 64y^4. \\ y_{01} = 1, y_{02} = \frac{1}{4}, (r_{11}, r_{12}) = (-3, 8), (r_{21}, r_{22}) = (2, 3), \end{cases}$$

$$\begin{cases} y'''y' = \frac{2}{3}y''^2 + \frac{5}{6}\frac{y''y'}{y} - \frac{7}{12}\frac{y'^3}{y^2} - \frac{40}{3}y''y^2 + \frac{170}{3}y'^2y - \frac{400}{3}y^4. \\ y_{01} = 1, y_{02} = \frac{1}{10}, (r_{11}, r_{12}) = (-4, 15), (r_{21}, r_{22}) = (2, 3), \end{cases}$$

$$\begin{cases} y'''y' = \frac{2}{3}y''^2 + \frac{5}{6}\frac{y''y'}{y} - \frac{7}{12}\frac{y'^3}{y^2} + \frac{80}{3}y''y^2 - \frac{130}{3}y'^2y + \frac{80}{3}y^4. \\ y_{01} = 1, y_{02} = -\frac{1}{2}, (r_{11}, r_{12}) = (-4, -5), (r_{21}, r_{22}) = (1, 10), \end{cases}$$

$$\begin{cases} y'''y' = \frac{2}{3}y''^2 + \frac{4}{3}\frac{y''y'}{y} - \frac{5}{6}\frac{y'^3}{y^2} + \frac{8}{3}y''y^2 + \frac{8}{3}y'^2y - \frac{64}{3}y^4. \\ y_{01} = 1, y_{02} = \frac{1}{4}, (r_{11}, r_{12}) = (-2, 4), (r_{21}, r_{22}) = (1, 2), \end{cases}$$

$$\begin{cases} y'''y' = \frac{7}{6}y''^2 + \frac{1}{3}\frac{y''y'}{y} - \frac{11}{24}\frac{y'^3}{y^2} + \frac{80}{3}y''y^2 - \frac{148}{3}y'^2y + \frac{128}{3}y^4. \\ y_{01} = 1, y_{02} = -\frac{1}{8}, (r_{11}, r_{12}) = (-3, -8), (r_{21}, r_{22}) = (1, 3), \end{cases}$$

$$\begin{cases} y'''y' = y''^2 + \frac{2}{3}\frac{y''y'}{y} - \frac{2}{3}\frac{y'^3}{y^2} + \frac{70}{3}y''y^2 - \frac{125}{3}y'^2y + \frac{100}{3}y^4. \\ y_{01} = 1, y_{02} = -\frac{1}{5}, (r_{11}, r_{12}) = (-4, -5), (r_{21}, r_{22}) = (1, 4), \end{cases}$$

$$\begin{cases} y'''y' = y''^2 + \frac{y''y'}{y} + 6y'^2y - 36y^4. \\ y_{01} = 1, y_{02} = -\frac{1}{3}, (r_{11}, r_{12}) = (-4, 6), (r_{21}, r_{22}) = (-2, 4), \end{cases}$$

$$\begin{cases} y'''y' = \frac{4}{3}y''^2 + \frac{1}{4}\frac{y''y'}{y} + \frac{9}{16}\frac{y'^3}{y^2} + 25y''y^2 - 60y'^2y + 75y^4. \\ y_{01} = 1, y_{02} = \frac{1}{5}, (r_{11}, r_{12}) = (-6, -5), (r_{21}, r_{22}) = (-3, 2), \end{cases}$$

22.3.3 Leading Order $k = -3$

In partiular if $a_i = 0$, $(i = 1, \ldots, 9)$ and $a_{10} \neq 0$. The ε-test,

$$x \to x_0 + \varepsilon x, \quad y \to \varepsilon^{-3}y, \ \varepsilon \to 0,$$

reduces equation (22.5) to

$$y'''y' = \beta y''^2 + c_1\frac{y''y'}{y} + c_2\frac{y'^3}{y^2} + a_{10}y'y^2. \tag{22.24}$$

In this case y_0 takes a single value i.e., there is only one branch. The resonances and the reduced equations for these cases are

$$\begin{cases} y'''y' = y''^2 + \frac{y''y'}{y} + 24y'y^2, \ y_{01} = 1, \ (r_{11}, r_{12}) = (-4, 6)\,, \\ y'''y' = \frac{7}{6}y''^2 + \frac{8}{9}\frac{y''y'}{y} - \frac{28}{27}\frac{y'^3}{y^2} + 56y'y^2, \ y_{01} = 1, (r_{11}, r_{12}) = (-7, 8)\,. \end{cases}$$

22.3.4 Leading Order $k = -4$

Suppose that $a_i = 0$, $(i = 1, \ldots, 10)$ and $a_{11} \neq 0$ hold. The dominant terms arise from $y'''y'$, y''^2, y^3. Then $k = -4$ and the corresponding equation is

$$y'''y' = \beta y''^2 + c_1\frac{y''y'}{y} + c_2\frac{y'^3}{y^2} + a_{11}y^3. \tag{22.25}$$

The resonances and the reduced equations for these cases are

$$y'''y' = \frac{1}{2}y''^2 + \frac{3}{2}\frac{y''y'}{y} - \frac{7}{8}\frac{y'^3}{y^2} + 24y^3, \ y_{01} = 1, (r_{11}, r_{12}) = (2, 3)\,,$$

$$y'''y' = \frac{1}{2}y''^2 + 2\frac{y''y'}{y} - \frac{9}{8}\frac{y'^3}{y^2} - 72y^3, \ y_{01} = 1, (r_{11}, r_{12}) = (-3, 6)\,,$$

$$y'''y' = \frac{1}{2}y''^2 + 3\frac{y''y'}{y} - 2\frac{y'^3}{y^2} - 168y^3, \ y_{01} = 1, (r_{11}, r_{12}) = (-7, 6)\,,$$

$$y'''y' = \frac{3}{2}y''^2 - \frac{3}{8}\frac{y'^3}{y^2} - 24y^3, \ y_{01} = 1, (r_{11}, r_{12}) = (-2, 3)\,,$$

$$y'''y' = \frac{11}{10}y''^2 + \frac{3}{2}\frac{y''y'}{y} - \frac{5}{4}\frac{y'^3}{y^2} - 120y^3, \ y_{01} = 1, (r_{11}, r_{12}) = (-6, 5)\,.$$

Bibliography

[1] M. J. Ablowitz, A. Ramani, and H. Segur, A connection between nonlinear evolution equations and ordinary differential equations of P-type I, II, *J. Math. Phys.*, **21**: 715–721, 1980; **21**: 1006–1015, 1980.

[2] Y. Adjabi, F. Jarad, A. Kessi, and U. Mugan, Third order differential equations with fixed critical points, *Applied Mathematics and Computation,* **208**(1): 238–248, 2009.

[3] Y. Adjabi and A. Kessi, Third order differential equations with fixed critical points, *Annals: The proceeding of the Institute of Math. (NASB) Belarus,* **12**(2): 12–17, 2004.

[4] F. J. Bureau, Differential equations with fixed critical points, *Ann. Mat. Pura. Appl.,* **64**: 29–364, 1964; **66**: 1–116, 1964.

[5] J. Chazy, Sur les équations différentielles du troisième et d'ordre supérieur dont l'intégrale générale à ses points critiques fixés, *Acta Math.,* **34**: 317–385, 1911.

[6] R. Conte and A. Fordy and A. Pickering, A perturbative Painlevé approach to nonlinear differential equations, *Physica D,* **69**(1–2): 33–58, 1993.

[7] C. M. Cosgrove, Chazy Classes IX–XII of third-order differential equations, *Stud. Appl. Math.,* **104**(3): 171–228, 2000.

[8] I. P. Martynov, Analytic properties of solutions of a third order equations, Translated from *Differntsial'nye Uravneniya,* **21**(5): 512–517, 1985; **21**(6): 937–946, 1985.

Author Information

Yasin Adjabi, Department of Mathematics, Faculty of Sciences, UMBB, Boumerdes, Algeria.

Arezki Kessi, Faculty of Mathematics, USTHB, BP 32 El Alia, 16 311 Bab Ezzouar, Algiers, Algeria.
E-mail: arkessi@yahoo.com

Painlevé Equations and Related Topics, 185–189

Chapter 23

Analytic Properties of Solutions of a Class of Third-Order Equations with an Irrational Right-Hand Side

Ivan P. Martynov, Vyacheslav A. Pronko and Tatsyana K. Andreeva

Abstract. In this chapter the Painlevé analysis of third order ordinary differential equations with irrational right-hand side is presented.

Keywords. Painlevé Classification, Differential Equation, Critical Point.

2010 Mathematics Subject Classification. 34M55.

Consider the differential equation

$$x''' = f_1(x, x', x'') + \frac{x'}{x^2} f_2(x, x', x'') + \frac{1}{x^2}\left(f_3(x, x', x'')\right)^{3/2}, \tag{23.1}$$

where $f_k(x, x', x'') = a_k x x'' + b_k x'^2 + c_k x^2 x' + d_k x^4, a_k, b_k, c_k, d_k, k = 1, 2, 3$ are constants; moreover, $f_3(x, x', x'')$ is not a complete square. We single out all classes of equations (23.1) with the Painleve property. We indicate which functions are integrated in the resulting equations. Equation (23.1) determines one of the components of the quadratic third-order system.

It is required that the solutions of equation

$$f_3(x, x', x'') = 0 \tag{23.2}$$

be the solutions of equation (23.1) [1]. If this requirement is performed, then equation (23.1) replaces the system

$$x'' + \frac{b_3}{a_3}\frac{x'^2}{x} + \frac{c_3}{a_3}xx' + \frac{d_3}{a_3}x^3 = \eta w^2 x, \quad w' = w^2 + \left(\alpha x + \beta\frac{x'}{x}\right)w, \tag{23.3}$$

where $\eta = 4/a_3^3, \alpha = (c_3/a_3 + a_1)/2, \beta = b_3/a_3 + (a_2 - 1)/2$. From (23.3) we have that $2w = f_3'(x, x', x'')/f_3(x, x', x'') - (\beta + 2)x'/x - \alpha x$. The following statement is valid.

Lemma 1. *Equation* (23.1) *has the Painlevé property if and only if the system* (23.3) *has the Painlevé property.*

Lemma 2. *If equation* (23.1) *has the Painleve property, then equation* (23.2) *has the Painleve property.*

Proof. With the setting in the system (23.3) $w = \varepsilon\omega$, we obtain the simplified system

$$x'' + \frac{b_3}{a_3}\frac{x'^2}{x} + \frac{c_3}{a_3}xx' + \frac{d_3}{a_3}x^3 = 0, \quad \omega' = \left(\alpha x + \beta\frac{x'}{x}\right)\omega,$$

where $\varepsilon = 0$ which should have the Painlevé property. It follows that Lemma 2 is valid.

With the setting in the system (23.3) $z = \varepsilon\tau$, at $\varepsilon = 0$ we obtain the simplified system

$$\ddot{x} + \frac{b_3}{a_3}\frac{\dot{x}^2}{x} = 0, \quad \dot{w} = \beta\frac{\dot{x}}{x}w, \tag{23.4}$$

as $\varepsilon = 0$, where $\dot{x} = dx/d\tau$, $\ddot{x} = d^2x/d\tau^2$, $\dot{\omega} = d\omega/d\tau$. The first equation of the system (23.4) has the Painlevé property if and only if

$$\frac{b_3}{a_3} = \frac{1}{n} - 1, \tag{23.5}$$

where $n \in \mathbb{Z} \setminus \{0\}$ or $n = \infty$. If $\beta \neq 0, a_3 + b_3 \neq 0$, then we find $w = C_1(\tau - \tau_0)^{\beta n}$, where C_1, τ_0 are arbitrary constants (in the following, τ_0 and $C_i, i = 1;2$, stand for arbitrary constants). Consequently, $\tau = \tau_0$ is not a critical point, only for

$$\beta = \frac{\mu}{n}, \tag{23.6}$$

where $\mu, n \in \mathbb{Z} \setminus \{0\}$. □

The following statement is valid.

Lemma 3. *If system* (23.3) *has the Painlevé property, then the condition* (23.5) *is true and at* $\beta \neq 0, n \in \mathbb{Z} \setminus \{0\}$ *condition* (23.6) *is true.*

Introducing in the system (23.3) the parameter ε by the formulas $w = \varepsilon^{-1}\omega, z = \varepsilon\tau$, we obtain the simplified system

$$\ddot{x} = \left(1 - \frac{1}{n}\right)\frac{\dot{x}^2}{x} + \eta w^2 x, \quad \dot{\omega} = \omega^2 + \beta\frac{\dot{x}}{x}\omega, \tag{23.7}$$

as $\varepsilon = 0$, where $\dot{x} = dx/d\tau$, $\ddot{x} = d^2x/d\tau^2$, $\dot{\omega} = d\omega/d\tau$. Let $\beta = 0$. The second equation in system (23.7) has a solution $\omega = -1/(\tau - \tau_0)$, then the first equations in system (23.7) acquires the form

$$\ddot{x} = \left(1 - \frac{1}{n}\right)\frac{\dot{x}^2}{x} + \eta\frac{x}{(\tau - \tau_0)^2}. \tag{23.8}$$

Equation (23.8) has the general solution $x = C_1 e^{C_2(\tau-\tau_0)}(\tau-\tau_0)^{-\eta}$, for $n = \infty$ and

$$x = (\tau-\tau_0)^{\frac{1}{2}(n-\sqrt{n(n+4\eta)})}\left(C_1 + C_2(\tau-\tau_0)^{\frac{1}{2}\sqrt{n(n+4\eta)}}\right)^n,$$

for $n \in \mathbb{Z}\setminus\{0\}$. If $\beta = 0, n = \infty$, then system (23.7) has the Painlevé property only if

$$\eta \in \mathbb{Z}\setminus\{0\}. \tag{23.9}$$

If $\beta = 0, n \in \mathbb{Z}\setminus\{0\}$, then system (23.7) has the Painlevé property if and only if one of the following conditions are present:

$$n = 2m, \qquad \eta = m(s^2-1)/2, \qquad m \in \mathbb{Z}\setminus\{0\}, \tag{23.10}$$

$$n = 2m+1, \qquad \eta = (2m+1)(l^2-l), \qquad m \in \mathbb{Z}, \tag{23.11}$$

where $s \in \mathbb{Z}\setminus\{-1;1\}, l \in \mathbb{Z}\setminus\{0;1\}$ is true. Consequently, the following statement is valid

Lemma 4. *Let* $\beta = 0$. *If system* (23.3) *has the Painlevé property, then the condition* (23.9) *for* $n = \infty$ *or one of the conditions* (23.10), (23.11) *is true.*

Let $\beta \neq 0$. From system (23.7), for ω, we have the equation

$$\ddot{\omega} = \left(1-\frac{1}{\mu}\right)\frac{\dot{\omega}^2}{\omega} + \left(1+\frac{2}{\mu}\right)\omega\dot{\omega} + \left(\mu\frac{\eta}{n}-\frac{1}{\mu}\right)\omega^3, \tag{23.12}$$

for $n \in \mathbb{Z}\setminus\{0\}$ and the equation

$$\ddot{\omega} = \frac{\dot{\omega}^2}{\omega} + \omega\dot{\omega} + \eta\beta\omega^3, \tag{23.13}$$

for $n = \infty$. Equation (23.12) has the Painleve property only if one of the following conditions [2,3] is true:

$$\eta = 10n \text{ or } \eta = n \text{ at } \mu = 1; \quad \eta = 5n/4 \text{ at } \mu = 2,$$

$$\eta = 3n/2 \text{ at } \mu = 3; \quad \eta = 2n \text{ at } \mu = 5; \ \mu = -2.$$

Equation (23.13), where $\beta \neq 0$, does not have the Painleve property [2,3]. From system (23.7) we find that

$$x = \omega^{\frac{1}{\beta}} e^{-\frac{1}{\beta}\int \omega dt}, \tag{23.14}$$

where ω is a solution to the of equation (23.12). The following statement is valid

Lemma 5. *If system* (23.3) *at* $\beta \neq 0$ *has the Painleve property, that one of the following conditions is true:*

(a) $n = (\mu+2)p, \eta = (\mu+2)(\mu+3)p/4$, *where* $p \in \mathbb{Z}\setminus\{0\}, \mu \in \{1;2;3;5\}$;

(b) $n = 3p, \eta = 30p$, *where* $p \in \mathbb{Z} \setminus \{0\}$, *for* $\mu = 1$;

(c) $\eta = pn(p+n)/(2p+n)^2$, *where* $p, n \in \mathbb{Z} \setminus \{0\}, p \neq -n, p \neq -\frac{n}{2}$, *for* $\mu = -2$;

(a) $\eta = n/4$, *where* $n \in \mathbb{Z} \setminus \{0\}$, *for* $\mu = -2$.

Using Lemmas 1–5, and the Painlevé analysis of system (23.3) we obtain the following statement

Theorem 1. *Equation* (23.1) *has the Painlevé property if and only if one of the following conditions is true:*

(a) $a_1 = c_3 = d_1 = d_3 = 0$, $a_2 = 3 - 2/n$, $b_1 = -c_2$, $b_2 = 2/n - 2$, $b_3 = (1/n - 1)a_3$, $c_1 = -d_2$, $a_3^3 = 8/(m(s^2-1))$, $n = 2m$, $s \in \mathbb{Z} \setminus \{-1; 1\}$, $m \in \mathbb{Z} \setminus \{0\}$ *or* $a_3^3 = 4/((2m+1)(l^2 - l))$, $n = 2m+1$, $l \in \mathbb{Z} \setminus \{0; 1\}$, $m \in \mathbb{Z}$;

(b) $a_1 = 1$, $a_2 = 3 - 2/n$, $a_3^3 = 16/(n(s^2-1))$, $b_1 = -c_2$, $b_2 = 2/n - 2$, $b_3 = (1/n - 1)a_3$, $c_3 = -a_3$, $d_1 = 0$, $d_2 = -c_1 - 2n/(n+2)^2$, $d_3 = na_3/(n+2)^2$, *where* $n \in \mathbb{Z} \setminus \{-2; -1; 0\}$, $s \in \mathbb{Z} \setminus \{-1; 1\}$, $(1-s)n/2 \in \mathbb{Z}$, $(1-s)n/2 + sk \neq -1$, *if* $n \in \mathbb{N}$, *then* $k = \overline{0, n}$; *if* $n \in \mathbb{Z}_- \setminus \{-2; -1\}$ *then* $k = 0, 1, 2, \ldots$;

(c) $a_1 = c_3 = d_1 = d_3 = 0$, $a_2 = 3$, $b_1 = -c_2$, $b_2 = -2$, $b_3 = -a_3$, $c_1 = -d_2$, $4/a_3^3 \in \mathbb{Z} \setminus \{0\}$;

(d) $a_1 = -c_3/a_3$, $a_2 = 3$, $b_1 = -c_2$, $b_2 = -2$, $b_3 = -a_3$, $c_1 = -d_2$, $c_3 \neq 0$, $d_1 = d_3 = 0$, $4/a_3^3 \in \mathbb{Z}_-$;

(e) $a_1 = c_3 = d_1 = d_3 = 0$, $a_2 = 3 + 2(\mu - 1)/n$, $b_1 = -c_2$, $b_2 = 2(1/n - 1)(1 + \mu/n)$, $b_3 = (1/n - 1)a_3$, $c_1 = -d_2$ *and one of the following conditions is true:*

1. $n = (\mu+2)p$, $a_3^3 = 16/((\mu+2)(\mu+3)p)$, *where* $\mu \in \{1; 2; 3; 5\}$, $p \in \mathbb{Z} \setminus \{0\}$;
2. $n = 3p$, $a_3^3 = 2/(15p)$, *where* $p \in \mathbb{Z} \setminus \{0\}$, *for* $\mu = 1$;
3. $a_3^3 = 16/n$ *or* $a_3^3 = 4(2p+n)^2/(pn(p+n))$ *where* $p, n \in \mathbb{Z} \setminus \{0\}$, $p \neq -n, p \neq -n/2$, *for* $\mu = -2$;

(f) $a_1 = 3$, $a_2 = 1$, $a_3^3 = 1/5$, $b_1 = -c_2$, $b_2 = b_3 = 0$, $c_1 = 6 - d_2$, $c_3 = a_3$, $d_1 = -4$, $d_3 = -a_3$;

(g) $a_1 = 1$, $a_2 = 2$, $a_3^3 = 8/15$, $b_1 = -1 - c_2$, $b_2 = -1$, $b_3 = -a_3/2$, $c_1 = 3 - d_2$, $c_3 = a_3$, $d_1 = -1$, $d_3 = -a_3/2$;

(h) $a_1 = 0$, $a_2 = 1$, $a_3^3 = -2/3$, $b_1 = -4/3 - c_2$, $b_2 = -4/9$, $b_3 = d_3 = -c_3 = -2a_3/3$, $c_1 = 8/3 - d_2$, $d_1 = -4/9$;

(i) $a_1 = 1 - 2\mu/(n+2)$, $a_2 = 3 + 2(\mu - 1)/n$, $b_1 = -c_2 - 6\mu/(n(n+2))$, $b_2 = 2(1-n)(n+\mu)/n^2$, $b_3 = (1/n - 1)a_3$, $c_1 = -d_2 + 2(\mu(n+3) - n)/(n+2)^2$, $c_3 = -a_3$, $d_1 = -2n\mu/(n+2)^3$, $d_3 = a_3 n/(n+2)^2$ *and one of the following conditions is true:*

1. $n = (\mu+2)p$, $a_3^3 = 16/((\mu+2)(\mu+3)p)$, *where* $\mu \in \{1;2;3;5\}$, $p \in \mathbb{Z}_-$;
2. $n = (\mu+2)p$, $a_3^3 = 16/((\mu+2)(\mu+3)p)$ *where* $\mu \in \{1;2;3;5\}$, $p \in \mathbb{N}\setminus\{1\}$ *or* $n = 3p$, $a_3^3 = 2/(15p)$, *where* $p \in \mathbb{N}$, *for* $\mu = 1$, *and there has been a correlation* $\sum_{k=1}^{m-1}(m-k)A_k a_{m-k} = -m a_m$, *where* $m = -a_0 - 1$, $A_0 = 1$, $A_k = \sum_{l=0}^{k-1}(m-l)A_l a_{k-l}/(k(m-k))$, $k = \overline{1, m-1}$, *where* a_k *are expansion coefficients* $(\mu+2)p(\dot{\omega}-\omega^2)/(\mu\omega) = a_0/(z-z_0) + a_1 + a_2(z-z_0) + \ldots$, *where* ω *is a solution to the equation* (23.12);
3. $n = 3p$, $a_3^3 = 2/(15p)$, *where* $p \in \mathbb{Z}_-$, *for* $\mu = 1$;
4. $a_3^3 = 16/n$, *at* $n \in \mathbb{Z}_- \setminus \{-2;-1\}$ *or* $a_3^3 = 4(n-2m)^2/(nm(m-n))$, *where* $m \neq \frac{n}{2}$, $n - m > 0$, $m, n \in \mathbb{N}$, *for* $\mu = -2$;

(j) $a_1 = 2$, $a_2 = 6$, $a_3^3 = 8(p-1)^2/(p(2-p))$, $p \in \mathbb{Z}\setminus\{0;1;2\}$, $b_1 = -3 - c_2$, $b_2 = -6$, $b_3 = -3a_3/2$, $c_1 = -d_2$, $c_3 = 0$, $d_1 = -1$, $d_3 = -a_3/2$;

(k) $a_1 = 2$, $a_2 = 6$, $a_3 = -2$, $b_1 = -3 - c_2$, $b_2 = -6$, $b_3 = 3$, $c_1 = -d_2$, $c_3 = 0$, $d_1 = -1$, $d_3 = 1$.

The solutions of these equations can be expressed via either elementary functions, elliptic functions, or solutions of linear equations.

Bibliography

[1] S. L. Sobolewski, *Movable singularities of solutions of ordinary differential equations*, BSU, Minsk, 2006, 119 p.

[2] E. L. Ince, *Ordinary differential equations*, ONTI, Kharkov, 1939, 719 p.

[3] F. Bureau, Differential equations with fixed critical points, *Ann. Math.* **64**: 229–364, 1964.

Author Information

Ivan P. Martynov, Popovicha str. 9 – 293, Grodno, Belarus.
E-mail: i.martynov@grsu.by

Vyacheslav A. Pronko, Tereshkovoy str. 13 – 6, Grodno, Belarus.
E-mail: v.a.pronko@gmail.com

Tatsyana K. Andreeva, Limozha str. 27/1 – 118, Grodno, Belarus.
E-mail: tatsyana.andreeva@gmail.com

Part V

Other Aspects

Chapter 24

The Sixth Painlevé Transcendent and Uniformizable Orbifolds

Yurii V. Brezhnev

Abstract. We show how the Picard–Hitchin solutions to the sixth Painleve transcendent generate the explicitly uniformizable (parametrizable) algebraic curves of nontrivial genera $g > 1$. As a byproduct of the theory we derive an example of the uniformizable representation to a holomorphic Abelian integral.

Keywords. Sixth Painleve Transcendent, Algebraic Solutions, Algebraic Curves of Higher Genera, Uniformization by Theta-functions.

2010 Mathematics Subject Classification. 30F10, 30F35, 33E17, 33E05, 11F03, 14H45, 14H55.

24.1 Algebraic Solutions of $\mathcal{P}_6$ and Uniformization Theory

The sixth Painlevé transcendent

$$\mathcal{P}_6: \quad y_{xx} = \frac{1}{2}\left(\frac{1}{y} + \frac{1}{y-1} + \frac{1}{y-x}\right)y_x^2 - \left(\frac{1}{x} + \frac{1}{x-1} + \frac{1}{y-x}\right)y_x$$
$$+ \frac{y(y-1)(y-x)}{x^2(x-1)^2}\left\{\alpha - \beta\frac{x}{y^2} + \gamma\frac{x-1}{(y-1)^2} - \left(\delta - \frac{1}{2}\right)\frac{x(x-1)}{(y-x)^2}\right\}$$

is known to be a rich source of nontrivial algebraic solutions $y = f(x)$ and genera of these solutions, as genera of corresponding algebraic curves $F(x, y) = 0$, may be made as great as is wished. The relation of such solutions to the uniformization theory is based on the $\wp$-representation of the $\mathcal{P}_6$:

$$-\frac{\pi^2}{4}\frac{d^2 z}{d\tau^2} = \alpha\,\wp'(z|\tau) + \beta\,\wp'(z-1|\tau) + \gamma\,\wp'(z-\tau|\tau) + \delta\,\wp'(z-1-\tau|\tau), \quad (24.1)$$

obtainable via the transcendental change $(x, y) \mapsto (z, \tau)$ (Painlevé 1906, Manin–Babich–Bordag 1996):

$$x = \frac{\vartheta_4^4(\tau)}{\vartheta_3^4(\tau)}, \qquad y = \frac{1}{3} + \frac{1}{3}\frac{\vartheta_4^4(\tau)}{\vartheta_3^4(\tau)} - \frac{4}{\pi^2}\frac{\wp(z|\tau)}{\vartheta_3^4(\tau)}. \quad (24.2)$$

Research supported by the Federal Targeted Program under contract 02.740.11.0238.

Thus, knowledge of $z(\tau)$-dependence leads to a parametric representation for solution $y = f(x)$ and, in particular, to parametric representation of algebraic solutions. In their full generality these dependencies are known for the Picard–Hitchin class of solutions. For example, Picard's case $\alpha = \beta = \gamma = \delta = 0$ corresponds to $z = A\tau + B$. In Hitchin's case $\alpha = \beta = \gamma = \delta = \frac{1}{8}$ the dependence $z(\tau)$ is more complicated (obtainable through Okamoto's transformations), but the parametric form of the solution is, however, found to be very compact;

$$y_{\text{Pic}} = -\frac{\vartheta_4^2(\tau)\,\theta_2^2}{\vartheta_3^2(\tau)\,\theta_1^2}, \qquad y_{\text{Hit}} = \frac{\vartheta_4^2(\tau)}{\vartheta_3^2(\tau)}\left\{\pi\,\frac{\vartheta_2^2(\tau)\cdot\theta_2\theta_3\theta_4}{\theta_1' + 2\pi A\,\theta_1} - \theta_2^2\right\}\frac{1}{\theta_1^2}, \tag{24.3}$$

where θ's are understood to be equal to $\theta_k(A\tau + B|\tau)$ with arbitrary constants A, B, and $\theta_1' := \theta_1'(A\tau + B|\tau)$. Purely algebraic solutions correspond to $A\tau + B = \frac{\nu}{N}\tau + \frac{\mu}{N}$ with integral ν, μ, and N.

Uniformizing functions are known to be determined in terms of the auxiliary second order linear Fuchsian ODEs $\Psi_{yy} = \frac{1}{2}\mathcal{Q}(x, y)\Psi$, where $\mathcal{Q}$, as a rational function of x and y, contains all the information about the corresponding Riemann surface $\mathcal{R}$ (or orbifold $\mathfrak{T}$). Since the function $x = \chi(\tau)$ in (24.2) is the very well-known one and its Fuchsian $\mathbf{\Gamma}(2)$-equation $\Psi'' = -\frac{1}{4}\frac{x^2-x+1}{x^2(x-1)^2}\Psi$ is also known, we obtain nontrivial (solvable) Fuchsian equations for the second uniformizing function $y(\tau)$. Manipulations with Fuchsian equations themselves are not convenient, because we constantly handle the multivalued function inversions: ratios like $\tau = \Psi_1(x)/\Psi_2(x)$. For this reason we invert the standard Schwarz derivative $\{\tau, x\}$ into the 'reverse' object $[x, \tau] = -\{\tau, x\}$ and work with the autonomous ODEs

$$[y, \tau] = \mathcal{Q}(x, y), \quad \text{where } [y, \tau] := \frac{\dddot{y}}{\dot{y}^3} - \frac{3}{2}\frac{\ddot{y}^2}{\dot{y}^4}, \tag{24.4}$$

defining uniformizing single-valued functions and other single-valued objects.

24.2 On the General Solution to Equation (24.1)

The complete structure of the analytic continuations (a connection problem) of arbitrary solutions to $\mathcal{P}_6$ is the subject matter of the series work by D. Guzzetti (see, e.g., [1]). Analyzing these results, it would appear reasonable that the ramification structure of all the (not necessarily algebraic) solutions to the $\mathcal{P}_6$-equation in the vicinity of critical points is described by a function series of the kind

$$y = A + R\big[(x-e)^a \ln^n(x-e)\big] + \cdots,$$

where $e = \{0, 1\}$, $a \in \mathbb{C}$, $n \in \mathbb{Z}$, and $R[\ldots]$ is a rational function of its argument. In the language of uniformizing Painlevé substitution (24.2) this point is self-suggested: in the upper (τ)-half-plain $\mathbb{H}^+$, the x-function has an exponential behavior in the neighborhood of the points $x = \{0, 1, \infty\}$:

$$x \overset{\tau\to 0}{=} 0 + 16\exp\Big(\frac{\pi}{\mathrm{i}\tau}\Big) + \cdots, \qquad x \overset{\tau\to\infty}{=} 1 - 16\mathrm{e}^{\pi\mathrm{i}\tau} + \cdots,$$

$$x \overset{\tau\to 1}{=} \frac{1}{16}\exp\Big(\frac{\pi\mathrm{i}}{\tau-1}\Big) + \cdots$$

(the uniformizing τ-parameter itself is defined up to a fraction-linear transformation). It follows (the conjecture) that the y-function also has the single-valued character about each of the branch-point preimages:

$$y(\tau) = A + B(\tau-\tau_\mathrm{o})^n \exp\Big(\frac{-a\pi\mathrm{i}}{\tau-\tau_\mathrm{o}}\Big) + \cdots, \quad y(\tau) = A + B\tau^n \mathrm{e}^{a\pi\mathrm{i}\tau} + \cdots, \tag{24.5}$$

as $\tau \to \tau_\mathrm{o} \in \mathbb{R}$ or, respectively, $\tau \to +\mathrm{i}\infty$. For example, all asymptotics appearing in [1] fit this behavior. We can therefore rewrite equations $\mathcal{P}_6$ and (24.1) in the form of the modification of purely 'algebraic' uniformizing Schwarz–Fuchs third-order ODE (24.4):

$$[y, \tau] = A\,y_x^{-4} + B\,y_x^{-3} + C\,y_x^{-2} + D\,y_x^{-1} + E\,, \tag{24.6}$$

where (A, B, C, D, E) are certain rational functions of x, y and quadratic polynomials in parameters $(\alpha, \beta, \gamma, \delta)$ (explicit expressions are too cumbersome to display here). Because of the outstanding character of $\mathcal{P}_6$, this equation may be treated as a generator of "infinite genus curves". In the case of algebraic solutions, the right hand side of equation (24.6) becomes a rational function $\mathcal{Q}(x, y)$, which is (24.4). We conjecture that all the Painlevé solutions to equation (24.6) are the globally single-valued analytic functions with the structure (24.5), and the domain of their existence is a half-plain (under suitable normalization of τ). It is known that solutions to the lower Painlevé equations $\mathcal{P}_{1\ldots 5}$ (under an appropriate modification [2]) are the single-valued functions on $\overline{\mathbb{C}}$. In this respect, the pass from $\mathcal{P}_6$-equation over $\overline{\mathbb{C}}\backslash\{0, 1, \infty\}$ to the $\mathbb{H}^+$ and uniformization theory related to the coverings of a three punctured $\mathbf{\Gamma}(2)$-orbifold becomes very natural.

24.3 Calculus: Abelian Integrals and Affine (Analytic) Connections

Insomuch as we have not only τ-representations for the scalar (i.e., automorphic) functions on $\mathcal{R}$'s but rules for differential computations with theta functions of arbitrary arguments [3], we can close the differential apparatus on orbifolds $\mathfrak{T}$ whose compactifications corresponding to Painlevé $\mathcal{R}$'s. This includes the additively automorphic functions (Abelian integrals), differentials, and covariant differentiation, say, of 1-differentials $\nabla = \partial_\tau - \mathbf{\Gamma}(\tau)$. The latter leads to the necessity of introducing the geometric connection object $\mathbf{\Gamma}(\tau)$, which transforms according to the standard rule

$\widetilde{\boldsymbol{\Gamma}}(\tilde{\tau})d\tilde{\tau} = \boldsymbol{\Gamma}(\tau)d\tau - d\ln\frac{d\tilde{\tau}}{d\tau}$ under $\mathrm{SL}_2(\mathbb{R})$-transformations and respects the factor topology of $\mathbb{H}^+/\pi_1(\mathfrak{T})$. The characteristic feature of the (complex) 1-dimensional case (orbifolds and Riemann surfaces) is that it is completely described by the invariant third-order ODE (24.4). Therefore, the closed collection of data for the theory is given by the set $\{y(\tau), \dot{y}(\tau), \ddot{y}(\tau)\}$ if, however, the automorphism group of the generator $y(\tau)$ coincides with $\pi_1(\mathfrak{T})$. In general, automorphisms of the field generators are not bound to coincide with $\pi_1(\mathfrak{T})$ since the choice of the pair (x, y) is not unique. It is found, however, that the set of Painlevé orbifolds coming from Picard–Hitchin's curves (24.3) is not the case: $\mathrm{Aut}\, y(\tau) \cong \pi_1(\mathfrak{T})$. In this regard many (we suggest all) Painlevé curves differ from the majority of classical modular equations originated from purely group-algebraic considerations related to the group $\mathrm{PSL}_2(\mathbb{Z})$ or some its subgroups. By this means the expression

$$\boldsymbol{\Gamma}(\tau) = \frac{d}{d\tau}\ln\dot{y}(\tau) + \text{arbitrary (Abelian) 1-differential}$$

provides a general form for the sought connection on Painlevé $\mathfrak{T}$. We can normalize this $\boldsymbol{\Gamma}(\tau)$ to have only first-order poles (residues) and, integrating the transformation law above, one can see that the sum of such residues is invariant.

$$\int_{\partial\mathcal{R}} \widetilde{\boldsymbol{\Gamma}}(\tilde{\tau})d\tilde{\tau} = \int_{\partial\mathcal{R}} \boldsymbol{\Gamma}(\tau)d\tau = (2g-2)\cdot 2\pi\mathrm{i};$$

it depends only on the genus and, in effect, is equal to the number of zeroes of a holomorphic differential $\dot{\mathfrak{u}}(\tau)$. Varying the holomorphic differentials $\dot{\mathfrak{u}}_k(\tau)$, we can impart the simpler form to the connection

$$\boldsymbol{\Gamma}(\tau) = \frac{d}{d\tau}\ln\dot{\mathfrak{u}}(\tau) + \sum_{k=1}^{g}\dot{\mathfrak{u}}_k(\tau)$$

and build the elementary $\boldsymbol{\Gamma}$ with a single pole (if genus $g > 1$ then the analytic connection does always have a singularity). So we have the set of invariant objects $\{y(\tau), \dot{y}(\tau), \boldsymbol{\Gamma}(\tau)\}$ since functions $x(\tau)$, $y(\tau)$ are completely at hand. The remarkable fact is that an *affine (analyitc) connection on an arbitrary* $\mathfrak{T}$ *satisfies an autonomous ODE* $\Xi(\dddot{\Gamma}, \ddot{\Gamma}, \dot{\Gamma}, \Gamma) = 0$ and there is an algorithm for how to derive it.

For completeness we should involve in the analysis the integrals of closed 1-forms on our $\mathfrak{T}$'s and $\mathcal{R}$'s, if only because there are exact 1-forms whose integrals lead to the scalar objects. On the other hand, uniformization of any higher genera curves is reduced to the uniformization of *zero* genus orbifolds and the latter form towers and hierarchies. In the Painlevé uniformizing theory, in one way or another, many classical and nonclassical known zero genus orbifolds appear [3]. In turn they are related to nonzero genus curves which may cover elliptic ones, i.e., tori. We thus obtain the possibility of constructing explicitly Abelian integrals if they come from an elliptic cover. The following is a good example along these lines.

The Chudnovsky orbifold defined by the Fuchsian equation $(z^3 - z)\Psi'' + (3z^2 - 1)\Psi' + z\Psi = 0$ is related, through the Halphen transformation (zero genus elliptic cover) $z = \wp(\mathrm{u})$, to the Fuchsian equation on the lemniscatic torus $\wp'^2 = 4\wp^3 - 4\wp$. Correlating these facts we derive the nice τ-representation for the everywhere finite object u and analog of (24.4), the uniformizing Schwarz equation:

$$[\mathrm{u}, \tau] = -2\wp(2\mathrm{u}), \qquad \mathrm{u}(\tau) = \frac{1}{2}\frac{\vartheta_3(\tau)}{\vartheta_2(\tau)} \cdot {}_2F_1\left(\frac{1}{2}, \frac{1}{4}; \frac{5}{4} \middle| \frac{\vartheta_3^4(\tau)}{\vartheta_2^4(\tau)}\right)$$

(the check is a good exercise). This is a first *explicit and analytic* τ-representation for an additively automorphic function (Abelian integral $\mathrm{u} = \wp^{-1}(z)$) on an orbifold (Riemann surface) of a *negative* curvature -1. Under suitable cover this $u(\tau)$ may produce the τ-representation for u-integrals on higher genus curves; examples of the analogous ODEs and their solutions can also be obtained. All of them can be related to the Painlevé curves.

Bibliography

[1] D. Guzzetti, The Elliptic Representation of the General Painlevé VI Equation, *Comm. Pure Appl. Math.*, **LV**(10): 1280–1363, 2002.

[2] V.I. Gromak, I. Laine, S. Shimomura, *Painlevé Differential Equations in the Complex Plane*, Walter de Gruyter: Berlin, 2002.

[3] Yu.V. Brezhnev, The sixth Painlevé transcendent and uniformization of algebraic curves, `http://arXiv.org/abs/1011.1645`.

[4] M.V. Babich and L.A. Bordag, The elliptic form of the sixth Painlevé equation, preprint NTZ 25/1997, pp. 1–7, Leipzig, 1997.

Author Information

Yurii V. Brezhnev, Tomsk State University, Lenina Av. 36, 634050, Tomsk, Russia.
E-mail: `brezhnev@mail.ru`

Chapter 25

On Uniformizable Representation for Abelian Integrals

Yurii V. Brezhnev

Abstract. We show how the unformizable representation for Abelian integrals of nontrivial genera arise. The technique makes use of the famous Chudnovsky's Fuchsian linear differential equations and their relation to the sixth Painlevé transcendent.

Keywords. Fuchsian Equations, Abelian Integrals, Hypergeometric Functions, Schwarz Equation, Theta-functions.

2010 Mathematics Subject Classification. 30F10, 30F35, 33E17, 33E05, 11F03, 14H45, 14H55.

25.1 Introduction

In this chapter we first exhibit examples of uniformizable τ-representation for functions having an additive automorphic property, i.e., Abelian integrals on 1-dimensional orbifolds of a negative constant curvature. Orbifolds are generalizations of a notion of Riemann surface for the case when fundamental group of this 'manifold' has elements of finite or formally infinite order. These are usually called conic singularities or punctures. Matrix representations for fundamental groups of 1-dimensional orbifolds are described by Fuchsian linear ordinary differential equations (ODEs) of the second order. Their singularities are precisely the conic points on these orbifolds.

In [1] we showed how the Fuchsian equations

$$p\Psi'' + p'\Psi' + (x + A)\Psi = 0, \qquad p := x(x-\alpha)(x-\beta) \tag{25.1}$$

arise in the theory of Painlevé equations and in [2] we announced a first explicit example of a solvable allied Schwarz ODE

$$[\mathrm{u}, \tau] = -2\wp(2\mathrm{u}), \qquad \mathrm{u}(\tau) = \frac{\vartheta_3(\tau)}{\vartheta_2(\tau)} \cdot {}_2F_1\!\left(\frac{1}{2}, \frac{1}{4}; \frac{5}{4}\,\middle|\, \frac{\vartheta_3^4(\tau)}{\vartheta_2^4(\tau)}\right),$$

where u is an elliptic holomorphic integral. Below are some details on these examples and the methods of getting explicit formulas.

Research supported by the Federal Targeted Program under contract 02.740.11.0238 and partially by the Royal Society/NATO.

25.2 Schwarz Equation and Equations on Tori

As is well known, Schwarz's equations are the third order ODEs coming from linear ODEs of the form $\Psi_{xx} = \frac{1}{2}\,Q(x)\,\Psi$. The ratio of its two linearly independent solutions

$$\tau = \frac{\Psi_1(x)}{\Psi_2(x)}$$

defines τ as function of x, and conversely. We may write down an autonomic ODE defining x as function of τ. If, as usual,

$$\{f, z\} := \frac{f_{zzz}}{f_z} - \frac{3}{2}\frac{f_{zz}^2}{f_z^2}$$

defines the standard Schwarz derivative, then the inverted function $x = \chi(\tau)$ satisfies the ODE

$$[x, \tau] = Q(x), \tag{25.2}$$

where the following notation has been adopted $[x, \tau] := -\{\tau, x\}$, i.e.,

$$[x, \tau] := \frac{x_{\tau\tau\tau}}{x_\tau^3} - \frac{3}{2}\frac{x_{\tau\tau}^2}{x_\tau^4}.$$

We shall deal with the two examples of Fuchsian equations of the class (25.1):

$$x(x-1)(x+1)\,\Psi'' + \qquad (3x^2-1)\,\Psi' + (x+0)\,\Psi = 0,$$
$$x(x^2+3x+3)\,\Psi'' + (3x^2+6x+3)\,\Psi' + (x+1)\,\Psi = 0,$$

They belong to the set of four equations known as Chudnovsky equations [3]. With the help of the well-known linear transformation $\Psi \mapsto \psi$ we can transform these equations into the normal form

$$\psi'' = -\frac{1}{4}\frac{(x^2+1)^2}{x^2(x-1)^2(x+1)^2}\,\psi, \tag{25.3}$$

$$\psi'' = -\frac{1}{4}\frac{(x+1)(x+3)(x^2+3)}{x^2(x^2+3x+3)^2}\,\psi. \tag{25.4}$$

It is known that each of these equations has the four parabolic singularities on the Riemann sphere and thereby define two punctured spheres: the simplest orbifolds of genus zero. Halphen used an original trick to relate these equations with elliptic functions corresponding to algebraic curves of the form $y^2 = x(x-1)(x+1)$ and $y^2 = x(x^2+3x+3)$. They have the standard Weierstrassian models

$$y^2 = 4x^3 - 4x \quad \text{and} \quad y^2 = 4x^3 - 4,$$

respectively, i.e., Gauss' lemniscate and the equi-anharmonic elliptic curve.

Let us make the substitutions $x = \wp(\mathrm{u}; g_2, g_3)$ in equations (25.3) and (25.4), where invariants g_2, g_3 are chosen according to the Weierstrassian models above. We shall obtain Fuchsian equations in variable u and, then, Schwarz's equation of the form (25.2). An easy computation gives

$$[\mathrm{u}, \tau] = -2\wp(2\mathrm{u}; g_2, g_3). \tag{25.5}$$

and this equation is equivalent to equations on tori considered for the first time in the classical work [5]. Thus, if we find solutions $\mathrm{u} = \mathrm{u}(\tau)$ to this equation we obtain nontrivial examples of a uniformizable representation for the object u which is an Abelian integral, because

$$\mathrm{u} = \wp^{-1}(x; g_2, g_3).$$

On the other hand, the above-mentioned Fuchsian and Schwarz equations correspond to orbifolds with punctures, and their curvature is therefore not equal to zero but is a negative constant. The theory of such orbifolds is very nontrivial. Automorphic functions on them (Klein's Hauptmodulen) are known, a few as they are, but up to now there have been no examples of additively automorphic objects.

25.3 Holomorphic Elliptic Integrals and Hypergeometric Functions

25.3.1 Lemniscate

We have in this case

$$\pm\mathrm{u} = \int_\infty^x \frac{du}{\sqrt{4u^3 - 4u}} = \cdots. \tag{25.6}$$

It follows that

$$\cdots = \frac{1}{2}\int_\infty^x u^{-\frac{1}{2}}(u^2-1)^{-\frac{1}{2}}\, du \tag{25.7}$$

which is a particular case of a consequence of the integral definition to the hypergeometric ${}_2F_1$-function. More precisely, the standard integral definition of the ${}_2F_1$-functions is through the definite integral [4, Sect. 2.1.3]

$${}_2F_1(\alpha, \beta; \gamma|z) := \frac{\Gamma(\gamma)}{\Gamma(\beta)\Gamma(\gamma-\beta)} \int_0^1 u^{\beta-1}(1-u)^{\gamma-\beta-1}(1-zu)^{-\alpha}\, du,$$

but in some cases this definition may be rewritten in terms of *indefinite* integrals. Changing the integration variable $u \mapsto zu$, the upper limit $u = 1$ gets mapped to $u = z$. Putting further $\gamma - \beta - 1 = 0$, we arrive at a definition of this particular case for the ${}_2F_1$-function through pure indefinite integral. Clearly, all the other cases are in fact the variations of this scheme[1]. We obtain (see also the tables in [7])

$$\int_0^z u^{\alpha-1}(u-1)^{-\beta}\,du = \frac{\mathrm{e}^{\pi i\beta}}{\alpha} z^{\alpha}\cdot{}_2F_1(\beta,\alpha;\alpha+1|z), \quad \operatorname{Re}(\alpha)>0, \tag{25.8}$$

and, therefore,

$$\int_\infty^z u^{\alpha-1}(u-1)^{-\beta}\,du = \frac{z^{\alpha-\beta}}{\alpha-\beta}\cdot{}_2F_1(\beta,\beta-\alpha;\beta-\alpha+1|z^{-1}), \quad \operatorname{Re}(\beta-\alpha)>0. \tag{25.9}$$

The lemniscate integral under question is thus transformed into

$$\mathfrak{u} = \frac{1}{\sqrt{x}}\cdot{}_2F_1\left(\frac{1}{2},\frac{1}{4};\frac{5}{4}\middle|\frac{1}{x^2}\right), \tag{25.10}$$

We also know that τ-representations for all the Chudnovsky equations are expressed through the standard elliptic modular functions, namely, Jacobi's ϑ-constants. By this means, correlating these two points, we obtain an explicit solution to the Schwarz equation (25.5).

Let us use the Hauptmodul $x=\chi(\tau)$ for the first Chudnovsky equation (25.3):

$$\chi(\tau) = \frac{\vartheta_2^2(\tau)}{\vartheta_3^2(\tau)},$$

where

$$\vartheta_2(\tau) := \mathrm{e}^{\frac{1}{4}\pi i\tau}\sum_{-\infty}^{\infty}{}_k\, \mathrm{e}^{(k^2+k)\pi i\tau}, \quad \vartheta_3(\tau) := \sum_{-\infty}^{\infty}{}_k\, \mathrm{e}^{k^2\pi i\tau},$$
$$\vartheta_4(\tau) := \sum_{-\infty}^{\infty}{}_k\,(-1)^k\,\mathrm{e}^{k^2\pi i\tau}.$$

To put it differently, this function solves equation

$$[x,\tau] = -\frac{1}{2}\frac{(x^2+1)^2}{(x^3-x)^2}.$$

Substituting this $\chi(\tau)$ into (25.10), we get that the function

$$\mathfrak{u}(\tau) = \frac{\vartheta_3(\tau)}{\vartheta_2(\tau)}\cdot{}_2F_1\left(\frac{1}{2},\frac{1}{4};\frac{5}{4}\middle|\frac{\vartheta_3^4(\tau)}{\vartheta_2^4(\tau)}\right)$$

solves equation (25.5) under $(g_2,g_3)=(4,0)$.

[1] To all appearances, integral formulae for definition of the ${}_2F_1$-functions with a variable upper limit were obtained for the first time in the 1876 dissertation by Tikhomandritskiĭ [8, p. 78].

25.3.2 Equi-Anharmonic Curve

First we shift x-variable $x = z - 1$ in equation (25.4) to obtain the canonical form $y^2 = 4z^3 - 4$ with $(g_2, g_3) = (0, 4)$. Halphen's transformation and Hauptmodul $x = \chi(\tau)$ [6] in this case have the form

$$z = \wp(\mathrm{u}; 0, 4), \quad z = 9\frac{\eta^3(9\tau)}{\eta^3(\tau)} + 1,$$

where $\eta(\tau) := \mathrm{e}^{\frac{\pi \mathrm{i}}{12}\tau} \prod_k (1 - \mathrm{e}^{2\pi \mathrm{i} k\tau})$ is the Dedekind eta-function. This Hauptmodul $z(\tau)$ satisfies the equation

$$[z, \tau] = -\frac{1}{2}\frac{z(z^3+8)}{(z^3-1)^2},$$

and one can show that the change $z \mapsto \mathrm{u}$ above transforms this Schwarz equation into equation (25.5). We have, according to (25.9),

$$\pm\mathrm{u} = \int_\infty^z \frac{du}{\sqrt{4u^3-4}} = \frac{1}{\sqrt{z}} \cdot {}_2F_1\!\left(\frac{1}{2}, \frac{1}{6}; \frac{7}{6} \,\middle|\, \frac{1}{z^3}\right)$$

and hence

$$\mathrm{u} = \left(9\frac{\eta^3(9\tau)}{\eta^3(\tau)} + 1\right)^{-\frac{1}{2}} \cdot {}_2F_1\!\left(\frac{1}{2}, \frac{1}{6}; \frac{7}{6} \,\middle|\, \left\{9\frac{\eta^3(9\tau)}{\eta^3(\tau)} + 1\right\}^{-3}\right).$$

It is a very nontrivial exercise to check directly that this function solves equation (25.5).

Remark 25.1. All the Chudnovsky Hauptmodule are single-valued functions and, hence, in spite of a square root in the last formula, this expression provides a single-valued object in the neighborhood of the point $z = \infty$. Indeed, making the change $\boldsymbol{z}^2 = z^{-1}$ in Schwarz's equation for z above, we get

$$[\boldsymbol{z}, \tau] = -\frac{1}{2}\frac{1}{\boldsymbol{z}^2} + \cdots$$

and hence $\boldsymbol{z} = \boldsymbol{z}(\tau)$ has an exponentially single-valued behavior in τ about point $z = \infty$,and ${}_2F_1$-function is, by definition, a single-valued Taylor series at the origin $\boldsymbol{z} = 0$.

We can, however, get an explicit root-free solution to this problem with the help of the following trick. Use formula (25.9) and the fact that the holomorphic integral is defined up to an additive constant:

$$\mathrm{u} = \int_\infty^x \frac{du}{\sqrt{4u^3-4}} = \int_\infty^0 \frac{du}{\sqrt{4u^3-4}} + \int_0^x \frac{du}{\sqrt{4u^3-4}}.$$

The first of these integrals is a transcendental constant $\mathrm{u}_\circ$ such that $\wp(\mathrm{u}_\circ) = 0$, i.e., zero of the Weierstrass $\wp(z; 0, 4)$-function. It is computed into elliptic integrals

$$\mathrm{u}_\circ = \frac{\mathrm{i}}{2\sqrt[4]{3}} F\Big(\sqrt[4]{3}(\sqrt{3}-1); \frac{1}{4}(\sqrt{6}+\sqrt{2})\Big) - \frac{1}{\sqrt[4]{3}} K\Big(\frac{\sqrt{3}-1}{2\sqrt{2}}\Big),$$

where F and K are the standard noncomplete and complete elliptic integrals [4]. The second integral, upon application of (25.8), becomes

$$\int_0^x \frac{du}{\sqrt{4u^3-4}} = \frac{\mathrm{i}}{2} x \cdot {}_2F_1\Big(\frac{1}{2}, \frac{1}{3}; \frac{4}{3}\Big| x^3\Big).$$

Turning to the Hauptmodul (variable) $z(\tau)$, we get finally

$$\pm\mathrm{u}(\tau) = \mathrm{u}_\circ + \frac{\mathrm{i}}{2}\Big(9\frac{\eta^3(9\tau)}{\eta^3(\tau)} + 1\Big) \cdot {}_2F_1\Big(\frac{1}{2}, \frac{1}{3}; \frac{4}{3}\Big| \Big\{9\frac{\eta^3(9\tau)}{\eta^3(\tau)} + 1\Big\}^3\Big).$$

The single-valuedness of this expression is now obvious; it solves equation (25.5).

25.4 Abelian Integrals for Genus $g > 1$

If α, β in formula (25.8) are the rational numbers then one has an Abelian integral belonging to a rationality that may have genus greater than unity. The integral may be holomorphic, meromorphic, or logarithmic. If, on the other hand, we have a single-valued τ-representation for z^α, then we automatically get an explicit τ-representation for this integral. This is a nontrivial consequence because, in general, Abelian integrals are not expressible in terms of any known functions and we thereby get analogs of Weierstrass' uniformization theory for elliptic curves.

Recall that Weierstrass' theory, in its complete description, includes closed and self-contained collections of the base functions and differentials

$$w^2 = 4z^3 - az - b, \quad \{z = \wp(\tau),\ w = \wp'(\tau)\}, \quad dz = \wp'(\tau)\,d\tau, \tag{25.11}$$

and the two principal integrals

$$\mathbf{II} := \int^z z\frac{dz}{w} = -\zeta(\mathrm{u}), \qquad \mathbf{III} := \frac{1}{2}\int^z \frac{w + \wp'(\alpha)}{z - \wp(\alpha)}\frac{dz}{w} = \ln\frac{\sigma(\mathrm{u}-\alpha)}{\sigma(\mathrm{u})} + \zeta(\alpha)\mathrm{u}, \tag{25.12}$$

as functions of the fundamental holomorphic object

$$\mathrm{u} := \int^z \frac{dz}{w}, \qquad \mathrm{u}(\tau) = \tau. \tag{25.13}$$

By this means, having the base holomorphic and meromorphic integrals we can manipulate with this integrals and functions in exactly the same manner as we do with Weierstrassian objects σ, ζ, $\wp$, and $\wp'$. In as much as no one explicit formula for such a theory was hitherto known, it is perhaps of some interest to exhibit examples.

25.4.1 Higher Genera. Examples

Consider the set of integrals

$$\int_0^z \frac{u^m\,du}{\sqrt[k]{u\,(u^4-1)}}.$$

Considering them as integrals for the algebraic irrationality $w^k = z\,(z^4-1)$, we find that they can be either holomorphic or meromorphic. For example, in the case $k = 2$ (hyperelliptic curves) we have two base holomorphic integrals when $m = 0, 1$ and two base meromorphic ones when $m = 2$ or 3. The only thing we need now is the formula for Hauptmodul $z = z(\tau)$. It is known explicitly in cases when z, as a solution to the corresponding Fuchs–Schwarz equation, has only parabolic singularities, i.e., the Fuchsian equation has punctures at all the points $z = \{0, \pm 1, \pm \mathrm{i}, \infty\}$:

$$z = \frac{\vartheta_2(\tau)}{\vartheta_3(\tau)}$$

We derive from (25.8)

$$\begin{aligned}\int_0^z u^{\alpha-1}(u^n-1)^{-\beta}\,du &= \frac{\mathrm{e}^{\pi \mathrm{i}\beta}}{\alpha} z^{\alpha}\cdot {}_2F_1\Big(\beta, \frac{\alpha}{n}; \frac{\alpha}{n}+1\Big| z^n\Big),\\ \int_\infty^z u^{\alpha-1}(u^n-1)^{-\beta}\,du &= \frac{z^{\alpha-n\beta}}{\alpha-n\beta}\cdot {}_2F_1\Big(\beta, \beta-\frac{\alpha}{n}; \beta-\frac{\alpha}{n}+1\Big|\frac{1}{z^n}\Big)\end{aligned} \qquad (25.14)$$

and $\alpha = m + 1 - \frac{1}{k}$. The value n does not affect the single-valuedness of the τ-representation, and under $k = 2$ we have to do the square root of $z(\tau)$. One can show that the ratio of any two Jacobi's ϑ-constant is a complete square. In particular,

$$\sqrt{\frac{\vartheta_2(\tau)}{\vartheta_3(\tau)}} = \sqrt{2}\,\frac{\vartheta_2(\tau)}{\vartheta_2(\frac{\tau}{2})}.$$

It follows that for the famous algebraic curve $w^2 = z^5 - z$ we can obtain a complete set of base Abelian integrals.

Theorem 25.2. *Every Abelian (homolomorphic, meromorphic, or logarithmic) integral belonging to the algebraic irrationality $w^2 = z^5 - z$ has a uniformizing τ-representation through the Jacobi's θ-functions of the two base holomorphic integrals.*

Proof. This curve admits a representation in the form of two isomorphic elliptic curves. Indeed, it is suffice to use the well-known formulas by Jacobi–Legendre for covering the curve

$$y^2 = x\,(x-1)(x-A)(x-B)(x-AB).$$

We have in this case the following cover of torus $(\mathfrak{u})$:

$$\begin{cases} \wp(\mathfrak{u}) = -\dfrac{(\sqrt{A} \pm \sqrt{B})^2}{(x-A)(x-B)}\, x - \dfrac{1}{3}(k^{\pm} + 1) \\[2ex] \wp'(\mathfrak{u}) = 2\dfrac{(\sqrt{A} \pm \sqrt{B})^2}{\sqrt{(1-A)(1-B)}}\, \dfrac{x \pm \sqrt{AB}}{(x-A)^2(x-B)^2}\, y \end{cases},$$

where

$$k^{\pm} = -\frac{(\sqrt{A} \pm \sqrt{B})^2}{(1-A)(1-B)}.$$

Then expression $\wp'(\mathfrak{u})^2 = 4\wp^3(\mathfrak{u}) - g_2\wp(\mathfrak{u}) - g_3$ is equivalent to the hyperelliptic curve above. Reduction of holomorphic differentials has the form

$$d\mathfrak{u} = \frac{1}{2}\sqrt{(1-A)(1-B)} \cdot (x \mp \sqrt{A}\sqrt{B})\frac{dx}{y}.$$

and the curve $w^2 = z^5 - z$ corresponds to the following parameters:

$$A = -1, \qquad B = \mathrm{i}, \qquad k^{\pm} = \frac{1}{2}(1 \pm \sqrt{2}),$$

$$\begin{aligned} \wp'^2 &= 4\wp^3 - \frac{5}{3}\wp \pm \frac{7}{27}\sqrt{2} \\ &= 4\Big\{\wp + \underbrace{\tfrac{1}{6}(3 \pm \sqrt{2})}_{-e}\Big\}\Big\{\wp - \underbrace{(\pm\tfrac{1}{3}\sqrt{2})}_{e'}\Big\}\Big\{\wp - \underbrace{\tfrac{1}{6}(3 \mp \sqrt{2})}_{e''}\Big\}. \end{aligned}$$

Using (25.14), we derive

$$\mathfrak{U} = \int_0^z \frac{u^m\, du}{\sqrt{u(u^4-1)}} = \frac{2\sqrt{2}\mathrm{i}}{2m+1}\, \frac{\vartheta_2^{m+1}(\tau)}{\vartheta_3^m(\tau)\vartheta_2(\frac{\tau}{2})} \cdot {}_2F_1\left(\frac{1}{2}, \frac{1}{4}m + \frac{1}{8}; \frac{1}{4}m + \frac{9}{8}\,\middle|\, \frac{\vartheta_2^4(\tau)}{\vartheta_3^4(\tau)}\right)$$

and these expressions provide not only two holomorphic integrals $\mathfrak{U}_1, \mathfrak{U}_2$ $(m = 0, 1)$ but the meromorphic ones as well. Inasmuch as we have holomorphic integrals as functions of τ we can compute any other Abelian integral in terms of 2-dimensional Θ-functions. On the other hand, by virtue of the cover above, all the Θ-functions are reducible to Jacobi's θ. It follows that all the Abelian integrals are expressed through such θ-functions having as their arguments the two holomorphic integrals above. These θ-formulas can be transformed into the Weierstrass $\wp$, $\wp'$, ζ, σ because all the integrals are reducible to the objects (25.11)–(25.13). □

Bibliography

[1] Yu. V. Brezhnev, The sixth Painlevé transcendent and uniformization of algebraic curves, http://arXiv.org/abs/1011.1645.

[2] Yu. V. Brezhnev, The sixth Painlevé transcendent and uniformizable orbifolds, *Chapter 24, this volume.*

[3] D. V. Chudnovsky and G. V. Chudnovsky, Transcendental Methods and Theta-Functions, *Proc. Sympos. Pure Math.,* **49**, Part 2: 167–232, 1989.

[4] A. Erdélyi, W. Magnus, F. Oberhettinger, and F. G. Tricomi, *Higher Transcendental Functions **I**. The Hypergeometric Function, Legendre Functions.* McGraw–Hill, New York, 1953.

[5] L. Keen, H. E. Rauch, and A. T. Vasquez, Moduli of punctured tori and the accessory parameter of Lamé's equation, *Trans. Amer. Math. Soc.,* **225**: 201–230, 1979.

[6] R. Maier, On rationally parametrized modular equations, *J. Ramanujan Math. Soc.,* **24**: 1–73, 2009.

[7] I. S. Gradshteyn, I. M. Ryzhik, *Table of Integrals, Series, and Products.* New York, Academic Press, 1980.

[8] M. A. Tikhomandritskiĭ, *On Hypergeometric Series,* Magister Thesis, St. Petersburg, 1876 (in Russian).

Author Information

Yurii V. Brezhnev, Tomsk State University, Lenina Av. 36, 634050, Tomsk, Russia.
E-mail: brezhnev@mail.ru

Chapter 26

Phase Shift for a Special Solution to the Korteweg–de Vries Equation in the Whitham Zone

Rustem N. Garifullin

Abstract. The aim of this work is to find the phase shift of a special solution of the Kortevege–de Vries equation in a Whitham zone. It is done without using any averaging method. We use a complimentary condition of the Kortevege–de Vries equation and a fifth-order ordinary equation.

Keywords. Nondissipative Shock Waves, Phase Shift, Korteweg–de Vries Equation.

2010 Mathematics Subject Classification. Primary: 37K10; Secondary: 76L05.

26.1 Introduction

In this chapter we investigate a special solution $u(t, x)$ of the Kortevege–de Vries equation

$$u_t + uu_x + u_{xxx} = 0. \tag{26.1}$$

This solution is at the same time a solution to the following ordinary differential equation (ODE)

$$\left(u_{xxxx} + \frac{5u_{xx}u}{3} + \frac{5u_x^2}{6} + \frac{5u^3}{18}\right)'_x - \frac{2u + xu_x - 3t(u_{xxx} + uu_x)}{6} = 0 \tag{26.2}$$

which is obtained as a combination of stationary parts of two symmetries of the KdV equation. One of them is the fifth-order higher (generalized) symmetry of the KdV equation,

$$u_{\tau_5} = \left(u_{xxxx} + \frac{5u_{xx}u}{3} + \frac{5u_x^2}{6} + \frac{5u^3}{18}\right)'_x \tag{26.3}$$

and the second one is classical dilation symmetry

$$u_{\tau_r} = 2u + xu_x - 3t(u_{xxx} + uu_x). \tag{26.4}$$

Equation (26.2) may be called the first high analog of Painleve I equation.

This work was completed with the support of RFBR, grants 10-01-91222 and 10-01-00186.

This solution plays the main role in the problem of nondissipating shock waves [1–3]. In [1] the asymptotic behavior of these simultaneous solutions in the zone of undamped oscillations is given by quasisimple wave solutions to Whitham equations. But the phase shift was not found there. In this chapter we find the asymptotic behavior of these solutions including the phase shift by a method suggested in [4].

26.2 Evaluation of the Phase Shift

Our aim is to construct asymptotics of the solution in the zone of undamped oscillations when $t \to \infty$. Following familiar techniques, we change the variables by

$$u = t\,U(t,s), \quad s = \frac{x}{t^2}.$$

Then equations (26.1) and (26.2) take the form

$$t^{-5}U_{sss} + tU_t - 2sU_s + UU_s + U = 0, \tag{26.5}$$

$$t^{-10}U_{sssss} + \frac{1}{6}t^{-5}(20U_sU_{ss} + (10U+3)U_{sss}) + \frac{1}{6}(5U^2 - s + 3U)U_s - \frac{1}{3}U = 0.$$

We now look for a solution U of the system in the form of the asymptotic series

$$U = U_0(\varphi, s) + t^{-7/4}U_1(\varphi, s) + t^{-7/2}U_2(\varphi, s) + \cdots, \tag{26.6}$$

where U_0, U_1 and U_2 are 2π-periodic in the fast variable φ. The latter is assumed to be of the form

$$\varphi = t^{5/2}f(s) + n(s),$$

where by $n(s)$ is meant precisely the phase shift.

For the unknown function U_0 we get the nonlinear system

$$(f')^2\partial_\varphi^3 U_0 + \left(\frac{5f}{2f'} - 2s + U_0\right)\partial_\varphi U_0 = 0, \tag{26.7}$$

$$(f')^4\partial_\varphi^5 U_0 + \frac{1}{6}\left(20\partial_\varphi U_0 \partial_\varphi^2 U_0 + (10U_0+3)\partial_\varphi^3 U_0\right)(f')^2$$
$$+ \frac{1}{6}(5U_0^2 - s + 3U_0)\partial_\varphi U_0 = 0,$$

and nonhomogeneous linear systems for unknown functions U_1, U_2, which are omitted.

From the compatibility condition of the equations for U_0 we obtain a second-order equation

$$(f')^2\partial_\varphi^2 U_0 + \frac{1}{2}U_0^2 - 2U_0 s + \frac{5}{2}s + 12s^2 + \frac{5}{4}(-24s + 2U_0 - 3)\frac{f}{f'} + \frac{75}{4}\frac{f^2}{(f')^2} = 0. \tag{26.8}$$

From this equation we can write:

$$(f'\partial_\varphi U_0)^2 = 2sU_0^2 - \frac{1}{3}U_0^3 - sU_0(24s+5) + (24s+3-U_0) \cdot \frac{5U_0 f}{2f'} - \frac{75U_0 f^2}{2(f')^2} + \frac{25}{4}b(s). \tag{26.9}$$

Here $b(s)$ is an arbitrary function (constant of integration).

From the compatibility condition of the equations for U_1 we obtain a system of the first-order nonlinear equations for functions

$$a(s) = 5/2f/f', b(s),$$

which are omitted. From the compatibility condition of the equations for U_2 we obtain the equation of the next form:

$$\partial_\varphi U_0(n''' + A_1 n'' + A_2 n') + \partial_s^3 U_0 + B_1 \partial_s^2 U_0 \partial_s U_0 + B_2 \partial_s^2 U_0 + B_3(\partial_s U_0)^3 + B_4(\partial_s U_0)^2 + B_5 \partial_s U_0 + B_6 = 0, \tag{26.10}$$

where

$$A_i = A_i(s, f, a, b), \quad B_i = B_i(U_0, s, f, a, b)$$

are some functions.

From (26.10) we can immediately get

$$n''' + A_1 n'' + A_2 n' = 0. \tag{26.11}$$

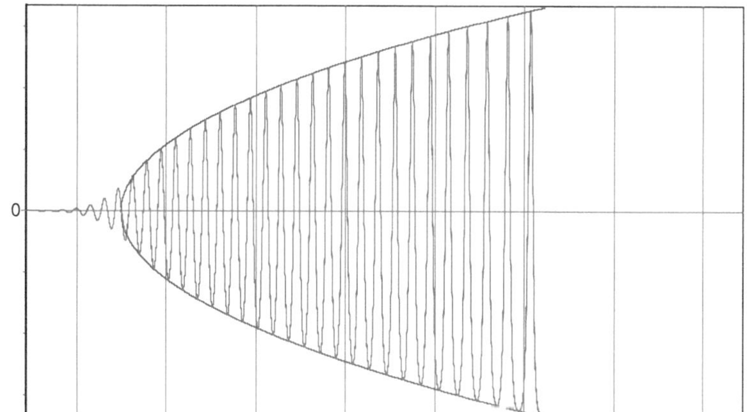

Figure 26.1. The numerical simulation for the function $U(t, z)$ corresponding to $t = 19$.

The common solution of (26.11) has the form:

$$n(s) = C_1 + C_2 n_1(s) + C_2 n_2(s). \tag{26.12}$$

By using numerical experiments we finally find

$$n(s) = \pi.$$

We observed the difference between numerical and asymptotic solutions and found that it decreases as $t^{-5/2}$ for this value of $n(s)$. In Figure 26.1 one can observe numerical solutions for the function $U(t, z)$ where $t = 19$.

Bibliography

[1] R. N. Garifullin and B. I. Suleimanov, From weak discontinuities to nondissipative shock waves, *JETP,* **110**: 133–146, 2010, DOI 10.1134/S1063776110010164.

[2] A. M. Kamchatnov and S. V. Korneev, Flow of a Bose–Einstein condensate in a quasi-one-dimensional channel under the action of a piston, *JETP,* **110**: 170–182, 2010, DOI 10.1134/S1063776110010206.

[3] G. A. El, V. V. Khodorovskii, and A. M. Leszczyszyn, Refraction of dispersive shock waves, `arXiv:1105.1920v1 [nlin.PS]`.

[4] R. Garifullin, B. Suleimanov, and N. Tarkhanov, Phase Shift in the Whitham Zone for the Gurevich-Pitaevskii Special Solution of the Korteweg–de Vries Equation, *Ph. Let. A*, **374**: 1420–1424, 2010, DOI 10.1016/j.physleta.2010.01.057.

Author Information

Rustem N. Garifullin, Chernushevskogo str., 112, 450008, Ufa, Russia.
E-mail: `rustem@matem.anrb.ru`

Chapter 27

Fuchsian Reduction of Differential Equations

Valentina A. Golubeva

Abstract. The historical review and some present results on the multidimensional theory of Fuchsian equations and also the exposition of some results concerning the Fuchsian reduction of certain linear and nonlinear equations are given.

Keywords. Griffits Method of Reduction of the Pole Order, Knizhnik–Zamolodchikov Equation, Fuchsian Reduction, Fuchsian System.

2010 Mathematics Subject Classification. Primary: 35Q40; Secondary: 35Q15.

At the end of 1950s, I. M. Gel'fand studied and delivered a course on quantum electrodynamics and the theory of elementary particles, and he formulated the problem of generalizing the well-known Riemann–Hilbert problem:

Construct a system of partial differential equations of the hypergeometric type for the Feynman integral which is a term of some finite order of the series for scattering the matrix of quantum electrodynamics.

The notion of a hypergeometric-type partial differential equation was not known. This statement was based on the ramification properties of such integrals on the Landau varieties investigated in some particular cases in physics publications.

The first publication of this problem was done by T. Regge in 1965 in the Proceedings of the Conference "Battelle Rencontres" (1965) [1]. T. Regge was known for this conjecture from O. Parasyuk who at that time was actively working in Moscow in the seminars of N. N. Bogolyubov and I. M. Gel'fand at the Moscow University.

However, since the beginning of the 20th century, hypergeometric functions of two variables had been made known by Appell and Kampé de Fériet [2]. R. Gérard tried to study their properties, which were very similar to the properties of the Gauss hypergeometric function, and even to solve the multidimensional local Riemann–Hilbert problem, but he could not realize this dream due to some errors.

This research was continued by V. A. Golubeva, who obtained the systems of partial differential equations of Fuchsian type for hypergeometric functions of two variables from Appell and Kampé de Fériet, and obtained the commutation relations between the coefficients of these equations [3]. The definitions of multidimensional Fuchsian-type equations or Pfaff systems of Fuchsian type were given.

This work was completed with the support of grants NSh-5998.2012.1

The partial differential equations for certain terms of the series for the S-matrix were obtained by V. Golubeva. These equations were of the Fuchsian type. Singular loci of these terms of the S-matrix appeared naturally connected with Landau's variety of the Feynman integral. The method of derivation of the equations was based on Ph. Griffiths algebraic-geometric theorem on reduction of the pole order of rational differential form. This method often does not give the Fuchsian system, but some Fuchsian reduction permits obtaining a global Fuchsian system.

The next step in the investigation of the ramification properties of Feynman integrals was published in a paper by V. Z. Enolskiiy, who obtained the local expansion of the Feynman integral in the neighborhood of the ordinary point of its Landau variety.

After construction of the first series of the multidimensional Fuchsian systems, the question was stated whether there are the other objects satisfying the partial differential equations of the Fuchsian type. The next problem was to find the physical models which can be described by equations of the Fuchsian type. Happily, the physicists prepared such models in conformal field theory. The well-known Knizhnik–Zamolodchikov equation [9] is the multidimensional equation of Fuchsian type whose singular locus is the set of hyperplanes in $\mathbb{C}^n$ turned out to be the reflection hyperplanes of Weyl group of A_n type and with the coefficients belonging to the tensor power of the corresponding Lie algebra A_n. This equation was investigated in detail (see e.g. Chr. Kassel, 1995 [11]). It was natural to state the problem of construction and investigation of the Knizhnik–Zamolodchikov-type equation with other symmetry algebras. For example, such an equation with symmetries corresponding to the Lie algebra B_n was presented in A. Leibman's paper (1994, [10]), who, using the Casimir element of the second order, obtained the solutions of the integrability relations for this equations. Leibman considered only the one-parameter case of the model. However, for the characterization of the models with symmetries described by the Lie algebra of B_n it is necessary to use two parameters. To date, the problem of constructing a two-parametric Knizhnik–Zamolodchikov equation of the B_n-type has not been solved. Using the higher Casimir elements, it is possible to construct some equations of the Knizhnik–Zamolodchikov type but they are not two-parametric.

It is necessary to note that the coefficients of Knizhnik–Zamolodchikov equations can be not only elements of (tensor products)of algebras, but also elements of other nature, e.g., knots and links. The application of central elements such as Casimir and Capelli elements of a higher order for generation of new Knizhnik–Zamolodchikov equations show promise.

The contemporary statement of the multidimensional problem analogues to the Riemann–Hilbert–Gelfand–Regge problem appears as follows:

In $\mathbb{C}^n$ the arrangement of hyperplanes is given as the reflection of hyperplanes of the Weyl group of one of the types B_n, C_n, D_n, G_2, F_4, E_6, etc. to construct the equations of the Knizhnik–Zamolodchikov type with singular locus on the given arrangement.

Some of these problems have now been solved.

Note. In the above statement of the problem, the monodromy is not prescribed (as in the one-dimensional case) because it depends on the symmetry algebra of the given arrangement. The choice of the coefficients sets of the equations is possible in rich bounds.

At present, in mathematical physics the method of Fuchsian reduction is applied to linear as well as nonlinear systems of equations. It consists of applying the series of local transformations as independent as well as dependent variables for deriving the local equations of Fuchsian type. Such a transformation permits the use of a rich technique of local developments for the solutions of Fuchsian and Painlevé equations. The method of Fuchsian reduction is used in astronomy, general relativity, nonlinear optics, soliton theory, differential geometry, etc. In particular, in the theory of solitons Fuchsian reduction enables us to answer the question of whether the formal solutions of the completely integrable equations represent actual solutions.

27.1 Fuchsian Reduction

In this section we briefly sketch the sufficiently new methods of investigation of the local properties of solutions of differential equation. These methods are called Fuchsian reduction.

The Fuchsian reduction is a representation of a differential equation in which the singularity locus can be of parameters. Usually the reduction is done in several steps.

Let the equation be of type $F(u) = 0$, where F depends on u, its derivatives and on an independent variable.

The first step is the analysis of the leading order. The aim is to find a function T and a pair (u_0, ν), such that $F(u_0 T^\nu) = 0$ to leading order. The hope is to find a solution such that $U \sim u_0 T^\nu$. The second step is a construction of the formal solution of the typical form

$$u = T^\nu \sum_{j=0}^{\infty} \sum_{p=0}^{j} u_{jp} T^j (\ln T)^p.$$

One has $u = T^\nu(u_0 + T^\epsilon v)$. Put $D = T\frac{d}{dT}$. If ϵ is small enough, v solves a system of the form $(D + A + \epsilon)v = T^\sigma f(T, v)$.

27.1.1 Two Simple Examples

(i) Consider the equation $u_{tt} = u^2$. Let $t = a$, $T = t-a$ and $u = u_0+T(u_1+v)$, here $u_0 = u(a)$, $u_1 = u'(a)$. We obtain $D(D + 1)v = T(u_0 + T(u_1 + Tv))$. This is a Fuchsian system.

(ii) Let us seek the manifold of solutions of $u_{tt} - u = 3u^2$. Let $T = exp(-T)$, we obtain $(D + 1)(D - 1)u = 6u^2$. Let $u = aT + T^2 v$, we obtain $(D + 1)(D + 3)v = 3(a + Tv)^2$. Further, let $v = a^2 + w$. We obtain the Fuchsian system $(D + 1)(D + 3)w = 3((a + Tv)^2 - a^2) = 3Tv(2a^2 + Tv)$.

27.1.2 A More Complex Example

Consider the equation

$$u_{tt} - 6u^2 - t = 0.$$

Let $T = t - a$. The equation becomes

$$u_{tt} - 6u^2 - T - a = 0.$$

We seek the leading term of the form

$$u_0 = T^\nu$$

with $u_0 \neq 0$. For $u_0 = 1$ and $\nu = -2$, upon multiplication by T^2 we obtain

$$D((D-1)u - 6(Tu)^2 - T^3 - aT^2 = 0.$$

Let

$$u = \frac{1}{T^2}(1 + T^\epsilon v(T)).$$

Multiplying that equation by $T^{2-\epsilon}$ and substituting u we obtain

$$(D+\epsilon-2)(D+\epsilon-3)v - 6T{-}\epsilon((1+T^\epsilon v)^2 - 1) - T^{2-\epsilon}(T^3 + aT^2) = 0.$$

Since

$$(D+\epsilon-2)(D+\epsilon-3) - 12 = (D+\epsilon+1)(D+\epsilon-6)$$

we obtain

$$(D+\epsilon-2)(D+\epsilon-6)v = 6T^\epsilon v^2 + T^{4-\epsilon}(T+a).$$

For $\epsilon = 4$ one gets

$$(D+5)(D-2)v = a + T + 6T4v^2.$$

Replacing v by $v - \frac{a}{10}$ one gets a Fuchsian system.

The general results yield the formal series solution

$$u = \frac{1}{T^2} - \frac{a}{10}T^2 - \frac{1}{6}T^3 + T^4(b + \ldots),$$

where b is arbitrary.

The second reduction: define a function w by

$$u = \frac{1}{T^2} - \frac{a}{10}T^2 - \frac{1}{6}T^3 + T^4 w(T).$$

This function solves $D(D+7)w = Tg(T, a, w)$, where $g = 6T^2(-1/10 - T/6 + T^2 w)^2$. The second reduction is now complete. The general result implies a unique solution u of the form

$$u = \frac{1}{(t-a)^2} - \frac{a}{10}(t-a)^2 - \frac{1}{6}(t-a)^3 + b(t-a)^4 + \cdots$$

with arbitrary b. The expansion converges in some disk $|t-a| < r(a,b)$. The singularity data are a, b. Solutions have a double pole at $t = a$.

27.2 Two Applications: Astronomy and Relativity Theory

In this section two examples of applications of Fuchsian reduction are considered. The examples are extracted from works by S. Kitchenassamy [12, 13].

27.2.1 Astronomy. A Model of Gaseous Stars

M is the mass of a star. The star lies inside a ball with radius R with a center in 0. S_r is a sphere of radius $r < R$.

The unknowns are:

- the mass $M(r)$ inside $S(r)$;
- the pressure $P(r)$;
- the density $\rho(r)$;
- the temperature $T(r)$ on $S(r)$;
- the luminosity $L(r)$, the outward flow of energy per second through $S(r)$.

Equations: the density is related to M by equation

$$\frac{dM}{dr} = 4\pi r^2 \rho, \tag{27.1}$$

$$\frac{dP}{dr} = \frac{GM(r)\rho(r)}{r^2}, \tag{27.2}$$

Here the constant G is defined from the Poisson equation $\operatorname{div}(\bar{g}) = -4\pi G\rho$, $\bar{g} = g(r)\bar{r}/r$ is a vector of acceleration of gravity with magnitude $g(r)$.

$$P = P_{gas} + P_{rad}.$$

In the case of the Sun one has put $P = P_{gas} = \frac{\rho}{\mu} RT$.

Assuming $dL = \epsilon dM$, where ϵ is total rate of energy outflow per second and per gramm. A possible choice $\epsilon \sim \rho T^{\nu}$. For luminosity we have

$$\frac{dL}{dr} = 4\pi r^2 \epsilon \rho,$$

for the temperature we have

$$\frac{dT}{dm} = -\frac{GM\rho T}{r^2 P}\nabla^*.$$

Here ∇^* equals $d \ln T / \ln P$.

Boundary conditions:

(i) at the surface $r = R$ $M = M_s$ is total mass, $P = 0$, $T = 0$, $L = L_s$ is total luminosity;

(ii) at the center $r = 0$ $M = 0$, $L = 0$, $P = P_c$, $T = T_c$ (P_c, T_c are unknown values).

The problem: define (M, P, T, L) as functions of r, or (r, P, T, L) as functions of M.

The Politropic Model

Setting $\rho = \lambda T^n$ ad P propotional to ρT, one finds $P = KT^{n+1}$. Substitute this to (27.2), and find that $M(r)$ is proportional to $r^2\frac{dT}{dr}$. Then equation (27.1) is written as the Lane–Emden–Fouler equation

$$\frac{d^2T}{dr^2} + \frac{1}{2}\frac{dT}{dr} + cT^n = 0.$$

27.2.2 Relativity. Gowdy Space-Time

In this section we transform the Gowdy equations to a second-order system for u and v which is then converted to a first-order Fuchsian system.

Gowdy space-time are Lorenzian manifolds with spacelike slices homeomorphic to the three-torus T^3 on which a two-dimensional Abelian isometry group G acts without fixed points.

The metric takes the form

$$ds^2 = e^{\lambda/2}t^{-1/2}(-dt^2 + dx^2) + t\big(e^{-Z}(dy + Xdz)^2 + e^Z dz^2\big),$$

where λ, X, Z are functions of t, x only and 2π-periodic with respect to x. A related metric occurs in the framework of axial symmetry, where a radial variable r plays the role of t. The following system of equations is equivalent to the vanishing of the Ricci tensor of the metric.

The mathematical problem is the analysis of the behavior X, Z as t tends to zero. X, Z are unknowns, and they depend on two real variables $t > 0$, x.

The system of equations is

$$\begin{aligned} D^2X - t^2X_{xx} &= 2(DXDZ - t^2X_xZ_x),\\ D^2Z - t^2Z_{xx} &= -e^{2Z}\big((DX)^2 - t^2X_x^2\big),\\ \lambda_x &= 2(Z_xDZ + e^{-2Z}X_xDX,\\ D\lambda &= (DZ)^2 + t^2Z_x^2 + e^{-2Z}\big((DX)^2 + t^2X_x^2\big). \end{aligned}$$

Here $D = t\partial_t$. This system is not Fuchsian.

At first, this system is studied in the case when $X = 0$. The equations for Z are reduced to the equation of Euler–Poisson–Darboux, which can be solved explicitly. This provides an asymptotic $Z = k\ln t + O(1)$, where k depends on x and can be arbitrary. Such spaces are called "polarized" Gowdy space-times. If we want to consider solutions depending only from x, they have behavior of the form $Z = k\ln t + O(1)$, $X = O(1)$.

Numerical computations for given nonlinear system enable us to characterize the velocity

$$v(x,t) = \sqrt{(DZ)^2 + exp(-2Z)(DX)^2},$$

where v should tend to $|k(x)|$ for $t \to 0$.

Solutions with $v < 1$ are said to be low-velocity and the others are called high-velocity. It is interesting that if one starts out with $v > 1$, for t tending to 0, velocity v become less then 1.

For $0 < k < 1$ the general solution of the system involves four arbitrary functions of x: k, X_0, ϕ and ψ.

$$Z(x,t) = k(x)\ln t + \varphi(x) + t^{\epsilon}u(x,t), \quad X(x,t) = X_0(x) + t^{2k(x)}\big(\psi(x) + v(x,t)\big).$$

If k is negative, and one starts from the expansions $Z = k(x)\ln t + \varphi(x)$ and $X = X_0 + t^{2k(x)}\psi(x)$, the generated solution will have positive k.

Let u, v be one of several local charts in considered (hyperbolic) space. The system of differential equations of the second order can be written in the form

$$\begin{aligned}
(D+\epsilon)^2 u &= t^{2-\epsilon}(k_{xx}\ln t + \varphi_{xx} + t^{\epsilon}u_{xx}) - \exp(-2\varphi - 2t^{\epsilon}u)\\
&\quad \times \Big\{t^{2k-\epsilon}\big((D+2k)(v+\psi)\big)^2 - t^{2-2k-\epsilon}\\
&\qquad\qquad \times \Big(X_{0x} + t^{2k}\big(v_x + \psi_x + 2k_x(v+\psi)\ln t\big)\Big)^2\Big\},\\
D(D+2k)v &= t^{2-2k}X_{0xx} + 2t^{\epsilon}(D+\epsilon)u(D+2k)(v+\psi) + t^2\\
&\quad \times \Big(\big((v+\psi)_{xx}4k_x(v_x+\psi_x)\ln t + \big(2k_{xx}\ln t + 4k_x^2(\ln t)^2\big)(v+\psi)\Big) - 2t^{2-2k}\\
&\quad \times \cdot\Big(X_{0x} + t^{2k}\big(v_x + \psi_x + k_x(v+\psi)\ln t\big)(k_x\ln t + \varphi_x + t^{\epsilon}u_x\Big).
\end{aligned}$$

If one lets u be $(u_0, u_1, u_2, v_0, v_1, v_2) = (u, Du, tu_x, v, Dv, tv_x)$, we obtain a Fuchsian system

$$\begin{aligned}
Du_0 &= u_1,\\
Du_1 &= -2\epsilon u_1 - \epsilon^2 u_0 + t^{2-\epsilon}(k_{xx}\ln t + \varphi_{xx}) + t\partial_x u_2 - \exp(-2\varphi - 2t^{\epsilon}u_0)\\
&\quad \times \Big\{t^{2k-\epsilon}(v_1 + 2kv_0 + 2k\psi)^2 - t^{2-2k-\epsilon}X_{0x}^2 - 2t^{1-\epsilon}X_{0x}\big(v_2 + t\psi_x\\
&\qquad\qquad + k_x(v_0+\psi)t\ln t\big) - t^{2k-\epsilon}\big(v_2 + t\psi_x + 2k_x(v_0+\psi)t\ln t\big)^2\Big\},\\
Du_2 &= t\partial_x(u_0 + u_1),\\
Dv_0 &= v_1,\\
Dv_1 &= -2kv_1 + t^{2-2k}X_{0xx} + t\partial_x(v_2 + t\psi_x) + 2k_x(v_2 + t\psi_x)t\ln t\\
&\quad + (v_0+\psi)\big(2k_{xx}t^2\ln t + 4(k_x t\ln t)^2\big) + 2t^{\epsilon}(v_1 + 2kv_0 + 2k\psi)(u_1 + \epsilon u_0)\\
&\quad - 2X_{0x}t^{2-2k}(k_x\ln t + \varphi_x + t^{\epsilon}\partial_x u_0) - 2t(\partial_x(v_0+\psi) + 2k_x(v_0+\psi)\\
&\quad + 2k_x(v_0+\psi)\ln t)(k_x t\ln t + t\varphi_x + t^{\epsilon}u_2),\\
Dv_2 &= t\partial_x(v_0 + v_1).
\end{aligned}$$

This system has the form

$$(D + A)u = g(t, x, u, u_x),$$

g involves various powers of t, possibly multiplies by logarithms. We choose ϵ so that all these terms tend to zero as t goes to zero. The low-velocity case is precisely the one in which it is possible to achieve the result without making any assumption on the singularity data k, X_0, φ, ψ.

The low-velocity case corresponds to the case $0 < k(x) < 1$.

Theorem 27.1. *Let $k(x)$, $X_0(x)$, $\varphi(x)$, $\psi(x)$ be real analytic and assume $0 < k(x) < 1$ for $0 < x < 2\pi$. Then there exists a unique solution of the form*

$$Z(x,t) = k(x)\ln t + \varphi(x) + t^{\epsilon}u(x,t), \quad X(x,t) = X_0(x) + t^{2k(x)}(\psi(x) + v(x,t)),$$

where u, v tend to zero as t tends to 0.

Thus, the obtained Fuchsian systems for physical models allow the application of all the powerful methods of analytic theory of differential equation developed during more then one-and-half centuries of investigation of physical objects.

27.3 Fuchsian Systems for Feynman Integrals

In this section the application of Griffiths' method for the derivation of the partial differential equations for Feynman integrals from the scattering theory of quantum electrodynamics is considered. Let us demonstrate this on the integral corresponding to the one-loop diagram with four vertices:

$$F(p) = \int_{-\infty}^{+\infty} \cdots \int_{-\infty}^{+\infty} \frac{\Pi_{i=1}^{n} d^4 k_i}{\Pi_{j=1}^{N} (q_j{}^2 - m_j{}^2 + i\epsilon)}$$

Using the following integral formula

$$\frac{1}{(x_1 - \xi_1)\dots(x_N - \xi_N)} = (N-1)! \int_{\xi_{a+1}=0} \frac{\omega^*(\alpha)}{(\alpha(x-\xi)^N)},$$

the Feymann integral of a one-loop diagram with four vertices can be written as

$$F(p) = (N-1)! \int_{-\infty}^{+\infty} \prod_{i=1}^{h} d^4 k_i \int_{h_0} \frac{\omega^*(\alpha)}{(\sum_{j=1}^{N} \alpha_j (q_j^2 - m_j^2 + i\epsilon))^N}.$$

For the analytic investigation of the Feymann integral consider the integral

$$I(y) = \int_{\Gamma} \frac{\omega^*(\alpha)}{Q^2},$$

where $Q = \sum_{1 \le i < j \le 4} \alpha_i \alpha_j y_{ij}$, and Γ is a hypersurface in the first quadrant surrounding the origin.

A derivation of a system of partial differential equation of the Fuchsian type for $I(y)$ is based on Griffiths' theorem on the reduction of pole of a rational differential form [4, 8]. Let us give a sketch of this reduction. One has

$$\frac{\partial I(y)}{\partial y_{ij}} = \int_\Gamma \frac{\alpha_i \alpha_j \omega^*(\alpha)}{Q^3}.$$

It is easy to see that if $\alpha_i = (1/2)\sum_{k=1}^4 y^{ik} Q_{\alpha_k}$, then $\alpha_i \alpha_j = \sum_{k=1}^4 \lambda_k(\alpha) Q_{\alpha_k}$.

And by Griffth's theorem $\partial I(y)/\partial y_{ij} = y^{ij} I + 2\int_\Gamma d\gamma_{ij}$, where

$$\gamma_{ij} = \frac{1}{Q^2}\sum_{\mu<\nu}(-1)^{\mu+\nu}(\alpha_\mu \lambda_\nu - \alpha_\nu \lambda_\mu).$$

Then

$$\int_\Gamma d\gamma_{ij} = \sum_{m=1}^{4}\int_{\Gamma_m} \frac{\lambda_m(\alpha)\alpha_j \omega_m}{Q_m^2(\alpha, y)},$$

where Γ_m is a part of the cycle boundary in hyperplane $\alpha_m = 0$, $Q_m = Q(\alpha, y)\mid_{\alpha_m} = 0$.

Further,

$$\text{const}\int_\Gamma d\gamma_{ij} = \sum_{m=1}^{4} y^{im}\int_{\Gamma^{im}}\int_{\Gamma_m} \frac{\alpha_j \omega_m^*}{Q_m^2(\alpha, y)} = \sum_{m=1}^{4}\ \sum_{k\in\{1,2,3,4\}} y^{im} y_m^{jk} \Phi_{mk}.$$

Here Γ_{mk} is a part of the cycle boundary in hyperplane $\alpha_m = 0$, $Q_m = Q(\alpha, y)\mid_{\alpha_m} = 0$, $\alpha_k = 0$, $Q_k = Q(\alpha, y)\mid_{\alpha_k} = 0$ and

$$\Phi_{mk} = \int_{\Gamma_{mk}} \frac{\omega_{mk}^*}{Q_{mk}(\alpha, y)}$$

is the Feynman integral for the reduced diagram of the second order (with reduced lines m and k). Let us write a system of partial differential equations for the Feynman integral $I(y)$ obtained using the Griffits method of pole reduction

$$\frac{\partial I}{\partial y_{ij}} + \frac{1}{2} y^{ij} I + \frac{1}{4}\sum_{m=1}^{4}\ \sum_{\substack{k\in\{1,2,3,4\}\\ -\{m\}}} y^{im} y_m^{jk} \Phi_{mk} = 0,$$

$$\frac{\partial \Phi_{mk}}{\partial y_{ij}} + \frac{1}{2} y_{mk}^{ij} \Phi_{mk} + \frac{1}{4}\sum_{\substack{p\in\{1,2,3,4\}\\ -\{m,k\}}}\ \sum_{\substack{q\in\{1,2,3,4\}\\ -\{m,k,p\}}} y_{mk}^{ip} y_{mkp}^{jq} \Phi_{mkpq} = 0,$$

$$\frac{\partial \Phi_{mkpq}}{\partial y_{ij}} = 0, \quad 1 \le i < j \le 4.$$

For details see [4, 5].

However, this system is not Fuchsian. Now we describe the Fuchsian reduction of this system [6, 7]. For its realization, it is necessary to introduce some new notations.

Let Y be a matrix with elements $y_{ij}, i, j = 1, 2, 3, 4$. By $Y_{p,q,r,t}$ denote the matrix Y with deleted rows and columns p, q, r, t. And let $Y\binom{i}{j}$ be the cofactor to y_{ij}. Further, let

$$y^{ij} = \frac{Y\binom{i}{j}}{|Y|}, \qquad y^{ij}_{pqrt} = \frac{Y_{pqrt}\binom{i}{j}}{|Y_{pqrt}|}$$

be the elements of the corresponding inverse matrices.

Introduce now the new functions

$$\psi_{kl} = y^{kl}\Phi_{kl}, \psi_{klmn} = y^{kl}Y^{mn}_{kl}\Phi_{klmn}.$$

Using the following evident relation

$$\frac{dy^{kl}}{y^{kl}} = \frac{dY\binom{k}{l}}{Y\binom{k}{l}} - \frac{d|Y|}{|Y|},$$

from the previous system of the partial Pfaffian system of the Fuchsian type

$$dI + \frac{1}{2}\frac{d|Y|}{|Y|}I + \frac{1}{4}\sum_{k<l}\left(\frac{d|Y_k|}{|Y_k|} + \frac{d|Y_l|}{|Y_l|}\right)\psi_{kl} = 0,$$

$$d\psi_{kl} + \left(\frac{1}{2}\frac{d|Y_{kl}|}{|Y_{kl}|} + \frac{d|Y|}{|Y|} - \frac{dY\binom{k}{l}}{Y\binom{k}{l}}\right)\psi_{kl}$$
$$+ \frac{1}{4}\sum_{m<n}\left(\frac{d|Y_{klm}|}{|Y_{klm}|} + \frac{d|Y_{kln}|}{|Y_{kln}|}\right)\psi_{klmn} = 0,$$

$$d\psi_{klmn} + \left(\frac{d|Y|}{|Y|} - \frac{dY\binom{k}{l}}{Y\binom{k}{l}} - \frac{d|Y_{kl}|}{|Y_{kl}|} - \frac{dY_{kl}\binom{m}{n}}{Y_{kl}\binom{m}{n}}\right)\psi_{klmn} = 0.$$

In Golubeva's paper [7], the Feynman integral of a self-energy diagram in a two -imensional moment space is investigated. Analogously, one can investigate the Feynman integral of the stairs diagram of the sixth order.

Bibliography

[1] T. Regge, *Algebraic topology methods in the theory of Feynman relativistic amplitudes*, Battelle rencontres, Benjamin, New York, 1968.

[2] P. Appell and Kampé de Fériet, *Fonction hypergeometriques et hyperspheriques*, Gauthier–Villars, Paris, 1926.

[3] V. A. Golubeva, Appell and Kampé de Fériet hypergeometric functions of two variables, *Siberian Mathematical Journal*, **20**(5): 705–717, 1979.

[4] V. A. Golubeva, Some problems in the analytic theory of Feynman integrals, *Russian Mathematical Surveys,* **31**(2): 139–207, 1976.

[5] V. A. Golubeva, V. Z. Enol'skii, The differential equations for the Feynman amplitude of a single-loop graph with four vertices, *Mat. Zametki,* **23**(1): 113–119, 1978.

[6] V. A. Golubeva, Pfaffian system of Fuchsian type for Feynman integral of one-loop diagram, *Voprosy teorii grupp i gomologicheskoi algebry,* Yaroslavl, 1982, 82–85.

[7] V. A. Golubeva, Differential equations for the Feynman integral of the self-energy diagram, *Differencial'nye Uravnenija,* **9**: 1298–1309, 1973.

[8] P. A. Griffiths, On the periods of certain rational integrals, *Ann. Math. Ser. 2,* **90**(3): 460–541, 1969.

[9] V. G. Knizhnik, A. B. Zamolodchikov, Current algebra and Wess-Zumino model in two dimensions, *Nuclear Physics*, **247**: 83–103, 1984.

[10] A. Leibman, Some monodromy representations of generalized braid groups, *Commun. Math. Phys.*, **164**(2): 293–304, 1994.

[11] Chr. Kassel, *Quantum groups,* Graduate Texts in Mathematics 155, Springer-Verlag, New York, 1995.

[12] S. Kichenassamy, *Fuchsian Reduction: Applications to Geometry, Cosmology and Mathematical Physics,* Birkhäuser, Boston, 2007.

[13] S. Kichenassamy, A. D. Rendall, Analytic description of singularities in Gowdy spacetime, *Clasas. Quantum Gravity,* **15**: 1339–1355, 1998.

Author Information

Valentina A. Golubeva, MAI, Volokolamskoe st., 4, 123060, Moscow, Russia.
E-mail: goloubeva@yahoo.com

Chapter 28

The Voros Coefficient and the Parametric Stokes Phenomenon for the Second Painlevé Equation

Kohei Iwaki

Abstract. The second Painlevé equation with a large parameter (P_2) is analyzed by using the exact WKB analysis. The purpose of this study is to investigate the problem of the degeneration of the P-Stokes geometry of (P_2), which is related to a kind of Stokes phenomena for asymptotic (formal) solutions of (P_2). We formulate the connection formula for this Stokes phenomenon, and confirm it in two ways: (i) by computing the "Voros coefficient" of (P_2), and (ii) by using the isomonodromic deformation theory. Our main claim is that the connection formulas derived by these two completely different methods coincide.

Keywords. Painlevé Equation, Exact WKB Analysis, Stokes Phenomenon.

2010 Mathematics Subject Classification. Primary: 34M55; Secondary: 34M40.

28.1 Introduction

We analyze the second Painlevé equation with a large parameter η

$$(P_2) : \frac{d^2\lambda}{dt^2} = \eta^2(2\lambda^3 + t\lambda + c)$$

by using the exact WKB analysis. (The general theory of the exact WKB analysis for Painlevé equations is presented in [1].) The purpose of this study is to investigate the problem of the degeneration of P-Stokes geometry (existence of P-Stokes curves connecting two turning points) of (P_2). (See Definition 28.1 for the definition of P-Stokes curve.) Here the following figures describe the P-Stokes curves of (P_2) near $\arg c = \pi/2$. The degeneration of P-Stokes curves observed when $\arg c = \pi/2$ suggests that a kind of Stokes phenomena occurs when c varies near $\arg c = \pi/2$, that is, the correspondence between asymptotic (formal) solutions and true solutions of (P_2) changes discontinuously before and after the degeneration. We call this the "parametric Stokes phenomenon" because this Stokes phenomenon (or the degeneration of Stokes geometry) occurs when the parameter c contained in (P_2) varies. In this case, asymptotic solutions mean the following 1-parameter family of transseries solutions (1-parameter solutions) of (P_2):

$$\lambda(t,c,\eta;\alpha) = \lambda^{(0)}(t,c,\eta) + \alpha\eta^{-\frac{1}{2}}\lambda^{(1)}(t,c,\eta)e^{\eta\phi_{\rm II}} + (\alpha\eta^{-\frac{1}{2}})^2\lambda^{(2)}(t,c,\eta)e^{2\eta\phi_{\rm II}} + \cdots .$$

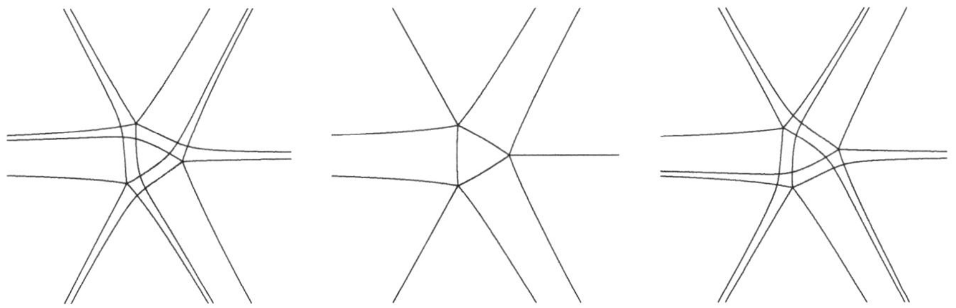

Here α is a free parameter, $\lambda^{(k)}(t,c,\eta) = \lambda_0^{(k)}(t,c) + \eta^{-1}\lambda_1^{(k)}(t,c) + \eta^{-2}\lambda_2^{(k)}(t,c) + \cdots$ $(k \geq 0)$ are formal power series of η^{-1} and $\phi_2 = \phi_2(t,c)$ is some function. (Note that $\lambda^{(0)}(t,c,\eta)$ itself is a formal power series solution of (P_2), called the 0-parameter solution.) We will formulate the connection formula for the parametric Stokes phenomenon for 1-parameter solutions.

28.2 Connection Formula for the Parametric Stokes Phenomenon

Before formulating the connection formula, we discuss normalizations of 1-parameter solutions. By letting $\tilde{\lambda}^{(1)}(t,c,\eta;\alpha) = \alpha\eta^{-\frac{1}{2}}\lambda^{(1)}(t,c,\eta)e^{\eta\phi_{\mathrm{II}}}$, we know that $\tilde{\lambda}^{(1)}$ satisfies the following second-order linear ordinary differential equation

$$\frac{d^2\tilde{\lambda}^{(1)}}{dt^2} = \eta^2\left(6\lambda^{(0)}(t,c,\eta)^2 + t\right)\tilde{\lambda}^{(1)}, \tag{28.1}$$

i.e., the Fréchet derivative of (P_2) at $\lambda = \lambda^{(0)}$. Thus $\tilde{\lambda}^{(1)}$ can be taken as a WKB solution [1] of (28.1) of the form

$$\begin{aligned}\tilde{\lambda}^{(1)}(t,c,\eta;\alpha) &= \alpha\frac{1}{\sqrt{R_{\mathrm{odd}}(t,c,\eta)}}\exp\left(\int^t R_{\mathrm{odd}}(t,c,\eta)\,dt\right)\\ &= \alpha\eta^{-\frac{1}{2}}\left(\lambda_0^{(1)}(t,c) + \eta^{-1}\lambda_1^{(1)}(t,c) + \eta^{-2}\lambda_2^{(1)}(t,c) + \cdots\right)e^{\eta\phi_{\mathrm{II}}},\end{aligned} \tag{28.2}$$

where R_{odd} is the odd part of a formal power series solution $R = \eta R_{-1} + R_0 + \eta^{-1}R_1 + \cdots$ of the Riccati equation

$$R^2 + \frac{dR}{dt} = \eta^2\left(6\lambda^{(0)}(t,c,\eta)^2 + t\right) \tag{28.3}$$

associated with (28.1). (Thus we have $\phi_2 = \int^t R_{-1}dt$.) We note that $\lambda^{(k)}$ $(k \geq 2)$ are determined uniquely once the normalization of $\tilde{\lambda}^{(1)}$ (i.e., the normalization of the integral $\int^t R_{\mathrm{odd}}(t,c,\eta)dt$ in (28.2)) is fixed.

Definition 28.1 ([1]). (i) A point t is called a *P-turning point of (P_2)* if t satisfies $6\lambda_0^{(0)\,2} + t = 0$.

(ii) For a P-turning point $t = \tau$, a real one-dimensional curve defined by

$$\operatorname{Im} \int_\tau^t \sqrt{6\lambda_0^{(0)\,2} + t}\, dt = 0$$

is said to be a *P-Stokes curve of (P_2)*.

Here we introduce two normalizations of $\tilde{\lambda}^{(1)}$. The first normalization of $\tilde{\lambda}^{(1)}$ is "the normalization at a P-turning point":

$$\tilde{\lambda}_\tau^{(1)}(t,c,\eta;\alpha) = \alpha \frac{1}{\sqrt{R_{\text{odd}}}} \exp\left(\int_\tau^t R_{\text{odd}}\, dt\right), \tag{28.4}$$

where τ is a P-turning point of (P_2). The second one is "the normalization at ∞":

$$\tilde{\lambda}_\infty^{(1)}(t,c,\eta;\alpha) = \alpha \frac{1}{\sqrt{R_{\text{odd}}}} \exp\left(\eta \int_\tau^t R_{-1}\, dt + \int_\infty^t \left(R_{\text{odd}} - \eta R_{-1}\right) dt\right). \tag{28.5}$$

We define the 1-parameter solution $\lambda_\tau(t,c,\eta;\alpha)$(resp. $\lambda_\infty(t,c,\eta;\alpha)$) by using $\tilde{\lambda}_\tau^{(1)}$ (resp. $\tilde{\lambda}_\infty^{(1)}$) for the normalization of $\tilde{\lambda}^{(1)}$. Then the connection formula for the parametric Stokes phenomena for 1-parameter solutions can be described as follows.

Connection Formula for the 1-parameter Solutions of (P_2). *Let ε be a sufficiently small positive number.*

(i) *If the true solutions represented by $\lambda_\infty(t,c,\eta;\alpha)$ for* $\arg c = \pi/2 - \varepsilon$ *and by $\lambda_\infty(t,c,\eta;\tilde{\alpha})$ for* $\arg c = \pi/2 + \varepsilon$ *coincide, then the following holds:*

$$\tilde{\alpha} = \alpha. \tag{28.6}$$

(ii) *If the true solutions represented by $\lambda_\tau(t,c,\eta;\alpha)$ for* $\arg c = \pi/2 - \varepsilon$ *and by $\lambda_\tau(t,c,\eta;\tilde{\alpha})$ for* $\arg c = \pi/2 + \varepsilon$ *coincide, then the following holds:*

$$\tilde{\alpha} = \left(1 + e^{2\pi i c\eta}\right)\alpha. \tag{28.7}$$

Assuming the Borel summability of the 1-parameter solutions, we derive the above connection formulas in two ways:

A: derivation through the analysis of "the Voros coefficient of (P_2)",

B: derivation by using the isomonodromic deformation.

Our main claim is that the connection formulas derived by these two completely different methods coincide. In this chapter, due to the lack of space, we explain only an outline of the derivation of the connection formula by **A**. The derivation by **B** will be discussed in the talk.

28.3 Derivation of the Connection Formulas Through the Analysis of the Voros Coefficient of (P_2)

We define the Voros coefficient of (P_2). It plays an important role in the analysis of the parametric Stokes phenomenon for the 1-parameter solutions of (P_2).

Definition 28.2. The Voros coefficient $W(c,\eta)$ of (P_{II}) is defined as follows:

$$W(c,\eta) = \int_{\tau}^{\infty} \big(R_{\mathrm{odd}}(t,c,\eta) - \eta R_{-1}(t,c)\big)dt. \tag{28.8}$$

$W(c,\eta)$ appears as the difference of the above two normalizations of $\tilde{\lambda}^{(1)}$:

$$\tilde{\lambda}^{(1)}_{\tau}(t,c,\eta;\alpha) = e^{W(c,\eta)}\, \tilde{\lambda}^{(1)}_{\infty}(t,c,\eta;\alpha). \tag{28.9}$$

Then we obtain the following theorem.

Theorem 28.3. *The Voros coefficient $W(c,\eta)$ of (P_2) is represented explicitly as follows:*

$$W(c,\eta) = -\sum_{n=1}^{\infty} \frac{2^{1-2n}-1}{2n(2n-1)} B_{2n} (c\,\eta)^{1-2n}. \tag{28.10}$$

Here B_{2n} is the 2n-th Bernoulli number defined by

$$\frac{w}{e^w - 1} = 1 - \frac{w}{2} + \sum_{n=1}^{\infty} \frac{B_{2n}}{(2n)!} w^{2n}. \tag{28.11}$$

A key to the proof of Theorem 28.3 is the calculation of the difference $W(c,\eta) - W(c-\eta^{-1},\eta)$, which is done by making use of the Bäcklund transformation that induces the translation of the parameter $c \mapsto c - \eta^{-1}$ of (P_2). Using the expression (28.10), we can analyze the parametric Stokes phenomenon for $W(c,\eta)$.

Corollary 28.4 ([2]). *By denoting the Borel resummation operator by $\mathcal{S}$, we obtain*

$$\mathcal{S}\left[e^{W(c,\eta)}\big|_{\arg c = \frac{\pi}{2}-\varepsilon}\right] = \left(1 + e^{2\pi i c \eta}\right) \mathcal{S}\left[e^{W(c,\eta)}\big|_{\arg c = \frac{\pi}{2}+\varepsilon}\right], \tag{28.12}$$

where ε is a sufficiently small positive number.

Assume that the connection formula (28.6) for λ_∞ is true. Then, combining Corollary 28.4, (28.6), and (28.9), we can derive the connection formula (28.7) for the 1-parameter solution $\lambda_\tau(t,c,\eta;\alpha)$.

Bibliography

[1] T. Kawai and Y. Takei, *Algebraic Analysis of Singular Perturbation Theory*, American Mathematical Society, Translations of Mathematical Monographs 227, 2005.

[2] Y. Takei, Sato's conjecture for the Weber equation and transformation theory for Schrödinger equations with a merging pair of turning points, *RIMS Kôkyûroku Bessatsu*, **B10**: 205–338, 2008.

Author Information

Kohei Iwaki, Research Institute for Mathematical Sciences, Kyoto University, Kyoto, 606-8502, Japan.
E-mail: `iwaki@kurims.kyoto-u.ac.jp`

Chapter 29

Integral Symmetry and the Deformed Hypergeometric Equation

Alexander Ya. Kazakov

Abstract. The deformed hypergeometric equation, a linear differential equation of the second order with three regular singularity and one apparent singularity, is under consideration. Euler integral symmetry for this equation is described. Analytic continuation of the corresponding contour integral makes it possible to calculate the monodromy group for this equation in explicit terms. Solutions of the deformed hypergeometric equation in integral form are also obtained.

Keywords. Euler Integral Symmetry, Monodromy, Connection Matrix, Deformed Hypergeometric Equation.

2010 Mathematics Subject Classification. Primary: 34M35; Secondary: 33C60, 34M03.

Linear special functions (like hypergeometric functions, Heun class functions, etc.) are connected with nonlinear special functions (solutions of the Painlevé equations) in two ways. Firstly, Painlevé equations govern isomonodromy deformations of the linear equations [1, 2]. Secondly, explicit information about monodromy groups can be exploited by the study of solutions of Painlevé equations (A. Its and colleagues). Here the monodromy group of the deformed hypergeometric equation

$$w''(z) + \left[\frac{\alpha}{z} + \frac{\beta}{z-1} - \frac{1}{z-\lambda}\right] w'(z) + \frac{1}{z(z-1)}\left[\frac{\chi}{z-\lambda} + \frac{\sigma}{\lambda(\lambda-1)}\right] w(z) = 0 \tag{29.1}$$

is under consideration. This equation has, like the hypergeometric one, three regular singular points $z = 0$, $z = 1$, $z = \infty$, and, moreover, an apparent (or "false") regular singular point of the first order at $z = \lambda$. It has four free parameters $\alpha, \beta, \lambda, \chi$, where the value of the parameter σ is determined by the condition that solutions do not have logarithmic behavior in the neighbourhood of the point $z = \lambda$. It can be shown that $\sigma = -(\beta\lambda - 2\lambda + \alpha\lambda + 1 + \chi - \alpha)\chi$. The order of the apparent singularity is equal to 1 in accordance with residue at $z = \lambda$ in the coefficient at $w'(z)$. This equation can be considered to be a particular case of the Heun equation [3] when the one of regular singularity is the apparent one. Our aim is to calculate the monodromy group for this equation; we apply here Euler integral symmetry and the procedure of analytic continuation. This approach has been previously discussed in [4, 5].

The initial point is the linear differential system

$$(Az + B)W'(z) = CW(z), \tag{29.2}$$

where A, B, C are constant $m{\times}m$-matrices. If $m = 2$, this system can be reduced to the deformed hypergeometric equation. The singular regular points of this system are the roots of the $det(Az + B)$ and, perhaps, the infinity.

Theorem 1. *Let $V(t)$ be the solution of the system*

$$(At + B)V'(t) = \big[C - (\mu + 1)A\big]V(t), \tag{29.3}$$

branching in the neighborhood of some regular singularity $t = t_*$ and contour L is a double loop on a complex plane around the point $t = z$ and $t = t_*$. Then $W(z)$ defined as

$$W(z) = \int_L (z - t)^{\mu} V(t)dt, \tag{29.4}$$

is the solution of the system (29.2) branching at the same point $z = t_*$.

Note that integral relation (29.4) is a scalar one, namely, it links corresponding components of the vector functions $V(t)$ and $W(z)$. Thus, if $m = 2$, reduction of systems (29.2) and (29.3) to differential equations leads to the integral relation between solutions of the deformed hypergeometric equations. In order to describe this link in explicit terms, it is necessary to express the parameters of (29.1) in terms of the parameters of (29.2). Then it is necessary to calculate the parameters of the deformed hypergeometric equation, which is a reduction of the system (29.3). These cumbersome calculations were realized with the help of Maple; we omit the details here. Results of this calculations are summarized in Proposition 2. We now need some convenient notions.

Let us denote by $w_h^{(0)}(z)$ normalized at $z = 0$ holomorphic solution of equation (29.1), $w_b^{(0)}(z)$ normalized branching at $z = 0$ solution of this equation, $w_h^{(0)}(z) \sim 1$, $w_b^{(0)}(z) \sim 1 \cdot z^{1-\alpha}$ at $z \sim 0$, $w_h^{(1)}(z)$, $w_b^{(1)}(z)$ normalized holomorphic and branching at $z = 1$ solutions, $w_h^{(1)}(z) \sim 1$, $w_b^{(1)}(z) \sim 1 \cdot (z - 1)^{1-\beta}$ at $z - 1 \sim 0$.

Theorem 2. *Let $v_b^{(0,1)}(t)$ be normalized branching solutions of the equation*

$$v''(t) + \left[\frac{\widetilde{\alpha}}{t} + \frac{\widetilde{\beta}}{t-1} - \frac{1}{t-\widetilde{\lambda}}\right]v'(t) + \frac{1}{t(t-1)}\left[\frac{\widetilde{\chi}}{t-\widetilde{\lambda}} + \frac{\widetilde{\sigma}}{\widetilde{\lambda}(\widetilde{\lambda}-1)}\right]v(t) = 0, \tag{29.5}$$

fixed by their behavior at the regular singular points $t = 0$ or $t = 1$, respectively, and let the contour $L = L_{0,1}$ be double loop including either $t = 0$ (for $L^{(0)}$), or $t = 1$ (for $L^{(1)}$), and point $t = z$, and

$$\widetilde{\alpha} = \alpha + \mu + 1, \widetilde{\beta} = \beta + \mu + 1, \tag{29.6}$$

$$\widetilde{\lambda} = \frac{\lambda(\alpha\lambda - \lambda - \alpha + \mu\lambda - \mu + \beta\lambda + \chi)}{-2\lambda + \beta\lambda + \alpha\lambda + 1 - \alpha + \chi} \tag{29.7}$$

$$\widetilde{\chi} = -\frac{1}{-2\lambda + \beta\lambda + \alpha\lambda + 1 - \alpha + \chi}(-\chi^2 - \alpha\mu\lambda - \beta\mu\lambda - \alpha\chi\lambda + \alpha\lambda^2 + \beta\lambda^2 - \lambda^2 - \beta\lambda + \lambda + \beta\mu\lambda^2 - \chi + \alpha\lambda^2\mu + \alpha\chi - \alpha\lambda + \mu^2\lambda^2 - \mu^2\lambda - \beta\lambda\chi + 2\chi\lambda). \tag{29.8}$$

Then function $w(z)$, defined by relation

$$w(z) = (\nu_{0,1})^{-1}\int_{L_{0,1}} (z-t)^{\mu} v_b^{(0,1)}(t)dt, \tag{29.9}$$

coincides with ñ $w_b^{(0,1)}(z)$. Here

$$\nu_{0,1} = \left[1 - \exp(2\pi i\mu)\right]\left[1 - \exp\left(2\pi i(\kappa_{0,1})\right)\right] B(\mu + 1, 1 + \kappa_{0,1}).$$

$\kappa_{0,1}$ are characteristic exponents for the solutions $v_b^{(0,1)}(t)$ at the corresponding singular point.

Let $W^{(0,1)}(z) = (w_b^{(0,1)}(z), w_h^{(0,1)}(z))^T$. Then, in accordance with the analytic theory of linear differential equations [1, 2], $W^{(0)}(z) = RW^{(1)}(z)$, where R is the corresponding connection matrix for equation (29.1). Considering the analytic continuation with the help of integral relation (29.9), one obtains

$$R_{11} = \frac{\exp(2\pi i\kappa_1) - 1}{\exp(-2\pi i\beta) - 1}\cdot\frac{\Gamma(\kappa_1 + 1)\Gamma(\kappa_0 + \mu + 2)}{\Gamma(\kappa_0 + 1)\Gamma(\kappa_1 + \mu + 2)}T_{11}, \tag{29.10}$$

where T is the corresponding connection matrix for equation (29.5). Note that integral transform (29.9) depends on the free parameter μ. Changing this parameter, one can change the parameters of equation (29.5) in accordance with (29.6)–(29.8). Let us fix parameter μ by the condition that the apparent singularity of equation (29.5) $t = \widetilde{\lambda}$ coincides with regular singularity $t = 0$. This means that $\mu = -(\alpha\lambda - \alpha + \beta\lambda - \lambda + \chi)/(\lambda - 1)$, and equation (29.5) reads

$$v''(t) + \left[\frac{-\alpha\lambda - 1 + \alpha - \beta + 2\lambda - \chi}{(t-1)(\lambda - 1)} + \frac{-\beta\lambda + \lambda - \chi}{t(\lambda - 1)}\right]v'(t) + \frac{(\beta\lambda - \lambda + \chi)(-\lambda^2 + t\lambda\alpha + t\beta\lambda + \lambda - 2t\lambda + t - t\alpha + t\chi)}{(\lambda - 1)^2\lambda t^2(t-1)}v(t) = 0. \tag{29.11}$$

Coefficient at $v(t)$ in equation (29.11) has s pole of the second order at $t = 0$. And so this equation can be transformed into a hypergeometric one by simple substitution.

Connection matrices for hypergeometric equation are described by Kummer relations [6]. Thus, one can calculate the value of T_{11} and, consequently, R_{11}. Let

$$\Delta = \sqrt{\frac{(\alpha-\beta)^2}{4} + \frac{(\beta\lambda - \lambda + \chi)(-\lambda + \alpha\lambda + \chi + 1 - \alpha)}{\lambda(\lambda - 1)}}, \tag{29.12}$$

then

$$R_{11} = -\frac{(\lambda - 1)\exp(i\pi\beta)\Gamma(\beta - 1)\Gamma(2-\alpha)}{(\beta\lambda + \chi - \lambda)\Gamma(\frac{\beta-\alpha}{2} + \Delta)\Gamma(\frac{\beta-\alpha}{2} - \Delta)}. \tag{29.13}$$

Relation (29.13) describes one entry of the connection matrix for equation (29.1). This connection matrix corresponds to the pair of singular points $z = 0, z = 1$. The other entries of this matrix can be evaluated with the help of teh elementary symmetries of the deformed hypergeometric equation. Namely, substitutions $w(z) = z^{1-\alpha}w_1(z)$, $w(z) = (z-1)^{1-\beta}w_2(z)$, and change of variable $z = 1 - s$ produce the same equation for functions $w_1(z), w_2(z), w(s)$ with different sets of parameters. In more detail, the first substitution transforms equation with parameters $[\alpha, \beta, \lambda, \chi]$ into equation with parameters $[2-\alpha, \beta, \lambda, 1 + \chi + \lambda\alpha - \lambda - \alpha]$, and exchanges branching and holomorphic at $z = 0$ solutions, solutions fixed by the behavior at $z = 1$ does not change its character. Thus,

$$R_{21}(\alpha, \beta, \lambda, \chi) = R_{11}(2-\alpha, \beta, \lambda, 1 + \chi + \lambda\alpha - \lambda - \alpha). \tag{29.14}$$

Analogously,

$$R_{12}(\alpha, \beta, \lambda, \chi) = R_{11}(\alpha, 2-\beta, \lambda, \chi - \lambda + \beta\lambda). \tag{29.15}$$

Combining

$$R_{22}(\alpha, \beta, \lambda, \chi) = R_{11}(2-\alpha, 2-\beta, \lambda\alpha + \beta\lambda - 2\lambda + 1 + \chi - \alpha). \tag{29.16}$$

These relations give us the connection matrix R, which corresponds to the pairs of singularities $z = 0, 1$. Equation (29.1) has once more regular singularity $z = \infty$. But change of the independent variable $z = s^{-1}$ (which exchanges points 0 and ∞)transforms equation (29.1) into equation, which differs from equation (29.1) by an elementary factor only (with a different set of parameters). This symmetry makes it possible to calculate the connection matrix, which corresponds to the pair of singularities $z = \infty, z = 1$. Solutions of the deformed hypergeometric equation in integral form are also obtained.

Bibliography

[1] K. Iwasaki, H. Kimura, S. Shimomura, and M. Yosida, *From Gauss to Painlevé: a modern theory of special functions,* Vieweg, Braunshweig, 1991.

[2] Y. Sibuya, *Linear differential equations in the complex domain: problem of analytic continuation,* AMS, Providence, 1985.

[3] S. Yu. Slavyanov, W. Lay, *Special functions: A unified theory based on singularities,* Oxford University Press, Oxford, New York, 2000.

[4] A. Ya. Kazakov, Integral symmetry, integral invariants and monodromy of ordinary differential equations, *Theoret. Math. Phys.*, **116**(3): 991–1000, 1998.

[5] A. Ya. Kazakov. The central two-point connection problem for the reduced confluent Heun equation, *J. Phys. A: Math.Gen.*, **39**: 2339–2347, 2006.

[6] H. Bateman, A. Erdelyi, *Higher Transcendental Functions*, vol. 1, McCraw-Hill Book Company Inc., New York, 1953.

Author Information

Alexander Ya. Kazakov, S.-Petersburg State University of Aerospace Instrumentation,
B. Morskaya 67, 190000, S.-Petersburg, Russia.
E-mail: a_kazak@mail.ru

Chapter 30

Integral Symmetries for Confluent Heun Equations and Symmetries of Painlevé Equation P_5

Alexander Ya. Kazakov and Sergey Yu. Slavyanov

Abstract. Integral relation between solutions of the confluent Heun equation with added apparent singularity produces corresponding symmetry of the Painlevé P_5 solutions.

Keywords. Heun Equation, Integral Symmetry, Monodromy.

2010 Mathematics Subject Classification. Primary: 33E17; Secondary: 33E99, 34M03.

A confluent Heun equation is a second-order equation with two regular singularities which are assumed to be $z_1 = 0$, $z_2 = 1$ and one irregular singularity at infinity of s-rank $R(\infty) = 2$ [1]. If the fourth singularity $z = q$, namely the apparent one, is added to this equation, we term the generated equation as the deformed confluent Heun equation *CHE1*

$$z(z-1)w''(z) + \left(-tz(z-1) + c(z-1) + dz - \frac{1}{z-q}\right) w'(z) + \left(-taz + t\lambda + \frac{pq}{z-q}\right) w(z) = 0. \quad (30.1)$$

It is known that the isomonodromy condition for equation (30.1) is equivalent to the Painleve equation P_5.

The Euler integral transforms for Heun equation have been studied before. They lead to symmetries for Heun equations and as a result to symmetries of the corresponding P_6 [2]. The idea that integral transforms lead to inner symmetries first appeared in the paper by Fock [3]

The problem of transforms for *CHE1* is rather sophisticated and needs computer algebra systems for particular calculations. Here we give an implicit solution of this problem using the ideas proposed in [2] based on study of an auxiliary linear 2×2-system for which the Euler transform and subsequent symmetries are much simpler. As a result, the symmetries of *CHE1* and resulting symmetries of P_5 are found. They generate a subset of Okamoto transforms for P_5.

The starting linear system is assumed to be of the form

$$(z^2A + zB + C)\frac{dW}{dz} = (-\alpha z A + E)W. \tag{30.2}$$

Here A, B, C, E are 2×2-matrices which do not depend on z, $W(z)$ is a 2-vector function. Particularly

$$A = \begin{pmatrix} 1 & 0 \\ 0 & 0 \end{pmatrix}, \qquad B = \begin{pmatrix} 0 & \tau \\ 1 & 0 \end{pmatrix},$$
$$C = \begin{pmatrix} c\tau + c + \tau & -c\tau - c - \tau \\ c & -c \end{pmatrix}, \qquad E = \begin{pmatrix} e_1 & e_2 \\ e_3 & e_4 \end{pmatrix}. \tag{30.3}$$

Here c and τ are parameters determining characteristic exponents at regular singularities. The determinant of $Az^2 + Bz + C$ can be easily computed:

$$\det(Az^2 + Bz + C) = -(c + \tau)z(z - 1). \tag{30.4}$$

The choice of matrices A, B, C was the crucial point in our studies. It is based on (30.4) but has much freedom, which can either simplify intermediate computations or make them more complicated. The roots of the polynomial $\det(z^2A + zB + C)$ are the regular singularities of *CHE1*.

Reduction of (30.2) to the second-order equation leads to the *CHE1*, and the corresponding relations between parameters of system (30.2) and equation (30.1) can be obtained. However, in several cases these relations are complicated (they include hundred of terms) and are presented only in the report.

Further the solution of the auxiliary system *CHE1* $\Phi(s)$ is introduced:

$$(s^2A + sB + C)\frac{d\Phi(s)}{ds} = \big((\alpha + 2)sA + E - (-\alpha + 1)B\big)\Phi(s) = 0. \tag{30.5}$$

It can be shown by integration by parts that if $\Phi(s)$ is the appropriate solution of (30.5) and L is the appropriate contour on a complex plane, then

$$W(z) = \int_L (z - s)^{\kappa}\Phi(s)ds \tag{30.6}$$

is a solution of the initial system (30.2). Reduction of the system (30.5) also leads to *CHE1*, but with the transformed coefficients. Due to the scalar nature of integral relation (30.6), the corresponding integral relation can be written for solutions of initial *CHE1* and transformed *CHE1*. Analytic continuation with the help of this integral makes it possible to link the monodromy matrices of the initial and transformed *CHE1*.

The rather boring computations where computer algebra systems can help allow us to find the explicit relations between parameters of initial *CHE1* and transformed *CHE1*.

Bibliography

[1] S. Yu. Slavyanov and W. Lay, *Special Functions: A Unified Theory Based on Singularities,* Oxford University Press, Oxford/New York, 2000.

[2] A. Ya. Kazakov and S. Yu. Slavyanov, *Theor. Math. Phys.*, **155**: 721–732, 2008.

[3] V. A. Fock, Zur Theorie des Wasserstoffatoms, *Zs. f. Phys.*, **98**: 145–154, 1935.

Author Information

Alexander Ya. Kazakov, S.-Petersburg State University of Aerospace Instrumentation, B. Morskaya 67, 190000, S.-Petersburg, Russia.
E-mail: `a_kazak@mail.ru`

Sergey Yu. Slavyanov, S.-Petersburg State University, Ulyanovskaya, 3, St. Peterhof, S.-Petersburg, Russia.
E-mail: `slav@SS2034.spb.edu`

Chapter 31

From the Tau Function of Painlevé P_6 Equation to Moduli Spaces

Dmitry Korotkin and Peter Zograf

Abstract. The isomonodromic tau-function of Hitchin's solution to the sixth Painlevé equation can be naturally generalized for an arbitrary Riemann–Hilbert problem with quasi-permutation monodromies (the corresponding tau-function also appears in the theory of Frobenius structures on Hurwitz spaces). Such a tau-function can be interpreted as a holomorphic section of the Hodge line bundle on a Hurwitz space. Analysing the asymptotic behavior of the tau-function near the boundary of the Hurwitz spaces, we obtain new relations between divisor classes on these spaces. Similar results hold for the spaces of holomorphic 1-differentials on Riemann surfaces.

Keywords. Painlevé VI Equation, Tau-function, Hurwitz Spaces.

2010 Mathematics Subject Classification. Primary: 34M55; Secondary: 32G15.

In his work [5], Hitchin solved the sixth Painlevé equation with coefficients $(1/8, -1/8, 1/8, 3/8)$, and his objective was to classify all $SU(2)$ invariant solutions of self-dual Einstein's equations. The isomonodromic tau-function corresponding to his solution was found later in [7] and is as follows:

$$\tau = \tau_B^{-1/2}\theta[p,q](0|\sigma); \tag{31.1}$$

here p, q are two complex valued constants, $\theta[p,q](u|\sigma)$ is the theta-function with characteristics, and σ is the period of elliptic curve $\nu^2 = z(z-1)(z-x)$. The function $\tau_B(x)$ (called the *Bergman tau-function* due to the reasons explained below) is given by

$$\tau_B = K(x(x-1))^{1/4}, \tag{31.2}$$

where $K(x)$ is the elliptic integral of the first kind. Formula (31.1) makes it possible to considerably simplify the original Hitchin's description of the corresponding Einstein manifolds [1].

We thank the Max Planck Institute for Mathematics in Bonn, where this chapter was written and a considerable part of the work described in it was done, for its support and hospitality. The work of DK was supported by the NSERC, FQRNT, and CURC programs. The work of PZ was partially supported by the RFBR, Grant 11-01-00677-a.

On the other hand, a straightforward generalization of (31.1) immediately yields the isomonodromic tau-function for an arbitrary Riemann–Hilbert problem with off-diagonal 2×2 monodromy matrices; in that case σ becomes the period matrix of the hyperelliptic curve of genus g given by $v^2 = \prod_{i=1}^{2g+2}(z - z_i)$; p and q are two arbitrary vectors from $\mathbf{C}^g$. The Bergman tau-function τ_B has in this case the form

$$\tau_B = \det A \prod_{i<j} (z_i - z_j)^{1/4} , \tag{31.3}$$

where A is the matrix of a-periods of nonnormalized differentials $z^{i-1} dz/v$, $i = 1, \ldots, g$. The expression (31.3) is also well known as the partition function of the Ashkin–Teller model [15].

Further generalization of the formulas (31.1) and (31.3) appears in the form of an isomonodromic tau-function of an arbitrary Riemann–Hilbert problem with quasi-permutation monodromy matrices [8]. Each Riemann–Hilbert problem of this kind in matrix dimension d is naturally associated to a d-sheeted covering of the complex plane, which defines a Riemann surface C of some genus g. Denoting the period matrix of this Riemann surface by σ as before, we find that the corresponding tau-function is given by the formula (31.1), where the vectors p and q are determined by the entries of monodromy matrices. To define the Bergman tau-function τ_B is this general case we introduce the canonical bimeromorphic differential $B(x, y) = d_x d_y \log E(x, y)$, where $E(x, y)$ is the prime form on C. Near the diagonal $x = y$ the bidifferential B behaves as follows: $B(x, y) = ((\xi(x) - \xi(y)^{-2} + S_B(x) + \cdots) d\xi(x) d\xi(y)$, where ξ is a local parameter. The term $S_B(x)$ transforms as a projective connection under the change of a local parameter; it is called the *Bergman projective connection* [14]. Another projective connection, denoted by S_{df}, is defined by the Schwarzian derivative of f: $S_{df} = \{f(x), \xi(x)\}$, where $f(x)$ is the meromorphic function on C, or the covering of the Riemann sphere. The space of pairs (C, f) for given genus g and degree d of f is called the Hurwitz space $\mathcal{H}_{g,d}$. We assume that all poles and critical values z_i of the function f are simple, C is considered up to an isomorphism, and f up to a linear fractional transformation $\tilde{f} = (af + b)(cf + d)^{-1}$; moreover, z_i serve as local coordinates on $\mathcal{H}_{g,d}$.

The difference of two projective connections $S_B - S_{df}$ is a meromorphic quadratic differential on the Riemann surface C. One can verify that the 1-form

$$\mathbf{q} = -\frac{1}{6} \sum_{i=1}^{n} \mathrm{Res}|_{x=x_i} \frac{S_B - S_{df}}{df} dz_i$$

is closed on the Hurwitz space; here x_i are the critical points of f, $i = 1, \ldots, n$ (i.e. $df(x_i) = 0$) and $z_i = f(x_i)$. Then the Bergman tau-function on the Hurwitz space (called so after the Bergman projective connection S_B) is defined as a local potential for the 1-form $\mathbf{q}$:

$$d \log \tau_B = \mathbf{q} \tag{31.4}$$

For the hyperelliptic coverings an explicit integration of (31.4) gives (31.3); for an arbitrary covering τ_B is given by a more complicated expression; cf. [9]. The function τ_B is a universal object appearing in different problems – from the theory of Frobenius manifolds [6] to Hermitian matrix models [2] and the spectral theory of Riemann surfaces [13].

It was shown in [11] that the product $\eta = \tau_B^{24(n-1)} V^{-6}$, where $V = \prod_{i<j}(z_i - z_j)$ denotes the Vandermonde determinant of critical values, is invariant under Möbius transformations of the function f. On the other hand, under symplectic transformations of the canonical basis on the Riemann surface given by a $2g \times 2g$-matrix $\begin{pmatrix} A & B \\ C & D \end{pmatrix}$, it transforms as follows: $\eta \to \eta\,[\det(C\sigma + D)]^{24(n-1)}$. Therefore, η is a section of the $24(n-1)$-th power of the Hodge line bundle on the Hurwitz space (moreover, it is holomorphic and nonvanishing as long as critical values of the function f remain finite and distinct). The Hurwitz space $\mathcal{H}_{g,d}$ can be compactified by means of admissible covers [4], and we deonote the boundary divisors of the compactification $\overline{\mathcal{H}}_{g,d}$ by $\Delta_\mu^{(k)}$, where $k = 2, \ldots, g+d-1$ and $\mu = [m_1, \ldots, m_r]$ is a partition of d describing the ramification over the double point (see [4,11] for details). The asymptotics of η near the boundary divisors can be explicitly computed, yielding the following relation [11]:

Theorem 1. *For the Hodge class $\lambda \in \mathrm{Pic}(\overline{\mathcal{H}}_{g,d}) \otimes \mathbf{Q}$ we have*

$$\lambda = \sum_{k=2}^{g+d-1} \sum_{\substack{\mu= \\ [m_1,\ldots,m_r]}} \prod_{i=1}^{r} m_i \left(\frac{k(n-k)}{8(n-1)} - \frac{1}{12}\left(d - \sum_{i=1}^{r} \frac{1}{m_i} \right) \right) \delta_\mu^{(k)} . \tag{31.5}$$

(here $\delta_\mu^{(k)}$ are the classes of the divisors $\Delta_\mu^{(k)}$).

Recently this theorem received a different proof s based on the Grothendieck–Riemann-Roch theorem (cf. [3]).

A natural version of the Bergman tau-function also exists on the moduli spaces of holomorphic differentials on Riemann surfaces [13]. These spaces were extensively studied in the theory of dynamical systems (see e.g. [10]). To each holomorphic 1-differential w on C one can naturally associate a flat metric with conical singularities $|w|^2$ that has trivial holonomy. As it was shown in [13], the regularized determinant of Laplace operator $\Delta^{|w|^2}$ for this metric is given by the formula

$$\det \Delta^{|w|^2} = \mathrm{Area}(C, |w|^2)\, \{\det \Im\sigma\} |\tau_B(C, w)|^2, \tag{31.6}$$

where the Bergman tau-function τ_B is defined by a system of equations similar to (31.4), and can also be computed in terms of theta-functions.

Analysing the behavior of τ_B near the boundary divisor of the projectivized moduli space $\mathbb{P}(\overline{H}_g)$ of holomorphic 1-differentials, we derive a formula for the Hodge class λ similar to (31.5) (cf. [12]):

Theorem 2. *In the rational Picard group* $\mathrm{Pic}(\mathbb{P}(\overline{H}_g)) \otimes \mathbf{Q}$ *of the space* $\mathbb{P}(\overline{H}_g)$

$$\lambda = \frac{g-1}{4}\psi + \frac{1}{24}\delta_{\deg} + \frac{1}{12}\delta_0 + \frac{1}{8}\sum_{j=1}^{[g/2]}\delta_j \,. \tag{31.7}$$

Here λ *is the pullback to* $\mathbb{P}(\overline{H}_g)$ *of the Hodge class on the compactified Riemann moduli space* $\overline{\mathcal{M}}_g$*,* ψ *is the tautological class of the projectivization,* δ_{deg} *is the class of the divisor of 1-differentials with multiple zeros, and* δ_j*,* $j = 0, \ldots, [g/2]$ *are the pullbacks of the classes of the boundary divisors of* $\overline{\mathcal{M}}_g$*.*

Bibliography

[1] M. Babich and D. Korotkin, Self-dual $SU(2)$ invariant Einstein metrics and modular dependence of theta-functions, *Lett. in Math. Phys.*, **46**: 323–337, 1998.

[2] B. Eynard, A. Kokotov, and D. Korotkin, Genus one contribution to free energy in Hermitian two-matrix model, *Nucl. Phys.*, **B694**: 443–472, 2004.

[3] G. van der Geer and A. Kouvidakis, The Hodge bundle on Hurwitz spaces, `arXiv:1010.3335`.

[4] J. Harris and D. Mumford, On the Kodaira dimension of the moduli space of curves, *Invent. Math.*, **67**: 23–86, 1982.

[5] N. Hitchin, Twistor spaces, Einstein metrics and isomonodromic deformations, *J. Diff. Geom.*, **42**(1): 30–122, 1995.

[6] B. Dubrovin, Painlevé transcendents in two-dimensional topological field theory, The Painlevé property, CRM Series Math. Phys., Springer, New York (1999), 287–412.

[7] A. Kitaev and D. Korotkin, On solutions of Schlesinger equations in terms of theta-functions, *IMRN*, **17**: 877–905, 1998.

[8] D. Korotkin, Solution of an arbitrary matrix Riemann-Hilbert problem with quasi-permutation monodromy matrices, *Math. Ann.*, **329**(2): 335–364, 2004.

[9] A. Kokotov and D. Korotkin, Isomonodromic tau function of Hurwitz–Frobenius manifolds and its applications, *IMRN,* **2006**: 1–34, 2006.

[10] M. Kontsevich and A. Zorich, Connected components of the moduli spaces of holomorphic differentials with prescribed singularities, *Invent. Math.*, **153**(3): 631–678, 2003.

[11] A. Kokotov, D. Korotkin and P. Zograf, Isomonodromic tau function and admissible covers, *Adv. Math.*, **227**(1): 586–600, 2011.

[12] D. Korotkin and P. Zograf, Tau function and moduli of differentials, *Math. Res. Lett.*, **18**(3), 2011.

[13] A. Kokotov, D. Korotkin, Tau functions on spaces of Abelian differentials and higher genus generalization of Ray–Singer formula, *J. Diff. Geom.*, **82**(1): 35–100, 2009.

[14] A. N. Tyurin, On periods of quadratic differentials, *Uspekhi Mat. Nauk.*, **33**(6): 169–221, 1978 (in Russian).

[15] Al. B. Zamolodchikov, Conformal scalar field on the hyperelliptic curve and critical Ashkin-Teller multipoint correlation functions, *Nucl. Phys.*, **B285**: 481–503, 1987.

Author Information

Dmitry Korotkin, Department of Mathematics and Statistics Concordia University, 1455 de Maisonneuve, Montréal, H3G 1M8 Québec, Canada.
E-mail: korotkin@mathstat.concordia.ca

Peter Zograf, Steklov Mathematical Institute, Fontanka 27, 191023, St. Petersburg, Russia.
E-mail: zograf@pdmi.ras.ru

Chapter 32

On particular Solutions of *q*-Painlevé Equations and *q*-Hypergeometric Equations

Yousuke Ohyama

Abstract. We study a degeneration diagram of linear q-difference equations of hypergeometric type, which are second order q-difference equations whose coefficients are linear functions. We obtain seven q-hypergeometric equations, including two types of q-Bessel equations and two types of q-Airy functions. We explain how our degeneration scheme corresponds to a degeneration diagram of q-Painlevé equations.

Keywords. q-Painlevé Equation, Basic Hypergeometric Series.

2010 Mathematics Subject Classification. Primary: 34M55; Secondary: 33E17.

32.1 Introduction

We give a unified theory for q-special functions which comes from the degeneration of the basic hypergeometric functions ${}_2\varphi_1(a, b, c; q; x)$. We obtain seven types of q-special functions. We have two different the q-Bessel functions. We also have two q-Airy equations, which are essentially equivalent.

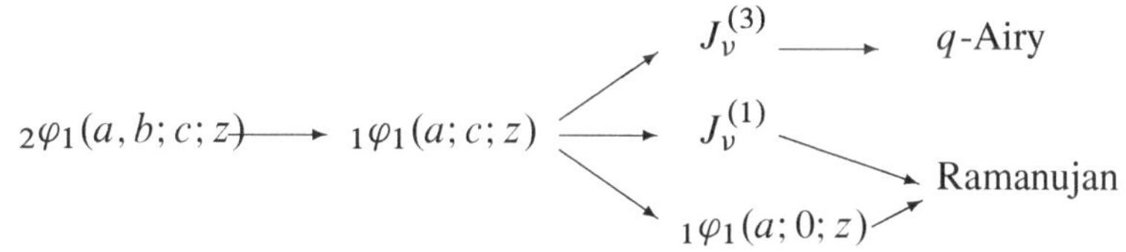

We also show that there is a relation to hypergeometric type q-Painlevé equations and our classification of q-special functions. See [6] for details.

32.2 *q*-Difference Equation of the Hypergeometric Type

We call a q-difference equation of the second order with the linear coefficients

$$(a_0 + b_0 x)u(xq^2) + (a_1 + b_1 x)u(xq) + (a_2 + b_2 x)u(x) = 0 \qquad (32.1)$$

This work was supported by the Mitsubishi foundation and the JSPS Grant-in-Aid for Scientific Research.

it the hypergeometric type. We denote the solution space of the above equation as

$$\Phi\begin{bmatrix} a_1 & a_2 & a_3 \\ b_1 & b_2 & b_3 \end{bmatrix}; x\Bigg]$$

following [1]. We set $\mathrm{lq}\, x = \log x / \log q$.

Theorem 32.1. *A q-difference equation of the hypergeometric type has transformations which remain hypergeometric;*

(A) *change $x \to cx$:*

$$\Phi\left[\begin{matrix} a_1 & a_2 & a_3 \\ b_1 & b_2 & b_3 \end{matrix}; cx\right] = \Phi\left[\begin{matrix} a_1 & a_2 & a_3 \\ cb_1 & cb_2 & cb_3 \end{matrix}; x\right];$$

(B) *change $u \to x^\gamma u$ $(c = q^\gamma)$*

$$x^\gamma\, \Phi\left[\begin{matrix} a_1 & a_2 & a_3 \\ b_1 & b_2 & b_3 \end{matrix}; x\right] = \Phi\left[\begin{matrix} c^2a_1 & ca_2 & a_3 \\ c^2b_1 & cb_2 & b_3 \end{matrix}; x\right];$$

(C) *change $x \to 1/x$*

$$\Phi\left[\begin{matrix} a_1 & a_2 & a_3 \\ b_1 & b_2 & b_3 \end{matrix}; \frac{1}{x}\right] = \Phi\left[\begin{matrix} b_3 & b_2 & b_1 \\ q^2a_3 & q^2a_2 & a_1 \end{matrix}; x\right];$$

(D) *change $u \to (ax;q)_\infty/(bx;q)_\infty u$:*

$$\Phi\left[\begin{matrix} a_1 & a_2 & a_3 \\ b_1 & b_2 & b_3 \end{matrix}; x\right] = x^s \frac{(-b_3x/a_3;q)_\infty}{(-b_1x/a_1q;q)_\infty}\, \Phi\left[\begin{matrix} a_3 & a_2 & a_1 \\ qb_3 & b_2 & q^{-1}b_1 \end{matrix}; x\right];$$

where $s = \mathrm{lq}(a_3/a_1)$.

We classify q-difference equations of the hypergeometric type up to the transformations in Theorem 1. Then we obtain seven classes of q-difference equations:

Theorem 32.2. *A q-difference equation* (32.1) *of the hypergeometric type reduces to one of the following equation by transforms in Theorem 32.1.* ($p = \sqrt{q}$).

(1) *When $a_1a_3b_1b_3 \neq 0$, Heine's hypergeometric ${}_2\varphi_1(a,b;c;q;x)$:*

$$(c - abqx)u(xq^2) - [c + q - (a+b)qx]u(qx) + q(1-x)u(x) = 0.$$

(2) *When $b_3 = 0$, $a_1a_3b_1b_2 \neq 0$, ${}_1\varphi_1(a;c;q;x)$:*

$$(c - aqx)u(xq^2) - (c + q - qx)u(qx) + qu(x) = 0.$$

(3-1) *When* $b_1 = b_2 = 0$, $a_3 \cdot a_2 a_1 b_3 \neq 0$, *Jackson's Bessel functions* $J_\nu^{(1)}(x;q)$:

$$u(xp^2) - (p^\nu + p^{-\nu})u(xp) + (1 + x/4)u(x) = 0.$$

(3-2) *When* $b_1 = b_3 = 0$, $a_2 \cdot a_3 a_1 b_2 \neq 0$, *Hahn-Exton's Bessel functions* $J_\nu^{(3)}(x;q)$:

$$u(xp^2) + [-(p^\nu + p^{-\nu}) + p^{2-\nu}x]u(xp) + u(x) = 0.$$

(3-3) *When* $b_3 = a_1 = 0$, $a_2 b_2 \cdot a_3 b_1 \neq 0$, *$q$-Hermite-Weber* ${}_1\varphi_0(a;-;q;x)$

$$axu(xq^2) + (1 - x)u(xq) - u(x) = 0.$$

(4-1) *When* $b_1 = a_2 = b_3 = 0$, *q-Airy* $Ai_q(x) = {}_1\varphi_1(0;-q;q;-x)$:

$$u(xq^2) + xu(xq) - u(x) = 0.$$

(4-2) *When* $a_1 = b_2 = b_3 = 0$, *the Ramanujan function* ${}_0\varphi_1(-;0;q;-tq)$:

$$qxu(xq^2) - u(xq) + u(x) = 0.$$

In the study of differential equations, shearing transformations are useful for studying irregular singular points whose Poincaré rank is noninteger.

Shearing transformations are also useful for q-differential equations when a slope of the Newton diagram is noninteger.

For a q-difference equation

$$a(x)u(xq^2) + b(x)u(xq) + c(x)u(x) = 0,$$

a *shearing transformation* is the following transformation:

$$x = t^2, \ p = \sqrt{q}, \ v(t) = u(x).$$

Then we have

$$a(t^2)v(tp^2) + b(t^2)v(tp) + c(t^2)v(t) = 0.$$

Lemma 32.3. *In Theorem 32.2,* (4-2) *is equivalent to* (4-1) *by shearing transformation.*

By the connection formula of the q-Airy function by T. Morita, we obtain a relation between the q-Airy function and the Ramanujan function [5]:

$$A_{q^2}\left(-\frac{q^3}{x^2}\right) = -\frac{1}{(q,-1;q)_\infty}\left\{\theta\left(\frac{x}{q}\right)\mathrm{Ai}_q(-x) + \theta\left(-\frac{x}{q}\right)\mathrm{Ai}_q(x)\right\}.$$

32.3 Hypergeometric Solutions of the q-Painlevé Equations

Just as the Painlevé differential equations have particular solutions represented by (confluent) hypergeometric functions, the q-Painlevé equations also have special solutions written by q-hypergeometric functions.

In [3], q-hypergeometric solutions of the q-Painlevé equations were examined. The degeneration diagram of the q-hypergeometric solutions of the q-Painlevé equations is as follows:

q-**P**	$q\text{-}P_{\mathrm{VI}}$	$\rightarrow$	$q\text{-}P_{\mathrm{V}}$	$\rightarrow$	$q\text{-}P_{\mathrm{IV}}$ $q\text{-}P_{\mathrm{III}}$	$\rightarrow$	$q\text{-}P_{\mathrm{II}}$	$\rightarrow$	$q\text{-}P_{\mathrm{I}}$
HG	${}_2\varphi_1$	$\rightarrow$	${}_1\varphi_1$	$\rightarrow$	${}_1\varphi_1\,(a;0;z)$ ${}_1\varphi_1\,(0;b;z)$	$\rightarrow$	${}_1\varphi_1\,(0;-q;z)$	$\rightarrow$	none
	(1)	$\rightarrow$	(2)	$\rightarrow$	(3-3) (3-2)	$\rightarrow$	(4-1)	$\rightarrow$	none

Comparing this with our list in Theorem 32.2, we do not have (3-1) and (4-2). Equation (4-2) is related to (4-1) by a shearing transformation. Equation (3-1) appears in another form of $q\text{-}P_{\mathrm{III}}$.

It is know that there are several types of q-Painlevé equations. For $q\text{-}P_{\mathrm{III}}$, one was called $q\text{-}P_{\mathrm{III}}(A_5^{(1)})$ by Sakai [8]:

$$\frac{y\bar{y}}{a_3a_4} = -\frac{\bar{z}(\bar{z}-b_2t)}{\bar{z}-b_3}, \qquad \frac{z\bar{z}}{b_3} = -\frac{y(y-a_1t)}{a_4(y-a_3)}.$$

Another one is known from Ramani, Grammaticos and Hietarinta [7]:

$$\frac{\overline{w}\underline{w}}{a_3a_4} = \frac{(w-a_1s)(w-sa_2)}{(w-a_1)(w-a_4)}, \tag{32.2}$$

which is a symmetric specialization of $q\text{-}P_{\mathrm{VI}}$ found by Jimbo and Sakai [2]. Also, $J_\nu^{(1)}(x;q)$, a solution of (3-1), is a special solution of (32.2), as shown in [4].

Bibliography

[1] W. Hahn, *Math. Nachr.*, **2**: 340–379, 1949.

[2] M. Jimbo and H. Sakai, *Lett. Math. Phys.*, **38**: 145–154, 1996.

[3] K. Kajiwara, T. Masuda, M. Noumi, Y. Ohta, and Y. Yamada, *Int. Math. Res. Not.*, **47**: 2497–2521, 2004.

[4] K. Kajiwara, Y. Ohta, and J. Satsuma, *J. Math. Phys.*, **36**: 4162, 1995.

[5] T. Morita, A connection formula between the Ramanujan function and the q-Airy function, arXiv:1104.0755.

[6] Y. Ohyama, A unified approach to q-special functions of the Laplace type, arXiv:1103.5232.

[7] A. Ramani, B. Grammaticos, and Hietarinta, *J. Phys. Rev. Lett.*, **67**: 1829–1832, 1991.

[8] H. Sakai, *Comm. Math. Phys.*, **220**: 165–229, 2001.

Author Information

Yousuke Ohyama, Department of Mathematics, Osaka University, 1-1 Machikaneyama-machi, Toyonaka, 560-0043, Japan.
E-mail: ohyama@ist.osaka-u.ac.jp

Chapter 33

Derivation of Painlevé Equations by Antiquantization

Sergey Yu. Slavyanov

Abstract. The procedure of antiquantization for derivation of Painlevé equation proposed as a recipe in earlier publications of the author is discussed in further detail.

Keywords. Antiquantization, Painlevé Equations.

2010 Mathematics Subject Classification. 34M55.

The Painlevé equations can be derived and studied by different means. One of the most fruitful approaches is the use of linear equations as a basic tool. The conventional ones are 2×2 first-order linear systems added by isomonodromie conditions [1] and second-order equations with apparent singularity. Quite straightforward from the analytical point of view, these approaches lead, however, to very sophisticated practical computations. The author recently proposed another approach which seems to be a tricky recipe, but leads to a much mor efficient result [2, 3]. This approach was later called the antiquantization procedure [5, 6]. It includes:

- choice of a quantum equation, which could be the only equation belonging to the Heun class;
- choise of the parameter which is termed as adiabatic and which plays the role of time;
- choice of "energy" or Hamiltonian;
- normalization of the hamiltonian; this step is equivalent to the introduction of an isomonodromie condition.

This is the content of a quantum model. Examples are discussed in [7]. Further, the quantum variables in the Hamiltonian coordinate and momentum are substituted for the classical coordinate and momentum. The classical Euler–Lagrange equations appear to be one of the Painlevé's equation.

In conventional antiquantization procedures, the role of the order of the quantum operators is important. In our procedure it is not. Normalization of the wave function is also not crucial.

The general Heun equation in essence can be studied as following [4]:

$$\sigma(z)y''(z) + \sum_{j=1}^{3}(1-b_j)\sigma_j(z)y'(z) + \Bigg[\sum_{j=1}^{3}\frac{a_j\sigma_j(z)}{(z-z_j)} + \delta(z-z_3) \\ - \left(\frac{\lambda\sigma_3(z_3)}{(z_2-z_1)} + \frac{1}{2}\sum_{j=1}^{2}(1-b_3)(1-b_j)\frac{\sigma_3(z_3)}{z_3-z_j}\right)\Bigg]y(z) = 0. \quad (33.1)$$

Here

$$b_j = (\rho_{1j}+\rho_{2j}), \quad a_j = \rho_{1j}\rho_{2j}, \quad j = 1,2,3, \quad a_\infty = \kappa_1\kappa_2,$$

$$\sigma(z) = \prod_{j=1}^{3}(z-z_j), \quad \sigma_j(z) = \frac{\sigma(z)}{z-z_j},$$

$$\delta = a_\infty - \sum_{j=1}^{3} a_j,$$

where ρ_{mj} are characteristic exponents at finite singular points and κ_1, κ_2 are characteristic exponents at infinity. It can be shown that the quantity λ stays invariant under linear transform of independent variable and s-homotopic transformation of the function y. The other invariants are squares of differences between characteristic exponents

$$\Delta_j = (\rho_{1j}-\rho_{2j})^2, \; j = 1,2,3 \quad \Delta_\infty = (\kappa_1-\kappa_2)^2.$$

Applying the antiquantization procedure, we arrive to the following equation [4]:

$$\frac{2\sigma_3(t)}{\sqrt{\sigma(q)}}\frac{d}{dt}\frac{\dot{q}\sigma_3(t)}{\sqrt{\sigma(q)}} + \frac{\dot{q}\sigma_3^2(t)}{\sigma(q)(q-t)} \\ + \left[-\Delta_\infty + \sum_{j=1}^{2}\frac{(\Delta_j+1-2b_j)\sigma_j(z_j)}{(q-z_j)^2} + \frac{(\Delta_3-1)\sigma_3(t)}{(q-t)^2}\right] = 0. \quad (33.2)$$

This is a general form of the Painlevé equation P^6 generated by the general form of the Heun equation. Two important features of equation (33.2) should be emphasized:

(i) the role of the singular point $z_3 = t$ is specific in (33.2) compared to the other points z_1, z_2;

(ii) the only influence of generalization due to s-homotopic transformation is a slight dependence on b_j $j = 1,2$ in (33.2).

The general approach to Hamiltonian structure for Painlevé equations is as follows. Each equation belonging to the Heun class in its canonical form may be presented in the form

$$\frac{1}{f(t)}\left[P_0(z,t)D^2 + P_1(z,t)D + P_2(z,t)\right]y(z) = \lambda y(z) \quad (33.3)$$

In equation (33.3) $P_0(z,t)$, $P_1(z,t)$, $P_2(z,t)$ are polynomials in z of not higher than third order, t is the scaling parameter, λ is the accessory parameter or the "energy' in the formulation above. The factor $f(t)$ is due to normalization of energy. If quantum observables $\hat{q}$, $\hat{p}$, ($\hat{q}$ the coordinate and $\hat{p}$ the momentum) are associated with z and D in equation (33.3) it becomes the Hamiltonian structure and can be rewritten as

$$H(\hat{q},\hat{p},t)y = \lambda y \tag{33.4}$$

In equation (33.4) the function H is the Hamiltonian adiabatically, depending on the parameter t, which can be considered as time, and λ is energy. The corresponding Hamiltonian in classical machanics is quadratic in the classical momentum p.

$$H(q,p,t) = \frac{1}{f(t)}\left[P_0(q,t)p^2 + P_1(q,t)p + P_2(q,t)\right] \tag{33.5}$$

The Legendre transform can be applied to this Hamiltonian turning from momentum p to velocity q_t. The following Euler–Lagrange equation of motion relates to this Lagrangian

$$\begin{aligned} q_{tt} &= \frac{1}{2}\frac{\partial}{\partial q}(\ln P_0(q,t))q_t^2 - \left(\frac{\partial}{\partial t}(\ln f(t)) - \frac{\partial}{\partial t}\big(\ln P_0(q,t)\big)\right)q_t \\ &+ \frac{P_0(q,t)}{f^2(t)}\left(\frac{\partial}{\partial q}\frac{P_1^2(q,t)}{2P_0(q,t)} + f(t)\frac{\partial}{\partial t}\frac{P_1(q,t)}{P_0(q,t)} - 2\frac{\partial P_2(q,t)}{\partial q}\right) \end{aligned} \tag{33.6}$$

The following theorem holds.

Theorem. *Each type of equation of the Heun class is generic for one of the equations of the Painlevé class in the sense that a Schrödinger-type equation corresponds via an antiquantization recipe to an equation of motion in classical dynamics. The inverse statement also holds. Each type of Painlevé equation may be generated by the corresponding Heun equation.*

Specially chosen solutions of Painleve equations constitute the class of special functions related to nonlinear mathematical physics – the so-called Painlevé transcendents.

Bibliography

[1] M. V. Babich, *Russian Math. Surv.*, **64**: 45–127, 2009.

[2] S. Slavyanov, *J. Phys. A.*, **29**: 7329–7335, 1996.

[3] S. Slavyanov, W. Lay, *Special functions: a unified theory based on singularities*, OUP, 2000.

[4] S. Slavyanov, *Operator Methods in Ordinary and Partial Differential Equations, series* **132**: 395–402, 2002.

[5] S. Yu. Slavyanov and F. R. Vukajlović, *Theor. Math. Phys.*, **150**(1): 123–131, 2007.

[6] S. Yu. Slavyanov, A. Ya. Kazakov, and F. R. Vukaylovic, *Albanian Journal of Mathematics*, **4**(4): 195–201, 2010.

[7] A. Müllary and S. Slavyanov, *Theor. Math. Phys.*, **166**(2): 224–227, 2011.

Author Information

Sergey Yu. Slavyanov, St. Petersburg State University, Ulyanovskaya, 3, St. Peterhof, St. Petersburg, Russia.
E-mail: `slav@SS2034.spb.edu`

Painlevé Equations and Related Topics, 257–261

Chapter 34

Integral Transformation of Heun's Equation and Apparent Singularity

Kouichi Takemura

Abstract. Kazakov and Slavyanov established that Heun's equation admits Euler's integral transformation which changes the parameters. In this article we investigate the transformation in detail. The existence of polynomial-type solutions corresponds to the apparency (nonbranching) of a singularity by the transformation, and we obtain integral representation of solutions of Heun's equation, which have an apparent singularity.

Keywords. Heun's Equation, Euler's Integral Transformation, Apparent Singularity.

2010 Mathematics Subject Classification. Primary: 34M35; Secondary: 33E10.

34.1 Heun's Equation and Integral Transformation

Heun's equation is a standard form of a second-order Fuchsian equation with four singularities, which is given by

$$\frac{d^2y}{dz^2}+\left(\frac{\gamma}{z}+\frac{\delta}{z-1}+\frac{\epsilon}{z-t}\right)\frac{dy}{dz}+\frac{\alpha\beta z-q}{z(z-1)(z-t)}y=0, \tag{34.1}$$

with the condition $\gamma+\delta+\epsilon=\alpha+\beta+1$ [2, 3]. The parameter q is independent from the local exponents and is called an accessory parameter. It is known that Heun's equation appears as a degenerate case of the linear differential equations which produce Painlevé VI equation by monodromy preserving deformation.

Kazakov and Slavyanov established that Heun's equation admits Euler's integral transformation which changes the parameters.

Proposition 34.1 ([1]). *Set*

$$\begin{aligned}&(\eta-\alpha)(\eta-\beta)=0, \quad \gamma'=\gamma-\eta+1, \quad \delta'=\delta-\eta+1, \quad \epsilon'=\epsilon-\eta+1,\\ &\{\alpha',\beta'\}=\{2-\eta,\alpha+\beta-2\eta+1\}, \quad q'=q+(1-\eta)\big(\epsilon+\delta t+(\gamma-\eta)(t+1)\big).\end{aligned} \tag{34.2}$$

The author was supported by the Grant-in-Aid for Young Scientists (B) (Nr. 22740107) from the Japan Society for the Promotion of Science.

Let $v(w)$ be a solution of

$$\frac{d^2v}{dw^2}+\left(\frac{\gamma'}{w}+\frac{\delta'}{w-1}+\frac{\epsilon'}{w-t}\right)\frac{dv}{dw}+\frac{\alpha'\beta' w-q'}{w(w-1)(w-t)}v=0. \tag{34.3}$$

Then the function

$$y(z)=\int_{[\gamma_z,\gamma_p]}v(w)(z-w)^{-\eta}dw \tag{34.4}$$

is a solution of equation (34.1) for $p\in\{0,1,t,\infty\}$, where $[\gamma_z,\gamma_p]=\gamma_z\gamma_p\gamma_z^{-1}\gamma_p^{-1}$ is the Pochhammer contour which turns around the points $w=z$ and $w=p$.

Note that Proposition 34.1 was rediscovered by considering middle convolutions of 2×2-Fuchsian differential equations with four singularities (see [6]).

34.2 Apparent Singularity and Integral Representation of Solutions

Let us consider local solutions of Heun's equation. The exponents of (34.1) about $z=t$ are 0 and $1-\epsilon$. If $\epsilon\notin\mathbb{Z}$; then we have a basis of local solutions as follows:

$$f(z)=\sum_{j=0}^{\infty}c_j(z-t)^j,\quad g(z)=(z-t)^{1-\epsilon}\sum_{j=0}^{\infty}\tilde{c}_j(z-t)^j,\quad (c_0\neq 0\neq\tilde{c}_0). \tag{34.5}$$

We now define an apparent singurality in the case $\epsilon\in\mathbb{Z}$. For simplicity, we only consider the case $\epsilon\in\mathbb{Z}_{\leq 0}$. If $\epsilon\in\mathbb{Z}_{\leq 0}$, then we have a basis of local solutions as

$$f(z)=\sum_{j=0}^{\infty}c_j(z-t)^j+Ag(z)\log(z-t),\quad g(z)=(z-t)^{1-\epsilon}\sum_{j=0}^{\infty}\tilde{c}_j(z-a)^j. \tag{34.6}$$

If the logarithmic term in equation (34.6) disappears, i.e., $A=0$, then the singularity $z=t$ is called apparent. Note that the apparency of a regular singularity is equivalent to that the monodromy about $z=t$ is trivial.

Set $n=1-\epsilon$. Then the condition that the singularity $z=t$ is apparent is written as $P^{\mathrm{app}}(q)=0$, where $P^{\mathrm{app}}(q)$ is a polynomial of the variable q of order n.

Example 34.2. If $\epsilon = -2$ ($n = 3$), then the condition that the regular singularity $z = t$ is apparent is written as

$$\begin{aligned} 0 = P^{\mathrm{app}}(q) &= q^3 + \{(-3\alpha\beta - 3\alpha - 3\beta - 1)t + (3\gamma - 4)\}q^2 \\ &+ \big\{(3\alpha^2\beta^2 + 6\alpha\beta(\alpha + \beta) + 10\alpha\beta + 2(\alpha^2 + \beta^2) + 2\alpha + 2\beta)t^2 \\ &+ \big((-6\alpha\beta - 4\alpha - 4\beta)\gamma + 4\alpha\beta + 4\alpha + 4\beta\big)t + 2(\gamma - 1)(\gamma - 2)\big\}q - \alpha \\ &\times \beta t\big\{(\alpha+1)(\alpha+2)(\beta+1)(\beta+2)t^2 - \gamma(3\alpha\beta+4\alpha+4\beta+4)t + 2\gamma(\gamma - 1)\big\}. \end{aligned} \tag{34.7}$$

By the integral transformation, polynomial-type solutions of Heun's equation corresponds to solutions which have an apparent singularity ([7]). Moreover we have the following proposition;

Theorem 34.3 ([7]). *If* $\epsilon \in \mathbb{Z}_{\le 0}$, $\alpha, \beta, \beta - \gamma, \beta - \delta \notin \mathbb{Z}$ *and the singularity* $z = t$ *of* (34.1) *is apparent, then there exists a nonzero solution of* (34.3) *which can be written as* $w^{\beta-\gamma}(w - 1)^{\beta-\delta}h(w)$ *where* $h(w)$ *is a polynomial of degree* $-\epsilon$ *and the functions*

$$\int_{[\gamma_z, \gamma_p]} w^{\beta-\gamma}(w - 1)^{\beta-\delta}h(w)(z - w)^{-\beta}dw, \tag{34.8}$$

$(p = 0, 1)$ *are nonzero solutions of* (34.1).

Example 34.4. Set $\epsilon = -2$. The condition that the singularity $z = t$ of equation (34.1) is apparent is written as equation (34.7). Then there exists a nonzero solution of equation (34.3) written as $w^{1-\gamma'}(w - 1)^{1-\delta'}h(w) = w^{\beta-\gamma}(w - 1)^{\beta-\delta}h(w)$ where

$$\begin{aligned} h(w) &= 2\alpha(\alpha + 1)w^2 + 2(\alpha + 1)\big\{q - \alpha(\beta + 2)t\big\}w + q^2 \\ &- \big\{2\alpha\beta + 3\alpha + \beta + 1)t - \gamma + 2\}q + \alpha t\{t(\alpha + 1)(\beta + 1)(\beta + 2) - \beta\gamma\big\}, \end{aligned} \tag{34.9}$$

and the functions in equation (34.8) $(p = 0, 1)$ are nonzero solutions of equation (34.1).

We have similar results for linear equations which produce Painlevé VI [7].

34.3 Elliptical Representation of Heun's Equation and Integral Transformation

It is known that Heun's equation has an elliptical representation. Let $\wp(x)$ be the Weierstrass doubly-periodic function with periods $(2\omega_1, 2\omega_3)$, $\omega_0(= 0)$, ω_1, ω_2 $(= -\omega_1 - \omega_3)$, ω_3 be the half-periods and $e_i = \wp(\omega_i)$ $(i = 1, 2, 3)$. Then Heun's equation (equation (34.1)) is transformed to

$$H^{(l_0,l_1,l_2,l_3)}f(x) = Ef(X), \quad H^{(l_0,l_1,l_2,l_3)} = -\frac{d^2}{dx^2} + \sum_{i=0}^{3} l_i(l_i + 1)\wp(x + \omega_i), \tag{34.10}$$

by setting $z = (\wp(x) - e_1)/(e_2 - e_1)$, $t = (e_3 - e_1)/(e_2 - e_1)$. The eigenvalue E corresponds to the accessory parameter q.

We rewrite the integral transformation of Heun's equation (i.e., Proposition 34.1) in elliptical representation form. It is remarkable that the eigenvalue E is unchanged by the integral transformation.

Proposition 34.5 ([7, 8]). *Let $\sigma(x)$, $\sigma_i(x)$ $(i = 1, 2, 3)$ be the Weierstrass sigma and cosigma functions and I be a suitable cycle. Let α'_i be a number such that $\alpha'_i = -l'_i$ or $\alpha'_i = l'_i + 1$ for each $i \in \{0, 1, 2, 3\}$. Set $d = -\sum_{i=0}^{3} \alpha'_i/2$ and $\eta = d + 2$. If $\tilde{f}(x)$ satisfies*

$$H^{(l'_0,l'_1,l'_2,l'_3)} \tilde{f}(x) = E \tilde{f}(x), \tag{34.11}$$

then the function

$$f(x) = \sigma(x)^{\alpha'_0+d+1}\sigma_1(x)^{\alpha'_1+d+1}\sigma_2(x)^{\alpha'_2+d+1}\sigma_3(x)^{\alpha'_3+d+1}. \tag{34.12}$$

$$\int_I \tilde{f}(\xi)\sigma(\xi)^{1-\alpha'_0}\sigma_1(\xi)^{1-\alpha'_1}\sigma_2(\xi)^{1-\alpha'_2}\sigma_3(\xi)^{1-\alpha'_3}\big(\sigma(x+\xi)\sigma(x-\xi)\big)^{-\eta} d\xi$$

satisfies

$$H^{(\alpha'_0+d,\alpha'_1+d,\alpha'_2+d,\alpha'_3+d)} f(x) = Ef(x). \tag{34.13}$$

Theorem 34.6 ([7, 8]). *Let $k \in \{1, 3\}$ and $M_{2\omega_k}^{(l_0,l_1,l_2,l_3)}(E)$ be the monodromy matrix by the shift of the period $x \to x + 2\omega_k$ with respect to a certain basis of solutions to $H^{(l_0,l_1,l_2,l_3)} f(x) = Ef(x)$. Then*

$$\operatorname{trace} M_{2\omega_k}^{(l'_0,l'_1,l'_2,l'_3)}(E) = \operatorname{trace} M_{2\omega_k}^{(\alpha'_0+d,\alpha'_1+d,\alpha'_2+d,\alpha'_3+d)}(E). \tag{34.14}$$

In other word, periodicity is preserved by the integral transformation.

We apply the integral transformation for the case where Heun's equation has the finite-gap property, i.e., the case where $l_0, l_1, l_2, l_3 \in \mathbb{Z}$ $(\gamma, \delta, \epsilon, \alpha - \beta \in \mathbb{Z} + 1/2)$. For the case $l_0, l_1, l_2, l_3 \in \mathbb{Z}$ we can calculate the monodomy in principle for all E by means of hyperelliptic integrals [4] and by the Hermite–Krichever Ansatz [5]. By applying monodromy invariance, we can calculate the monodromy of Heun's equation for the case $l_0, l_1, l_2, l_3 \in \mathbb{Z}+1/2$ and $l_0+l_1+l_2+l_3 \in 2\mathbb{Z}+1$ $(\gamma, \delta, \epsilon, \alpha+1/2, \beta+1/2 \in \mathbb{Z})$, which have not been studied so far.

Bibliography

[1] A. Kazakov and S. Slavyanov, Integral relations for special functions of the Heun class, *Theoret. and Math. Phys.*, **107**: 733–739, 1996.

[2] A. Ronveaux (ed.), *Heun's differential equations,* Oxford Science Publications, 1995.

[3] S. Slavyanov and W. Lay, *Special Functions,* Oxford University Press, 2000.

[4] K. Takemura, The Heun equation and the Calogero–Moser–Sutherland system III: the finite gap property and the monodromy, *J. Nonlinear Math. Phys.* **11**: 21–46, 2004.

[5] K. Takemura, The Heun equation and the Calogero–Moser–Sutherland system IV: the Hermite–Krichever Ansatz, *Comm. Math. Phys.* **258**: 367–403, 2005.

[6] K. Takemura, Middle convolution and Heun's equation, *SIGMA*, **5**: 040, 22 pages, 2009.

[7] K. Takemura, *Integral transformation of Heun's equation and some applications*, preprint, `arXiv:1008.4007[math.CA]`.

[8] K. Takemura, Integral transformation and Darboux transformation of Heun's differential equation, in: *Nonlinear and modern mathematical physics,* AIP Conference Proceedings **1212**: 58–65, 2010.

Author Information

Kouichi Takemura, Department of Mathematics, Faculty of Science and Technology, Chuo University, 1-13-27 Kasuga, Bunkyo-ku Tokyo 112-8551, Japan.
E-mail: `takemura@math.chuo-u.ac.jp`

Painlevé Equations and Related Topics, 263–266

Chapter 35

Painlevé Analysis of Solutions to Some Nonlinear Differential Equations and their Systems Associated with Models of the Random-Matrix Type

Vladimir Tsegel'nik

Abstract. Certain analytical properties of solutions to some differential equations and their systems associated with models of random-matrix type are investigated.

Keywords. Painlevé Test, Painlevé Property, Model of Random-Matrix Type.

2010 Mathematics Subject Classification. Primary: 34M55; Secondary: 35Q51.

35.1 Introduction

The past three decades have witnessed an increasing interest in studying certain classes of continuous and discrete probability models by the name "models of the random-matrix type". The origins of such models can be very different [1].

One of the most important characteristics of these models is the emptiness formation probability, i.e., the probability that particles are absent from a given interval or union of intervals.

The emptiness formation probabilities, as a rule, can be represented in the form of the Fredholm determinant $\det(\mathbf{1}-K)_J$, where K is an integral operator and J is the set inside which no particles must be present. The kernel of the operator K usually has the form

$$K(x,y) = \frac{A(x)B(y) - B(x)A(y)}{x-y}\sqrt{\psi(x)\psi(y)} \tag{35.1}$$

with appropriate functions ψ, A, and B. The only currently known way to calculate such Fredholm determinants is to present them as solutions of an ordinary differential equation or a system of partial differential equations.

35.2 Model of the Random-Matrix Type with Airy Kernal

This chapter deals with the Hamilton system

$$q' = p - qu + \alpha s, \qquad v' = -pq - \alpha sq, \tag{35.2}$$

$$p' = sq - 2qv + pu + \alpha su, \qquad u' = -q^2, \tag{35.3}$$

where s is an independent variable, α is a parameter. The Hamiltonian of system (35.2) has the form

$$H = \frac{p^2}{2} - \frac{sq^2}{2} + q^2 v - pqu + \alpha sp - \alpha squ.$$

In case of $\alpha = 0$ system (35.2), (35.3) corresponds to a model of the random-matrix type with the Airy kernel [2]. We have carried out Painlevé analysis of solutions to system (35.2), (35.3) and have proved the following theorem:

Theorem 35.1. *System* (35.2), (35.3) *is a Painlevé-type system. Its solutions are expressed in terms of solutions to the second Painlevé equation*

$$q'' = 2q^3 + (s + C)q + \alpha, \tag{35.4}$$

where C is an arbitrary constant of integration.

The correctness of this theorem follows from the existence of the first integral $u^2 - 2v - q^2 = C$ of system (2), (3) and the next formulas $u(s) = -\int q^2(s)ds$, $p = q' + qu - 2s$, $2v = u^2 - q^2 - C$.

Example. As equation (35.4) in the case $\alpha = -1$ has a solution $q = \frac{1}{s+C}$, the system (35.2), (35.3) has a solution

$$u = \frac{1}{s+C} + C_1, \quad v = \frac{C_1}{s+C} + \frac{C_1^2 - C}{2}, \quad p = \frac{C_1}{s+C} + s,$$

where C_1 is an arbitrary constant.

35.3 System of Differential Equations Associated with the Dyson Process

We have carried out Painlevé-analysis of solutions to the system of differential equations

$$\begin{aligned} q'' &= \left(s^2 - 2n - 1\right) q + 2q^2 p, \\ p'' &= \left(s^2 - 2n + 1\right) p + 2p^2 q, \end{aligned} \tag{35.5}$$

associated with the so-called Dyson process [3].

Theorem 35.2. *System* (35.5) *satisfies the Painlevé test.*

35.4 Solutions of the Traveling-Wave Form of a Partial Differential Equation

We have studied a class of selfsimilar solutions to the partial differential equations

$$\left(A_1^3 - 4\left(A_3 - \frac{1}{2}\right)\right) f + 6(A_1 f)^2 = 0, \tag{35.6}$$

where $A_n = \sum_{i=1}^{2r} x_i^{\frac{n-1}{2}} \cdot \frac{\partial}{\partial x_i}$, $n = 1, 3$; f is an unknown function of independent variables $x_1, x_2, \ldots x_{2r}$.

Equation (35.6) is also associated with a model [4] of the random-matrix type.

The set $E = \bigcup_{i=1}^{r} [x_{2i-1}, x_{2i}] \subset R$ is associated with equation (35.6). The transformation $w_1 = f(\tau), \tau = x_1 + x_2 + \ldots + x_{2r}$ reduces equation (35.6) to

$$r w_1''' - 2r w_1' + w_1 + 6r w_1'^2 = 0, \quad w_1' = \frac{dw_1}{d\tau}, \quad w_1''' = \frac{d^3 w_1}{d\tau^3} \tag{35.7}$$

that has the first integral

$$w_1''^2 + 4w_1'^3 - \frac{2}{r}\tau w_1'^2 + \frac{2}{r} w_1 w_1' = K_1, \tag{35.8}$$

where K_1 ia an arbitrary constant. The substitution $w_1 = \lambda_1 y$, $\tau = \mu_1 s$, $\lambda_1 \mu_1 = 1$, $\mu_1^3 = r$ transforms [5] equation (35.8) into the equation

$$y''^2 + 4y'^3 - 2sy'^2 + 2yy' - \left(\alpha - \frac{\varepsilon}{2}\right)^2 = 0, \tag{35.9}$$

where α is an arbitrary parameter; $\varepsilon^2 = 1$.

In case of $C = 0$ there is a correspondence between solutions of equation (35.9) and solutions of equation (35.4) given by the next formulas

$$y = w'^2 - \left(w^2 + \frac{s}{2}\right)^2 - 2\left(\alpha - \frac{\varepsilon}{2}\right) w, \tag{35.10}$$

$$w = \left(y'' + \frac{1}{2} - \alpha\varepsilon\right) \left(-2\varepsilon y'\right)^{-1}. \tag{35.11}$$

Theorem 35.3. *Let* $y = y(s)$ *be a solution of equation* (35.9) *with fixed parameter values* α *and* ε. *Then the function* $f(\tau) = \lambda_1 y\left(\frac{\tau}{\mu_1}\right)$, $\tau = x_1 + x_2 + \cdots + x_{2r}$, $\lambda_1 \mu_1 = 1$, $\mu_1^3 = 1$ *is a solution of equation* (35.1).

In case $E = (x, +\infty)$ the function $f(x)$ satisfies the following equation:

$$f''' - 4xf'' + 2f + 6f'^2 = 0,$$

which has the first integral

$$f''^2 + 4f'^3 - 4xf'^2 + 4ff' = K_2, \tag{35.12}$$

where K_2 is an arbitrary constant.

A scale transformation of unknown function f and independent variable x reduces equation (35.12) to the form (35.9).

Equation (35.12) for K_2 is obtained in [2].

Bibliography

[1] A. M. Borodin, *Isomonodromic deformations and Painlevé equations in the random-matrix-type problems,* preprint POMI Nr 04-2005, St. Peterburg Dept. Steklov Inst. Math., St. Peterburg (2005), http://www.pdmi.ras.ru/preprint/2005/05-04.html (in Russian).

[2] C. Tracy and H. Widom, Level-spacing distributions and Airy kernel, *Commun. Math. Phys.,* **159**: 151–174, 1994.

[3] C. Tracy and H. Widom, Differential equations for Dyson processes, http://xxx.lanl.gov/math.PR/0309082.

[4] M. Adler and P. van Moerbeke, Random matrices, Virasoro algebra and noncommutative KP, *Duke Math. J.,* **94**: 379–431, 1998.

[5] F. J. Bureau, Equations différentielles du second ordré en y et du second ordre degre en y'' dont l'intégrale générale est á points critiques fixes, *Annali di Mathematica,* **91**: 163–281, 1972.

Author Information

Vladimir Tsegel'nik, Belarusian State University of Informatics and Radioelectronics,
P. Browka Str., 6, 220013, Minsk, Belarus.
E-mail: tsegvv@bsuir.by

Chapter 36

Reductions on the Lattice and Painlevé Equations P_2, P_5, P_6

Pavlos Xenitidis

Abstract. The symmetry analysis of the discrete integrable systems of the Adler–Bobenko–Suris classification is reviewed, and symmetry reductions of the latter are discussed. In particular, reductions of the discrete potential KdV equation to discrete Painlevé equations are presented, and continuous symmetric reductions leading to integrable systems of partial differential equations are considered. As a by-product, an interpretation of solutions of continuous Painlevé equations as particular solutions of the discrete potential KdV and a discrete Painlevé II equations is demonstrated.

Keywords. Integrable Equations, Symmetries, Reductions, Painlevé Equations.

2010 Mathematics Subject Classification. Primary: 35B06, 39A14; Secondary: 34M55.

36.1 Introduction

Adler, Bobenko, and Suris (ABS) [2] recently classified all the integrable scalar difference equations which are defined on an elementary quadrilateral of the lattice and possess the following two properties: i) they are affine linear, and ii) they are multidimensionally consistent[1]. Their classification led to the following list of seven equations, which includes some new cases (H2, $\mathrm{H3}_{\delta=1}$, $\mathrm{Q1}_{\delta=1}$, Q2, $\mathrm{Q3}_{\delta=1}$) as well as some well-known equations (H1, $\mathrm{H3}_{\delta=0}$, $\mathrm{Q1}_{\delta=0}$, $\mathrm{Q3}_{\delta=0}$, Q4); see e.g. [1,3,6]:

H1 $\quad (u-z)(x-y)-\alpha+\beta=0;$

H2 $\quad (u-z)(x-y)+(\beta-\alpha)(u+x+y+z)-\alpha^2+\beta^2=0;$

H3 $\quad \alpha(ux+yz)-\beta(uy+xz)+\delta(\alpha^2-\beta^2)=0;$

Q1 $\quad \alpha(u-y)(x-z)-\beta(u-x)(y-z)+\delta^2\alpha\beta(\alpha-\beta)=0;$

Q2 $\quad \alpha(u-y)(x-z)-\beta(u-x)(y-z)$
$\qquad +\alpha\beta(\alpha-\beta)(u+x+y+z)-\alpha\beta(\alpha-\beta)(\alpha^2-\alpha\beta+\beta^2)=0;$

[1] This means that these equations can be extended to a multidimensional lattice in a self-consistent way. This property serves as an integrability criterion which played a central role in the ABS classification.

$$\text{Q3}\quad (\beta^2-\alpha^2)(uz+xy)+\beta(\alpha^2-1)(ux+yz)$$
$$-\alpha(\beta^2-1)(uy+xz)-\delta^2(\alpha^2-\beta^2)(\alpha^2-1)(\beta^2-1)/(4\alpha\beta)=0;$$
$$\text{Q4}\quad a_0uxyz+a_1(uxy+xyz+yzu+zux)+a_2(uz+xy)$$
$$+\bar{a}_2(ux+yz)+\tilde{a}_2(uy+xz)+a_3(u+x+y+z)+a_4=0.$$

In the above list we have used the following shorthand notation for the values of the function $u:\mathbb{Z}^2\to\mathbb{R}$:

$$u:=u_{n,m},\quad x:=u_{n+1,m},\quad y:=u_{n,m+1},\quad z:=u_{n+1,m+1},$$

and α, β denote the *lattice parameters* assigned to the n- and the m-directions of the lattice, respectively. In Adler's equation Q4 the parameters are functions of the lattice parameters given in terms of the Weierstrass $\wp$-function [1].

The study of equations H1, $\mathrm{H3}_0$, and $\mathrm{Q1}_0$ led to the derivation of a special type of solutions, namely similarity solutions [5], which have been shown to be related to discrete Painlevé equations [4, 7]. Moreover, a new type of connection among integrable discrete and continuous equations was demonstrated in [4], where a system of partial differential equations was constructed in relation with the discrete potential KdV equation, i.e., equation H1.

In this chapter we will formulate this analysis in terms of symmetries and symmetry reductions, and consequently we will derive these results in a systematic and simple manner which can be applied to any equation of the ABS classification. In particular, the symmetries of the above equations have been studied [10] and it has been demonstrated how they can be used effectively in the construction of group invariant solutions [10] and continuously symmetric solutions [11]. The former are related to solutions of discrete Painlevé equations, and the latter to solutions of the continuous Painlevé equations. Finally, relations among the solutions of particular discrete and continuous Painlevé equations follow from these considerations.

36.2 Symmetries of the ABS Equations

All of the ABS equations

$$Q(u_{n,m},u_{n+1,m},u_{n,m+1},u_{n+1,m+1},\alpha,\beta)\;=\;0$$

admit a pair of symmetries generated by the vector fields

$$V_1=\left(\frac{f(u_{n,m},u_{n+1,m},\alpha)}{u_{n+1,m}-u_{n-1,m}}-\frac{1}{2}\partial_{u_{n+1,m}}f(u_{n,m},u_{n+1,m},\alpha)\right)\partial_{u_{n,m}}$$
$$G_1=n\,V_1\;+\;\xi(\alpha)\,\partial_\alpha\,,$$

where the symmetric and biquadratic polynomial f is completely determined by the polynomial Q and its derivatives. Moreover, the invariance of the equation under the interchange of the lattice directions, i.e., the mutual interchange $(n, m, u_{n+i,m+j}, \alpha, \beta) \longrightarrow (m, n, u_{n+j,m+i}, \beta, \alpha)$, implies the existence of two similar symmetries V_2 and G_2 in the m-direction [10].

Symmetries generated by V_i are referred to as generalized symmetries, and they will be used in the derivation of similarity solutions. The other two symmetries G_i are referred to as extended generalized symmetries, and the reduction of the ABS equations under the action of both of these symmetries leads to systems of partial differential equations, also known as generating PDEs.

36.3 Reduction on the Lattice and Discrete Painlevé Equations

The solutions of an equation which remain invariant under the action of a symmetry of the equation are called group invariant solutions [8]. These particular solutions of the equation satisfy not only the equation, but also certain *constraints* following from their invariance.

In the case of the ABS equations which admit a generalized symmetry of the form $X = P\partial_{u_{n,m}} := c_1 V_1 + c_2 V_2$, $c_i \in \mathbb{R}$, such group invariant solutions can be derived systematically. More precisely, they are solutions of the equation which also satisfy the constraint $P = 0$. From the symmetry analysis of the ABS equations [10], it is known that the numerator and the denominator of the rational function P depends linearly on $u_{n\pm1,m}$ and $u_{n,m\pm1}$, which implies that the equation $P = 0$ can be solved uniquely with respect to any of these values of u. Using this observation, the system constituted from the original equation and the above constraint leads, in general, to a fourth-order dimensional map, the derivation of which is straightforward. If the equation admits point symmetries which commute with X, then the order of the map can be further reduced. In the case of H1 and a specific form of X, one finds that these group invariant solutions are determined by solutions of the asymmetric, alternate discrete Painlevé II equation [10].

36.4 Continuous Symmetry Reductions

In the same fashion, the solutions of the ABS equation, which remain invariant under the action of the extended generalized symmetries G_1 and G_2, are the ones which additionally satisfy the system of differential-difference equations

$$\xi(\alpha)\partial_\alpha u_{n,m} = nR(u_{n,m}, u_{n+1,m}, u_{n-1,m}, \alpha),$$

$$\xi(\beta)\partial_\beta u_{n,m} = mR(u_{n,m}, u_{n,m+1}, u_{n,m-1}, \beta).$$

From the compatible system of the discrete equation and the above two differential-difference equations, one can systematically derive a system of partial differential equations involving only $u_{n,m}$, $u_{n+1,m}$, and $u_{n,m+1}$ [11].

The continuous system corresponding to equation H1 generalizes the Ernst equation [9], and some of its similarity solutions are determined by solutions of Painlevé V and VI equations. From the derivation of this system follows that solutions of the continuous Painlevé V equation can be regarded as the solution of the asymmetric, alternate discrete Painlevé II equation [11].

Bibliography

[1] V. E. Adler, Bäcklund Transformation for the Krichever–Novikov equation, *Int. Math. Res. Notices*, **1**, 1–4, 1998.

[2] V. E. Adler, A. I. Bobenko, amd Yu. B. Suris, Classification of Integrable Equations on Quad-Graphs. The Consistency Approach, *Commun. Math. Phys.*, **233**: 513–543, 2003; Discrete nonlinear hyperbolic equations. Classification of integrable cases, *Funct. Anal. Appl.*, **43**: 3–21, 2009.

[3] F. W. Nijhoff and H. Capel, The discrete Korteweg-de Vries equations, *Acta Appl. Math.*, **39**: 133–158, 1995.

[4] F. Nijhoff, N. Joshi, and A. Hone, On a Schwarzian PDE associated with the KdV hierarchy, *Phys. Lett. A*, **267**: 147, 2000.

[5] F. W. Nijhoff and V. Papageorgiou, Similarity reductions of integrable lattices and discrete analogues of the Painlevé II equation, *Phys. Lett. A*, **153**: 337–344, 1991.

[6] F. W. Nijhoff, G. R. W. Quispel, and H. W. Capel, Direct linearization of nonlinear difference-difference equations, *Phys. Let. A*, **97**: 125–128, 1983.

[7] F.(1983)W. Nijhoff, A. Ramani, B. Grammaticos, and Y. Ohta, On discrete Painlevé equations associated with the lattice KdV systems and the Painlevé VI equation, *Stud. Appl. Math.*, **106**: 261–314, 2001.

[8] P. J. Olver, *Applications of Lie Groups to Differential Equations*, Graduate Texts in Mathematics 107, Springer Verlag, 1993.

[9] A. Tongas, D. Tsoubelis, and P. Xenitidis, A family of integrable nonlinear equations of the hyperbolic type, *J. Math. Phys.*, **42**: 5762–5784, 2001.

[10] A. Tongas, D. Tsoubelis, and P. Xenitidis, Affine linear and D_4 symmetric lattice equations: symmetry analysis and reductions, *J. Phys. A: Math. Theor.*, **40**: 13353–13384, 2007.

[11] D. Tsoubelis and P. Xenitidis, Continuous symmetric reductions of the Adler–Bobenko–Suris equations, *J. Phys. A: Math. Theor.*, **42**: 165–203, 2009.

Author Information

Pavlos Xenitidis, School of Mathematics, University of Leeds, LS2 9JT, Leeds, UK.
E-mail: `P.Xenitidis@leeds.ac.uk`

Comments

Chapter 18. The full list of asymptotic expansions of solutions to P_6 can be found in reference [5] of this chapter. A detailed comparison of results (18.4) and (18.5) with the full list would be highly interesting, especially for exotic and comlicated expansions.

Chapter 19. Systems considered in this chapter can be studied by methods of the multidimensional generalization of Plane Power Geometry developed for systems of ODEs. See references [14, Chap. 3], [18], and [19] of Chapter 1.

Chapter 20. "The Painlevé formal test" finds power expansions of solutions near nonsingular points of an ODE. It consists of 3 steps. Indeed, Plane Power Geometry (Chapter 1) gives algorithms for these 3 steps, which are (i)–(iii). Moreover, the so-called "resonance equation" $Q(s) = 0$ is exactly $\nu(s + r) = 0$, where $\nu(k)$ is the characteristic polynomial (1.3) in Chapter 1. But Plane Power Geometry also allows us to find logarithmic and essential singularities in any expansion and can thus help to verify the existence of the whole Painlevé property. Reduction of an ODE to a system of ODEs was considered in References [4–7] in Chapter 4. But now it has been found that the approach of Space Power Geometry (see Chapters 6 – 8) is more convenient. For the convergency of power expansions of solutions see Chapter 3.

Chapter 22. Let us have ODE $f(x, y) = 0$, where $f(x, y)$ is a differential sum (see Chapter 1) with unknown coefficients. Put $x = x_0 + t$ and $g(x_0, t, y) = f(x_0 + t, y)$, where $x_0 =$ const differs from singular points of $f(x, y)$ and will be considered as a new parameter. Using Plane Power Geometry (see Chapter 1), one can algorithmically find all the expansions of the solutions $y = \varphi(x_0, t)$ of the equation $g(x_0, t, y)$ for nonsingular x_0 and $t \to 0$. If all the expansions do not have branching and essential singularity, then the initial equation has the Painlevé property. It was made for P_6 (see [2] in Chapter 4) and for P_5 (see [5] in Chapter 5). Using Space Power Geometry (see Part II) one can algorithmically check the absence of "bad" solutions as $t \to 0$.

This approach can be applied to equation (22.5) considered in Chapter 22. Its support S is shown below in Figure 1. In the generic case its polygon has on the left side one vertex $(-4, 2)$ with the truncated equation (22.6) and one edge, marked by $r = -1$, with the truncated equation (22.15). In degenerated cases, the vertex is the same $(-4, 2)$, but the edges are marked in Figure 1 by $r = -2$ with the truncated equation (22.22), $r = -3$ with the truncated equation (22.24), and $r = -4$ with the truncated equation (22.25).

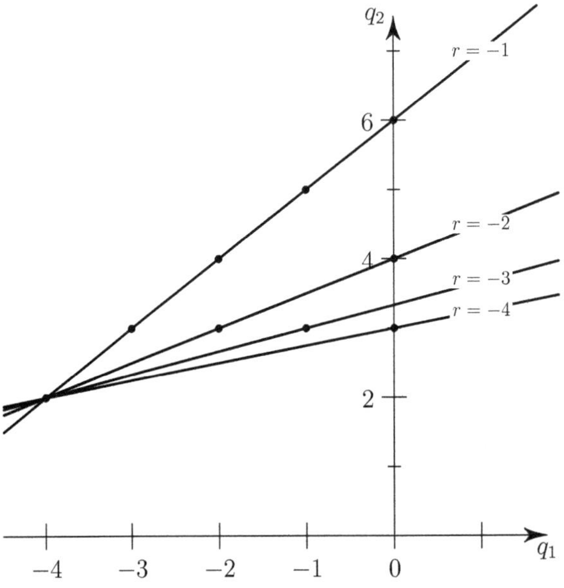

Figure 1. Plane support and upper lines for polygon in different cases.

Chapter 23. Equation (23.1) can be written in the form

$$\left(x^2x''' - x^2 f_1 - x' f_2\right)^2 - f_3^3 = 0.$$

Its left hand side is the differential sum (see Chapter 1) with unknown coefficients. So the Painlevé property of the equation can be studied by the methods of Plane and Space Power Geometry as described in the previous comment.

Alexander D. Bruno

Bei Fragen zur Produktsicherheit wenden Sie sich bitte an:
If you have any questions regarding product safety,
please contact:

Walter de Gruyter GmbH
Genthiner Straße 13
10785 Berlin
productsafety@degruyterbrill.com